BACTERIAL
ADHESION

WILEY SERIES IN
ECOLOGICAL AND APPLIED MICROBIOLOGY

SERIES EDITOR, Ralph Mitchell, Division of Applied Sciences,
Harvard University

BACTERIAL

ADHESION

Molecular and Ecological Diversity

Edited by

MADILYN FLETCHER

Belle W. Baruch Institute for Marine Biology and Coastal Research
University of South Carolina
Columbia, South Carolina

A JOHN WILEY & SONS, INC., PUBLICATION
New York • Chichester • Brisbane • Toronto • Singapore

Address All Inquiries to the Publisher
Wiley-Liss, Inc., 605 Third Avenue, New York, NY 10158-0012

Copyright © 1996 Wiley-Liss, Inc.

Printed in the United States of America

Library of Congress Cataloging-in-Publication Data
Bacterial adhesion : molecular and ecological diversity /
  edited by Madilyn Fletcher.
     p.   cm. — (Wiley series in ecological and applied
  microbiology)
   Includes bibliographical references and index.
   ISBN 0-471-02185-7 (alk. paper)
   1. Bacteria—Adhesion. 2. Bacterial diversity. 3. Molecular
microbiology. I. Fletcher, Madilyn. II. Series.
QR96.8.M65 1996
589.9'0875—dc20   96-15427
                                        CIP

CONTENTS

CONTRIBUTORS

EDWARD A. BAYER, Department of Membrane Research and Biophysics, The Weizmann Institute of Science, Rehovot, 76100 Israel [155]

ROBERT BELAS, Center of Marine Biotechnology, University of Maryland Biotechnology Institute, Baltimore, MD 21202 [281]

ROBERT A. BURNE, Department of Dental Research, University of Rochester, Rochester, NY 14652 [201]

RICHARD P. ELLEN, Department of Periodontics, University of Toronto, Toronto, Ontario M5G 1G6, Canada [201]

MADILYN FLETCHER, Belle W. Baruch Institute for Marine Biology and Coastal Research, Department of Biological Sciences, University of South Carolina, Columbia, SC 29208; formerly of Center of Marine Biotechnology, University of Maryland Biotechnology Institute, Baltimore, MD 21202 [1]

PAUL E. KOLENBRANDER, Laboratory of Microbial Ecology, National Institute of Dental Research, National Institutes of Health, Bethesda, MD 20892-4350 [249]

RAPHAEL LAMED, Department of Molecular Microbiology and Biotechnology, Tel Aviv University, Ramat Aviv, 69978 Israel [155]

JACK LONDON, Laboratory of Microbial Ecology, National Institute of Dental Research, National Institutes of Health, Bethesda, MD 20892-4350 [249]

KEVIN C. MARSHALL, School of Microbiology and Immunology, The University of New South Wales, Sydney, New South Wales 2052, Australia [59]

ANN G. MATTHYSSE, Department of Biology, University of North Carolina, Chapel Hill, NC 27599-3280 [129]

AARON L. MILLS, Laboratory of Microbial Ecology, Department of Environmental Sciences, University of Virginia, Charlottesville, VA 22903 [25]

The numbers in brackets are the opening page numbers of the contributors' articles.

Marc W. Mittelman, Departments of Surgery and Microbiology, University of Toronto, Toronto, Ontario M5G 2C4, Canada [89]

Ely Morag, Department of Membrane Research and Biophysics, The Weizmann Institute of Science, Reovot, 76100 Israel [155]

David K. Powelson, Laboratory of Microbial Ecology, Department of Environmental Sciences, University of Virginia, Charlottesville, VA 22903 [25]

Alice Prince, Department of Pediatrics, College of Physicians & Surgeons, Columbia University, New York, NY 10032 [183]

Lawrence J. Shimkets, Department of Microbiology, University of Georgia, Athens, GA 30602 [333]

Yuval Shoham, Department of Food Engineering and Biotechnology, Technion—Israel Institute of Technology, Haifa, 3200 Israel [155]

José Tormo, Department of Molecular Biophysics and Biochemistry, Bass Center for Molecular and Structural Biology, Yale University, New Haven, CT 06520 [155]

PREFACE

Many different types of bacteria attach to a wide range of solid surfaces, and bacterial attachment to surfaces often has a significant effect on their ecology, physiology, and behavior. Our understanding of this phenomenon began with the classical studies of Claude ZoBell and coworkers in the late 1930s and early 1940s. Interest in bacterial adhesion increased in the early 1970s and has become a focus of researchers cutting across the spectrum of clinical, ecological, physiological, and genetic microbiology.

Twenty years ago, many researchers had the common goal of elucidating the mechanism(s) of attachment to surfaces. Many also aimed to develop an understanding of the influence of attachment on the bacterium, as well as on its host when the substratum is typically a plant or animal surface. As research progressed, however, it became clear that there were many mechanisms for attachment, and that this is the inevitable result of the diversity of organisms involved and the equally diverse natures of their habitats and ecological requirements. Certain physical and chemical principles may apply to these various forms of adhesion, and certain cell surface structures and polymers are often involved. However, there is an enormous diversity in the specific cell components that act as adhesives, the different behaviorial strategies, and the genetic determinants that underly the adhesion phenomena. The purpose of this book is to illustrate this diversity and the ways in which different environmental situations have resulted in the evolution of a variety of attachment strategies and mechanisms.

Attachment mechanisms have frequently been classified as nonspecific and specific types of interactions. Nonspecific mechanisms are those in which the bacterium can attach to a range of substratum chemistries, even though there is usually a preference for particular chemistries over others. Specific mechanisms, on the other hand, are those that involve sterochemical complementarity between the bacterial "adhesin" and the substratum receptor. Bacteria that are capable of specific attachment are also generally capable of nonspecific attachment, but at much lower levels. As the chapters in this volume will illustrate, nonspecific attachment is typical of colonization of nonbiological surfaces in aquatic and soil environments. Specific mechanisms, however, have evolved to enable adhesion to plant or animal hosts or to other microorganisms.

Nonspecific adhesion in aquatic and soil environments is often fortuitous, that is the result of physicochemical interactions that lead to adsorption (Chapters 1 and 2). However, conditions on a surface may provide an advantage to the attached microorganism, particularly because they may be nutrient-enriched (Chapter 3). Bacteria in these environments may have evolved mechanisms to facilitate or stabilize adhesion, which may even involve specific genetic responses. Many of these bacteria also demonstrate mechanisms for detaching from surfaces (Chapter 3), which enhances the flexibility of these organisms in fluctuating and unpredictable environments. Although the nonspecific adhesion strategies are typical of colonization in aquatic or soil environments, they are also observed in animals and humans with the colonization of foreign bodies, such as the biomaterials used as implants or prosthetic devices (Chapter 4). Although biomaterials are relatively new in evolutionary time, many bacteria are able to take advantage of their presence and become entrenched in the host through surface colonization.

Specific attachment mechanisms evolved in situations where there was an advantage to being on a surface that was sufficiently common to be encountered by the microorganism and had predictable surface components that could act as receptors. Such adhesive interactions facilitated both symbiotic and pathogenic relationships in plants (Chapter 5) and have ensured recognition and binding to certain food sources (Chapter 6). In humans, there is a considerable diversity in bacterial adhesins and receptors. Adhesins have evolved to ensure establishment on mucosal surfaces (Chapters 7 and 8) or interactions with other organisms that participate in commensal relationships (Chapter 9). Because of the natural focus of research on human systems and the rapid advances in genetic approaches, there have been enormous strides in our understanding of bacterial attachment mechanisms in humans. The diversity of bacterial adhesion mechanisms has been fostered by an equally diverse array of receptors, as well as the need to evade host defense systems.

It has also become clear that attachment to surfaces may be only a component or phase in complex behavioral or life cycle processes. Such systems may require recognition of the surface (Chapter 10), or at least the ability to move across a surface (Chapter 11). Swarming behavior in microorganisms, for example, occurs in a variety of eubacteria and involves complex signaling and response behaviors, as well as attachment to a suitable substratum.

What began as a desire to understand "bacterial attachment" has resulted in the recognition of the complexity of the phenomenon—that there are many attachment mechanisms, often several on the same organism; that there have been numerous selection pressures that have led to the evolution of different attachment mechanisms; and that attachment may be part of a larger behavior strategy that involves complex signals and responses. The diversity of bacterial environments has fostered the molecular and ecological diversity of bacterial adhesion that we have come to appreciate today. My hope is that these chapters serve to illustrate that diversity by focusing on a selection of well-described attachment phenomena. The literature contains many additional examples, particularly in human systems, which the reader may want to pursue.

My sincere thanks to the contributors, whose commitment and efforts made this book possible, to Ralph Mitchell for his long-standing encouragement, and to David Ades, for his support and advice.

MADILYN FLETCHER

BACTERIAL ATTACHMENT IN AQUATIC ENVIRONMENTS: A DIVERSITY OF SURFACES AND ADHESION STRATEGIES

MADILYN FLETCHER

Belle W. Baruch Institute for Marine Biology and Coastal Research, Department of Biological Sciences, University of South Carolina, Columbia, South Carolina 29208

1.1. INTRODUCTION

The attachment of bacteria to surfaces in aquatic environments takes a variety of forms and occurs in a wide range of habitats. Attachment occurs both to surfaces of nonbiological materials and to surfaces of organisms in marine, freshwater, and groundwater systems. Attached bacteria may be sparsely distributed on the surface with no visible means of attachment, or they may be embedded in a hydrated polymer matrix along with other micro- and macroorganisms.

The colonization of nonbiological surfaces, such as rocks, sediment particles, and man-made structures, generally follows a sequence in which bacteria are the first colonizers and are followed by attachment and colonization by microalgae, algal spores, and invertebrate larvae. The eventual result is a complex community of micro- and macroorganisms, referred to as a *biofilm*. Bacteria also attach to and colonize the surfaces of other organisms. However, surfaces of many plants and animals remain free of bacteria because their chemistries are unsuitable for adhesive interactions (Rosowski, 1992), they produce toxins or repellants (e.g., Fletcher, 1975; Keifer and Rinehart, 1986;

Bacterial Adhesion: Molecular and Ecological Diversity, pages 1–24
© *1996 Wiley-Liss, Inc.*

Gerhardt et al., 1988; Todd et al., 1993), or they continually slough surface layers (e.g., Johnson and Mann, 1986) (see also Chapter 4, section 4.5.3, this volume). In some cases, bacteria readily colonize particular animal surfaces (e.g., Boyle and Mitchell, 1978), and attachment mechanisms may be "specific," that is, they have stereochemical specificity similar to that observed in bacterial–animal interactions in other environments (see Chapters 8 and 9, this volume).

Bacteria in aquatic environments also attach to and colonize detritus, such as dead phytoplankton and invertebrates (Montgomery and Kirchman, 1993, 1994). They are a component of detrital aggregates, such as marine snow, which constitute a complex microhabitat (Alldredge and Silver, 1988) and appear to select for particular types of microorganisms. Analyses of polymerase chain reaction (PCR)–amplified and cloned ribosomal RNA (DeLong et al., 1993) and low-molecular-weight (transfer and 5S) RNA (Bidle and Fletcher, 1995) have demonstrated that particle-associated communities are unlike those suspended in the water column. With aggregated detritus, it is not clear to what extent association with these particles involves attachment mechanisms or whether the bacteria are essentially entrapped in a complex mixture of aggregated material.

Microbial attachment plays a critical role in the formation of microbial mats, the thick laminated, microbial communities, which form at solid–liquid interfaces. Because of the thickness of mats and the metabolism of stratified community members, gradients in environmental nutrients and low-molecular-weight substances are often established (Cohen and Rosenberg, 1989). Colonization of mat communities may involve changes in cell surface properties that modulate adhesiveness, such as changes in hydrophobicity that alter the adhesion of benthic cyanobacteria (Shilo, 1989). Bacteria also colonize surfaces at hydrothermal vents and include the hyperthermophiles that are adapted to temperatures near the boiling point of water (Belkin and Jannasch, 1989). It is not known whether surface colonization in this extreme habitat requires attachment polymers or structures or whether the bacteria are retained on surfaces by incorporation within mineral deposits.

In many of these examples of bacterial attachment to surfaces, the ability to colonize the surface provides the bacteria with an advantage, usually by enhancing their access to nutrients (see Chapter 3, this volume) or protecting against toxic conditions, e.g., biocides (see Chapter 4, section 4.5.1, this volume). Thus, it is clear that there are many different environmental situations and selection pressures that may have guided the evolution of bacterial attachment properties. Because of the diversity of attachment substrata and associated environmental conditions, e.g., adsorbed nutrients, electrolytes, and attached macroorganisms, it is likely that there is also considerable diversity and/or flexibility in the adhesion abilities and mechanisms of aquatic bacteria. There can also be advantages in avoiding attachment or in detaching from a surface if conditions deteriorate, and accordingly many bacteria—and in some situations most—do not attach and remain free in the water column.

In the past 25 years, many studies have focused on bacterial attachment in aquatic environments. A persistent goal has been to identify and elucidate general principles that apply to mechanisms of adhesion. Progress has been made in understanding how bacteria attach, but what has become more clear is the enormous diversity and flexibility in bacterial attachment mechanisms and strategies. The purpose of this chapter is to describe the factors that influence adhesion in aquatic environments and to examine the properties of bacteria that determine their ability to adhere. An important question to address is whether life on a surface provides bacteria with sufficient advantages to foster the evolution of mechanisms that ensure colonization of surfaces. Moreover, we can ask whether the diversity in environmental conditions and attachment substrata has led to multiple adhesion mechanisms.

1.2. FACTORS THAT CONTROL ADHESION

1.2.1. Attractive and Repulsive Forces

As a bacterium moves toward a solid surface, numerous factors come into play and affect the forces that determine whether adhesion will actually occur (Fig. 1.1). Moreover, these forces differ in strength and in the separation distance at which they influence the interaction between the bacterium and the potential attachment surface. At separation distances of tens of nanometers, the interaction is a balance of attractive and repulsive forces between the two surfaces. van der Waals interactions are usually attractive, occur between all adjacent phases regardless of composition, and act over relatively long separation distances (>50 nm), but they are relatively weak (Busscher et al., 1990). Electrostatic forces become significant at closer separation distances (10–20 nm). If the opposing surfaces have opposite net surface charges, then electrostatic interactions are attractive. On the other hand, most often with bacteria and potential attachment surfaces, the net surface charges are negative, and therefore electrostatic repulsion occurs and may prevent the bacterium from getting close enough to the surface for adhesion to occur. However, repulsion forces decrease with an increase in ionic strength of the medium, and many natural environments, such as seawater, have sufficient electrolyte concentrations to eliminate the electrostatic repulsion barrier.

As the bacterium moves closer to the surface, another potential barrier to attachment is water that is adsorbed to the bacterial or substratum surfaces. Displacement of adsorbed water to allow closer approach of the two surfaces is energetically unfavorable. However, if either surface has nonpolar groups or patches, these can assist the exclusion of water by hydrophobic interactions (Busscher et al., 1990). There are many examples of bacteria attaching preferentially to hydrophobic surfaces (cf. Pringle and Fletcher, 1983; Rosenberg and Kjelleberg, 1986), and hydrophobic interactions are likely to be important

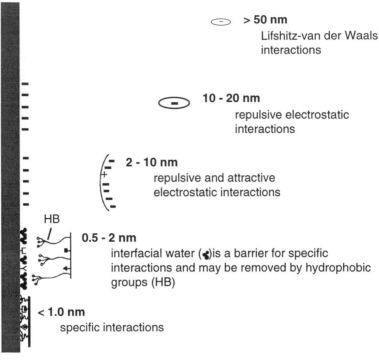

Fig. 1.1. Forces involved in attractive and repulsive interactions between a bacterium and a surface and approximate distances at which forces are significant. HB indicates hydrophobic functional groups on the cell surface or cell surface appendages that may assist in removing the layer of water adsorbed onto the substratum surface. Objects not drawn to scale. (Reproduced from Wiencek, 1995, as adapted from Busscher et al., 1990.)

in many cases, either as the primary mechanism of adhesion or by facilitating close approach, so that other adhesion interactions can occur.

Once the bacterium overcomes any water barrier and the separation distance is on the order of <1.0 nm (Busscher et al., 1990), additional adhesion interactions can come into play, such as hydrogen bonding and cation bridging, as well as the receptor–ligand interactions that are characteristic of specific interactions between bacteria and plant or animal hosts.

1.2.2. Theoretical Models of Bacterial Adhesion

Although the various types of interactions (e.g., electrostatic, hydrophobic) that can determine attachment are known, it is extremely difficult to evaluate these experimentally. However, two theoretical approaches derived from colloid and surface chemistry that incorporate attractive and repulsive interactions have been applied to studies of bacterial attachment in an attempt to evaluate and understand the interactions that control adhesion. If such models prove to describe or predict experimental observations of bacterial adhesion, then attachment may be dominated by the same forces that control adhesion of nonbiological particles. These theoretical approaches are (1) the DLVO theory, which takes into account the attractive van der Waals interactions and repulsive electrostatic interactions; and (2) thermodynamic models, in which the adhesive interaction is treated as an equilibrium process and is described in terms of the surface free energies of the bacterium, substratum, and separating liquid.

The DLVO theory (cf. Rutter and Vincent, 1980; van Loosdrecht et al., 1989) (see Chapter 2, sections 2.5.2.1, 2.5.2.2, and Chapter 4, section 4.3.3.2, this volume) predicts two distances of separation at which there is net attraction between the surface and an approaching bacterium. First, there is the "secondary minimum," the larger separation distance, at which the bacterium is weakly held near the surface and easily removed by shear forces. Closer approach is inhibited by electrostatic repulsion; however, if the bacterium overcomes this barrier, it may be bound at the closer separation distance, the "primary minimum," where attractive forces are strong and attachment is presumably irreversible.

A number of laboratory studies have investigated the significance of DLVO theory in bacterial adhesion by measuring the effects of net surface charges of bacteria and substrata and by determining the influence of electrolyte concentration. Some studies have shown attachment numbers to increase with increase in ionic strength of the medium (Weerkamp et al., 1988; van Loosdrecht et al., 1989). It is likely that in situations where bacteria and surfaces bear significant negative surface charges that attachment is inhibited. However, many experimental results cannot be exclusively explained by DLVO theory, and additional theoretical models have been explored.

In a thermodynamic approach, attachment is viewed as a spontaneous change, which is accompanied by a decrease in free energy of the system. To test this theory experimentally, it is necessary to determine values for interfacial free energies at the bacterium and substratum surfaces, and these are generally estimated indirectly by measuring the contact angles of liquids on the test substrata and on lawns of bacterial cells (Fletcher and Marshall, 1982; van der Mei et al., 1991) (see Chapter 2, section 2.5.2.2, and Chapter 3, section 3.2.2.2, this volume). Some studies have found that bacterial attachment data are consistent with thermodynamic predictions, but in other cases results cannot be explained by the model (Bellon-Fontaine et al., 1990).

There are several reasons why thermodynamic parameters would not accurately describe the bacterial attachment process. First, thermodynamic theory assumes an equilibrium situation, whereas bacteria may be directing metabolic energy into initiation or stabilization of attachment, e.g., through the production of adhesive polymers. Second, it is difficult to identify all components of the process because of the chemical complexity of bacterial attachment surfaces. For example, there may be unidentified entropy-gaining processes that drive adhesion, such as conformational changes in polymers, alterations in solvation of molecules, or water–bacterial polymer interactions. (See Chapter 3, sections 3.2 and 3.3, this volume, for a discussion of ways in which adsorption of molecules on surfaces affects bacterial adhesion.) Finally, it is also extremely difficult to determine values accurately for the relevant parameters, particularly bacterium surface free energy, which are used for calculating predictions (van der Mei et al., 1991).

The bacterial surface is extremely complex—both structurally and chemically—which is difficult to incorporate into a reductionist, theoretical approach. Bacterial surfaces, and often substrata, have no finite boundaries, but comprise long chain polymers, e.g., extracellular polysaccharide, or appendages, e.g., fimbriae, flagella, extending out into the medium. Furthermore, attachment may involve several stages of interaction, each involving different molecular components of the bacterial, or substratum, surface. For example, hydrophobic groups, such as nonpolar groups on fimbriae, lipopolysaccharide, or outer membrane proteins, may allow the bacterium to approach the substratum closely. This may be followed by conformational changes in surface polymers, allowing other functional groups to come close enough to the surface for short-range attractive interactions. Attachment could be strengthened even further by the synthesis *in situ* of additional adhesive polymers.

Thus, although the application of theoretical models to experimental observation may provide some insights into which physicochemical parameters should be significant and under what circumstances, it is clear that adhesion processes are determined by a combination of complex chemical, biological, and temporal factors. To elucidate the interactions that determine bacterial adhesion, it is therefore necessary to define better the chemistries of substrata and bacterial surfaces and to understand the ways in which these can influence the attachment process.

1.3. SUBSTRATUM SURFACE PROPERTIES

Many studies on mechanisms of attachment of aquatic bacteria have focused on the properties of the attachment substrata, primarily because of the desire to identify a "nonstick" surface. Microbial fouling is a serious problem on surfaces ranging from ship hulls, to heat exchangers, to drinking water pipelines. To inhibit microbial colonization by the use of material that resists adhesion would reduce the need for costly cleaning or replacement of heavily fouled structures.

Numerous studies have attempted to relate bacterial attachment to physicochemical properties of substrata, such as net surface charge (Feldner et al., 1983; MacRae and Evans, 1983), surface free energy (van Pelt et al., 1985; Busscher et al., 1986a,b), critical surface tension (Dexter, 1979; Becker and Wahl, 1991), or work of adhesion for water (W_A) (Pringle and Fletcher, 1983). Values for most of these parameters are derived from the measurement of contact angles (Θ) of liquids on the surfaces (see Chapter 3, section 3.2.2, this volume). The angle formed by the edge of the drop when placed on the surface is the advancing contact angle (Θ_A). When liquid is removed from the drop, the angle may change or remain the same; this angle is termed the receding contact angle (Θ_R) and provides information on chemical and physical interactions between the surface and liquid.

The critical surface tension (γ_c) is determined from the contact angles of a series of liquids with a range of surface tensions. It is equivalent to the surface tension of a liquid in that series where surface wetting first occurs—that is, where an advancing contact angle (Θ_A) of zero is first achieved (Baier et al., 1968). W_A is calculated from water contact angle and is inversely related to surface hydrophobicity. Bacterial attachment has also been related to surface hydrophobicity (Gerson and Scheer, 1980; Kawabta et al., 1983; Hogt et al., 1983; Ludwicka et al., 1984), which is generally inversely related to surface free energy. Surface hydrophobicity can be directly measured by determination of contact angles of water on test surfaces, whereas thermodynamic parameters must be indirectly assessed from contact angles of several liquids.

A number of studies have found that attachment was directly related to substratum hydrophobicity (Fletcher and Loeb, 1979; Fletcher and Marshall, 1982; Paul and Loeb, 1983) with respect to numbers of attached bacteria (Fletcher and Loeb, 1979; van Pelt et al., 1985), rate of attachment (Paul and Loeb, 1983; Sjollema et al., 1990), or strength of binding (Rijnaarts et al., 1993). However, other workers have observed that attachment to hydrophilic and hydrophobic surfaces is similar (Pedersen et al., 1986; Busscher et al., 1990) or that maximum adhesion occurs in a midrange between the most hydrophilic (e.g., glass) and most hydrophobic (e.g., teflon) substrata (Dexter et al., 1975; Pringle and Fletcher, 1983).

It is clearly impossible to generalize about the effect of substratum hydrophobicity on bacterial attachment. This is illustrated in Table 1.1, which details attachment numbers of groundwater isolates of pseudomonads to two types of polystyrene, one that is hydrophobic (Θ_A of water = 90°) and one that has been treated commercially to make it more hydrophilic (Θ_A of water = ca. 66°). Although most isolates (nine) attach in higher numbers to the hydrophobic surface, three have a greater affinity for the more hydrophilic substratum. This is a single illustration, but similar variations in attachment abilities of isolates could be expected from many, possibly most, environmental samples. Hydrophilic substrata might be expected to be poor attachment substrata because of adsorbed water that must be displaced for adhesion to occur (see section 1.2.1). Highly hydrated surfaces, such as hydrogels, have been found to be quite resistant to bacterial attachment (Pringle and Fletcher, 1986). Nevertheless, hydrophilic surfaces, such as glass, are often extensively colonized by many bacteria over time. Thus, the suitability of a particular surface for bacterial attachment is strongly dependent on the composition of the bacterial surface and the presence of complementary chemistries on the bacterium and substratum surfaces that result in net attraction forces.

TABLE 1.1. Adhesion of Gram-Negative Subsurface Isolates to Hydrophobic Petri Dishes (PD) and Hydrophilic Tissue Culture Dishes (TCD)[a]

Isolate designation	A590 $\times$ 10^3 (standard deviation)	
	PD	TCD
G884	357.8 (43.4)	61.6 (8.0)
G946	132.3 (14.8)	75.3 (10.8)
G953	150.1 (17.4)	72.7 (7.0)
G985	107.6 (20.5)	68.7 (29.0)
Zat10	149.6 (21.6)	49.8 (8.8)
Zat36	92.7 (22.6)	129.2 (18.2)
Zat37	169.1 (42.0)	49.6 (3.9)
Zat40	45.9 (4.7)	60.2 (10.9)
Zat43	51.9 (4.9)	66.0 (5.9)
Zat361	187.3 (19.8)	79.2 (10.9)
Zat408	195.8 (9.8)	72.9 (9.1)
Zat408	164.8 (60.4)	63.5 (11.1)

[a] Attachment is expressed as the absorbance at 590 nm of crystal violet–stained attached bacteria, measured in a spectrophotometer. Before the assay, log phase cultures of each strain were harvested and starved for 48 h in a minimal salt solution (Williams and Fletcher, 1996).

Most investigations of the effects of substratum properties on bacterial attachment have involved measurement of attached cells after the substrata have been removed from the bacterial suspension or aquatic environment. Such measurements give information on only one time point and may be affected by passing the surface through the air–water interface or rinsing procedures. Additional insights into the attachment process can be gained by observing living bacteria in real time by microscopy, particularly when combined with computer enhancement and analysis (Caldwell and Germida, 1985; Caldwell and Lawrence, 1986; Lawrence and Caldwell, 1987; Lawrence et al., 1987). With such systems, not only is the number of attached cells quantified, but also the kinetics of attachment (both attachment and detachment) and movement of cells on the surface (Lawrence et al., 1987; Lawrence and Caldwell, 1987) can be assessed. For example, in a study utilizing image analysis of adhesion of an estuarine pseudomonad and the marine *Pseudomonas* sp. NCIMB 2021 to surfaces in a flow chamber, rates of attachment and detachment and residence times of cells on the surface were quantified (Wiencek and Fletcher, 1995; Wiencek, 1995). The substrata were homogeneous hydrophobic and hydrophilic surfaces that were constructed from alkanethiol self-assembled monolayers (SAMs) (Lopez et al., 1993; Prime and Whitesides, 1993; DiMilla et al., 1994). A series of substrata with different functional groups making up each surface was constructed. The SAMs ranged from surfaces with only hydroxyl groups exposed at the surface to those with only methyl groups exposed at the surface and included a series of mixtures of hydroxyl-terminated and methyl-terminated alkanethiols. This produced surfaces with a range of wettabilities, determined by contact angles of both water and hexadecane (Fig. 1.2).

Both organisms showed the highest numbers of attachment on the hydrophobic, methyl-terminated SAM, although for *Pseudomonas* sp. NCIMB 2021 the difference in numbers attached on the methyl-terminated and hydroxyl-terminated SAMs was small (Fig. 1.3).

In addition to total numbers of attached cells, levels of adsorption and desorption were measured. For both organisms, desorption was considerably greater on the hydrophilic hydroxyl-terminated SAM, whereas desorption from the hydrophobic, methyl-terminated SAM was almost negligible (Fig. 1.3). The high levels of desorption from the hydroxyl-terminated SAM suggest that strength of binding was considerably less than that achieved on the hydrophobic surface. This raises the question as to how the high levels of attachment of *Pseudomonas* sp. NCIMB 2021 on this surface was achieved.

Microscopic observations of attaching cells were intriguing, as they suggested that different mechanisms were involved in attachment to the hydro-

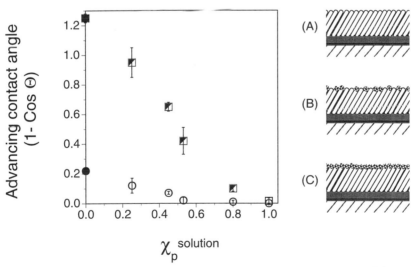

Fig. 1.2. Wettability of single- and mixed-component SAMs constructed from solutions of mercapto-undecanol and undecanethiol. Wettability is expressed as $1 -$ the cos of the advancing contact angle of water (squares) or hexadecane (circles). Solution concentrations are given as the mole fraction (χ_p of the polar alkanethiol, i.e., mercapto-undecanol. Surface hydrophobicity decreases from left to right on the χ axis. Schematic illustrations of single-component SAMs of undecanethiol **(A)**, mixed SAMs of mercapto-undecanol and undecanethiol **(B)**, and single-component SAMs of mercapto-undecanol **(C)**. (Wiencek, 1995.)

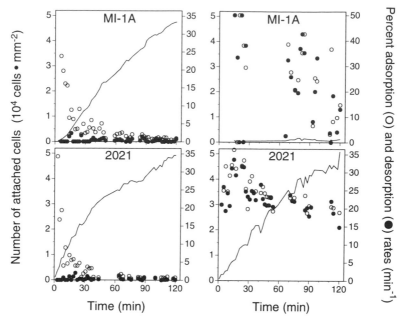

Fig. 1.3. Comparison of attachment kinetics of *Pseudomonas* sp. MI-1A **(top)** and marine *Pseudomonas* sp. NCIMB 2021 **(bottom)** to hydrophobic and hydrophilic SAMs. Net numbers of attached cells (solid line) and percent adsorption (open circles) and desorption (closed circles) rates with **(left)** methyl-terminated SAMs and **(right)** hydroxyl-terminated SAMs. (Wiencek, 1995.)

phobic and hydrophilic substrata (Fig. 1.4). With hydrophobic SAMs, the bacteria attached to the substrata in a random orientation and did not exhibit any flipping or rotational movements at the surface. The cell bodies appeared to be firmly attached to the surface. In contrast, cells consistently attached to the hydrophilic SAMs in one direction, parallel to the direction of flow, and exhibited rotating and flipping motions while anchored to the substratum. If the rate of flow was increased, the cells appeared to be forced down onto the surface, all in the direction of flow. It was as though the cells were tethered by their polar flagella, with their cell bodies hovering above the surface downstream of the point of contact. Then with the increased flow, the force pushed the cell bodies onto the surface.

The observation that the bacteria attached to hydroxyl-terminated SAMs by their flagella, but to methyl-terminated SAMs by the cell body, is consistent with a previous study that indicated that this organism may attach by different mechanisms to hydrophobic and hydrophilic surfaces (Fletcher and Marshall, 1982). Furthermore, other studies of pseudomonads have demonstrated that flagella are involved in attachment. Real-time observations of *P. fluorescens* attaching to glass in a flow chamber indicated that initial (reversible) attach-

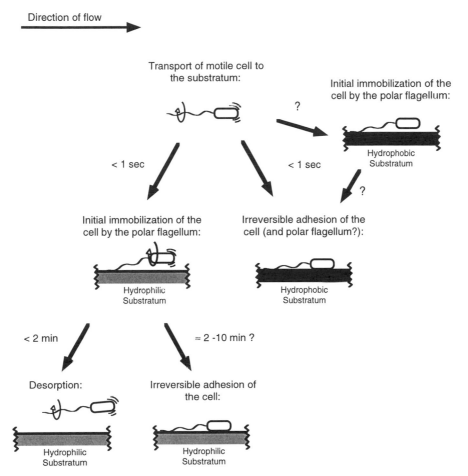

Fig. 1.4. Schematic representation of proposed model for attachment of *Pseudomonas* sp. NCIMB 2021 to hydrophobic and hydrophilic substrata. Attachment to the hydrophobic surface **(right)** is rapid and irreversible. Attachment to the hydrophilic surface **(left)** occurs initially by the flagellum and may proceed to either desorption or irreversible attachment. (Reproduced from Wiencek, 1995.)

ment was by the flagellum (Lawrence et al., 1987). Studies of another *P. fluorescens* strain (pf0–1) demonstrated that mutations that resulted in modifications in quantity of flagellin (either deletion or overexpression) resulted in a decrease in adhesion to quartz sand, soil, or a variety of plant seeds (DeFlaun et al., 1990,1994).

These results suggest that two basic attachment processes may occur with hydrophobic and hydrophilic surfaces (Fig. 1.4). Adsorption to hydrophobic surfaces appears to be rapid, and binding may be stronger than that on hydrophilic surfaces (Rijnaarts et al., 1993). In contrast, adhesion to hydrophilic surfaces may more closely follow the model of "reversible" and "irreversible" adhesion described by Marshall et al. (1971) in which a bacterium that encounters the surface is first weakly held at the surface at a separation distance of several nanometers, the "secondary minimum" described by DLVO theory (see section 1.2.2). This weak stage of adhesion is reversible in that the bacteria are easily removed by shear forces, and many also can be observed by microscopy to desorb spontaneously. This stage may progress into "irreversible" adhesion, presumably by the synthesis of extracellular adhesive or possibly by stabilizing conformational changes in existing polymer (Robb, 1984). Such polymer may be able to bridge the narrow separation distance enforced by adsorbed water or electrostatic repulsion on hydrophilic surfaces.

1.4. BACTERIAL SURFACE PROPERTIES

The bacterial surface is a heterogeneous, three-dimensional structure with a complex chemical composition. Numerous macromolecules interface with the surrounding medium, and these vary in composition and quantity from species to species and from strain to strain and with environmental conditions and physiological processes. There can be a great deal of diversity, variability, and flexibility in the attachment properties of bacteria (Table 1.1).

It is likely that different polymers are active with different substratum chemistries and medium compositions (Fletcher and Marshall, 1982; Paul and Jeffrey, 1985). Furthermore, it is theoretically possible that different functional groups on a given polymer may function in adhesion at different times, depending on the substratum chemistry or solution properties. For example, polymers with nonpolar sites, e.g., fimbriae, lipopolysaccharide, may dominate in binding to hydrophobic surfaces, whereas polymers capable of hydrogen bonding or electrostatic interactions, e.g., polysaccharides, may function with hydrophilic surfaces. Moreover, different polymer types may act cooperatively in binding to the surface to stabilize the adhesive interaction (Doyle et al., 1982). Also important are bacterial surface polymers that tend to prevent adhesion by binding water or through other steric effects (Pringle et al., 1983; Robb, 1984) and thus may mask adhesive polymers and keep the bacterium suspended in solution.

1.4.1. Extracellular Polysaccharides

There is evidence that extracellular polysaccharides (EPS), proteins, and lipopolysaccharide can play a role in attachment, possibly at different stages in the attachment of process. EPS is synthesized by attached bacteria (Allison and Sutherland, 1987), strengthens their binding to surfaces (Costerton et al., 1985), and largely comprises the intercellular matrices in biofilms. However, it is not clear whether, or to what extent, EPS functions as the initial adhesive, as there is evidence that proteins (section 1.4.2) or lipopolysaccharides (section 1.4.3) may be involved.

Composition of polysaccharides produced by aquatic microorganisms varies considerably. These are generally heteropolysaccharides, containing largely neutral sugars and various amounts of uronic acids and acetate or pyruvate substitutions (Sutherland, 1983; Boyle and Reade, 1983; Read and Costerton, 1987). A survey of exopolysaccharides produced by plant-associated fluorescent pseudomonads demonstrated that these organisms produced a variety of polysaccharides, including acetylated alginates (Fett et al., 1989), and the same diversity may occur in aquatic strains. The various characteristics of bacterial alginate from different organisms is a good illustration of how chemical structure can influence the ultimate adhesive function of the polymer. Alginate is a linear polymer of β-1,4-linked D-mannuronic and L-guluronic acids. Alginate from *P. aeruginosa*, which is expressed in the lungs of cystic fibrosis patients, is a highly viscous polymer, binding cells together and to epithelial cells of the respiratory tract (Ramphal et al., 1987; Pedersen, 1992) (see also Chapter 7, this volume). In contrast, the alginate produced by a river isolate of *P. fluorescens* was highly soluble in aqueous solution and strongly inhibited adhesion (Pringle et al., 1983). The degree of acetylation of bacterial alginates and ratio of mannuronic to guluronic acid vary considerably (Fett et al., 1989; Conti et al., 1994) and may be responsible for different physical properties such as ion binding and gel formation (Sengha et al., 1989), as well as adhesiveness.

Although polysaccharides are significant subsequent to attachment (Allison and Sutherland, 1987) and in cell accumulation in biofilms, there is little evidence that they are involved in initial adhesion. An exception is a study of a marine *Hyphomonas* sp. Attachment of this organism appeared to be mediated by an EPS capsule that bound lectins specific for *N*-acetyl-D-galactosamine, as well as the dye Calcofluor (Quintero and Weiner, 1995).

1.4.2. Proteins

Investigations on the effects of proteases on attachment indicate that protein (e.g., outer membrane proteins, fimbriae, flagella, or enzymes) may be required for the early stages of adhesion to surfaces (Danielsson et al., 1977; Fletcher and Marshall, 1982; Paul and Jeffrey, 1985). Observations of continuous flow slide cultures have shown that lateral flagella are important for initial attachment of vibrios (Lawrence et al., 1992) and that polar flagella can be involved in initial attachment of *P. fluorescens* (Korber et al., 1989). Furthermore, by generation of transposon mutants that were deficient in adhesion, DeFlaun et al. (1990, 1994) demonstrated that flagella are involved in attachment of *P. fluorescens* to soil and a variety of plant seeds. Proteins often function as adhesins in specific attachment mechanisms, i.e., those that involve stereochemical specificity (see Chapters 7, 8, and 9, this volume) but the extent of their involvement in nonspecific attachment is still not clear.

1.4.3. Lipopolysaccharide

Most culturable bacteria in aquatic environments are gram negative. Because of its dominance in the outer leaflet of the outer membrane of gram-negative bacteria, lipopolysaccharide should have a significant effect on adhesive interactions. It is anchored in the outer membrane by the lipid A moiety, while the polysaccharide (O-antigen) is at the cell surface. Electron microscopic observations of antibody bound to O-antigen indicate that the latter is flexible and can be stretched to more than 20 nm from the lipid component (Shands et al., 1967). However, the preferred conformation of the lipopolysaccharide is thought to be "bent," so that the O-antigen lies flat on top of the head groups of other outer membrane molecules (Kastowsky et al., 1992). In many strains or phenotypes, the O-antigen is attenuated or missing, producing the "rough" colony form, which can also result in an increase in cell surface hydrophobicity (de Maagd et al., 1989), hence possibly adhesion. There is some indication that marine strains may have largely homogeneous polysaccharide, or only oligosaccharides, on the lipopolysaccharide (Sledjeski and Weiner, 1991). A study of the lipopolysaccharide of 19 marine strains by sodium dodecyl sulfate-polyacrylamide gel electrophoresis (SDS-PAGE) demonstrated that most (15) had a low molecular weight lipopolysaccharide without the ladder pattern that typically represents size heterogeneities in the O-antigen.

In a separate study of a freshwater *P. fluorescens* (strain H2), transposon mutagenesis was used to generate mutants with altered adhesion (Williams and Fletcher, 1996). Southern analysis of transposon insertion indicated that two types of adhesion mutant were obtained, and both attached in higher numbers to hydrophobic polystyrene, and in lower numbers to hydrophilic, tissue culture polystyrene. All mutants demonstrated increased adhesion when suspensions were passed through glass columns containing silica sand. These changes in adhesion appeared to be related to alterations in lipopolysaccharide, as SDS-PAGE of the adhesion mutants demonstrated that the O-antigen on the lipopolysaccharide was either absent or shorter and less heterogeneous than that on the wild type. Absence or attentuation of the O-antigen would explain the increase in adhesion to hydrophobic surfaces, as this would result in greater exposure of the lipid moiety of the lipopolysaccharide (and possibly nonpolar sites on surface proteins).

1.4.4. Parameters That Approximate Surface Properties

Numerous workers have utilized parameters that are evaluated more easily than surface biochemistry and are an average expression of cell surface chemistry, e.g., surface hydrophobicity (Rosenberg et al., 1980; Gannon et al., 1991), surface free energy (or related parameters) (Busscher et al., 1984), and surface charge (James, 1991; Gannon et al., 1991). For example, attachment to soil or sediment particles may be favored by microbial surface hydrophobicity

(Stenström, 1989). However, gross measurements of surface properties do not consistently correlate with attachment and transport through porous media (e.g., Gannon et al., 1991). Relationships between attachment and physico-chemical parameters may be improved by consideration of multiple parameters. For example, an investigation of the relationship between cell surface hydrophobicity and electrokinetic potential of 23 strains and their adhesion to negatively charged polystyrene was demonstrated (van Loosdrecht et al., 1987). The greatest attachment was found with the most hydrophobic cells, irrespective of electrokinetic potential. However, with more hydrophilic bacterial surfaces, attachment decreased with increase in negative surface potential. Thus, measurements of gross bacterial surface characteristics can have value in describing bacterial adhesiveness, particularly when more than one parameter is considered.

1.4.5. Influence of Physiological Status on Attachment

The composition and quantity of bacterial surface polymers, hence cell surface and attachment properties, can be considerably influenced by the physiology of the organisms. Factors that influence production of surface polymers, hence adhesion, include carbon source, carbon : nitrogen ratio, carbon flux, and nutrient concentration (Wardell et al., 1980; Molin et al., 1982; McEldowney and Fletcher, 1986). Since many aquatic habitats are oligotrophic and resident bacteria are nutrient deficient, the influence of starvation on adhesiveness could be of considerable significance in aquatic environments. Wrangstadh et al. (1990) observed that the marine bacterium *Pseudomonas* sp. S9 produced an EPS that was closely associated with the cell surface during growth, but when the cells were starved a second loosely associated polymer was produced. The first polymer appeared to mediate attachment; however, as attached cells underwent starvation and the second polymer was produced, bacteria detached. Detachment in this case may have been a response to nutrient deprivation, allowing the organism to be dispersed to a different, possibly more suitable environment (see also Chapter 3, section 3.5, this volume). Polymer properties are also directly influenced by environmental factors. For example, pH, temperature (Annaka and Tanaka, 1992), and cations (Marshall et al., 1989) can affect cross-linking and gel characteristics of polymers, which in turn would influence their function as adhesives.

1.4.6. Evolution of Specific Responses to Attachment Substrata

An unanswered question is to what extent bacterial surface polymers have evolved to function as adhesives in aquatic environments. Extracellular polysaccharides bind cells together in biofilms, and the selection advantage of surface colonization has probably been strong enough to drive the evolution of biofilm polymer production. Furthermore, these polymers also appear to provide some protection from toxins or environmental perturbations (Nickel et al., 1985; Anwar et al., 1989; Brown et al., 1988) (see also Chapter 4, section 4.5.1), which would further enhance their selection. This suggests that adhesive polymer synthesis might be induced by attachment or the proximity of the attachment substratum. The first evidence that surfaces could provide a proximate stimulus for gene expression was described by Belas et al. (1986) in their elegant studies of *Vibrio parahaemolyticus* surface colonization (see Chapter 10, this volume). This organism, which commonly attaches to chitin particles in aquatic environments, is polarly flagellated when free swimming, but produces lateral flagella and adopts a swarming behavior after attachment to surfaces. By using *lux* fusions to investigate expression transcription of lateral flagella genes *(laf)*, these researchers demonstrated that factors that interfere with motion of the polar flagellum (e.g., medium viscosity, attachment) triggered *laf* expression.

More recently, studies have shown that conditions at surfaces can induce expression of additional genes (Dagostino et al., 1991), morphology changes (Dalton et al., 1994), and synthesis of macromolecules, including proteins (Brözel et al., 1995) and the extracellular polymer alginate. Although *P. aeruginosa* is found in many natural environments, the main reason for its being the subject of intense study is its role as an opportunistic pathogen, particularly in the lungs of cystic fibrosis patients, where its production of alginate affords it protection from host defense mechanisms (Pedersen, 1992) (see also Chapter 7, this volume). Alginate is also produced by many plant-associated pseudomonads (Fett et al., 1986; Conti et al., 1994) and by a river isolate of *P. fluorescens* (Pringle et al., 1983), although the adhesiveness of these alginates can differ significantly. Synthesis of alginate by *P. aeruginosa* is induced by environmental conditions such as dehydration, high osmolarity, and starvation, and its production is also enhanced by attachment to surfaces (Davies et al., 1993; Davies and Geesey, 1995). Expression of *algC* in biofilms was assessed by using *P. aeruginosa* with plasmids carrying an *algC–lacZ* transcriptional fusion and measuring β-galactosidase activity (Davies and Geesey, 1995). The specific activity of the *algC* reporter gene product was 19-fold higher in biofilm bacteria than in those remaining suspended in continuous culture. Initial attachment appeared to be independent of *algC* promoter activity, but after attachment to a glass surface for 15 minutes *algC* up-expression could be detected. Moreover, bacteria that did not remain attached had an expression level between that of the biofilm and suspended populations, indicating that *algC* expression was in some way related to stabilization of attachment. The gene *algC* encodes a bifunctional enzyme, which is not only involved in alginate synthesis but is also required for synthesis of O-antigen in lipopolysaccharide (Goldberg et al., 1993). Thus, *algC* up-expression could indicate enhanced synthesis of alginate or of lipopolysaccharide.

A study of groundwater organisms indicated that attachment could induce EPS synthesis. When sand was added to pure cultures of isolates from groundwater, there was increased synthesis of extracellular polysaccharide compared with free-living bacteria (Vandevivere and Kirchman, 1993). Furthermore, when attached bacteria were redispersed in fresh medium, extracellular polysaccharide synthesis decreased to the level previously observed for unattached cells, indicating that enhanced synthesis was indeed a phenotypic change and not the result of selection of a subpopulation.

Most aquatic bacteria appear to attach by "nonspecific" mechanisms; that is, there is not the stereochemical specificity that occurs in receptor–ligand interactions. The evolution of "specific" adhesion mechanisms in aquatic environments has been rarely observed, possibly because it would be strategically disadvantageous for bacteria to be restricted to specific types of substrata in environments where diversity is the norm. (Specific adhesion interactions, however, may be more common on plant or animal hosts, a subject that has received little attention.) In cases where a particular type of substratum is relatively abundant and also a potential nutrient, it would be an advantage to be able to attach and utilize the resource. Such a surface is chitin (poly-N-acetylglucosamine), which forms the exoskeletons of many aquatic invertebrates and is a constituent of some algae. In marine systems, common colonizers of chitinaceous particles are vibrios, and *Vibrio harveyi* has been shown to attach specifically to chitin, apparently by means of two chitin-binding peptides (Montgomery and Kirchman, 1993). A 53 kDa peptide is produced constitutively and appears to be involved in the initial attachment to the surface, whereas synthesis of a second 150 kDa chitin-binding peptide is induced rapidly after attachment and may serve to strengthen binding (Montgomery and Kirchman, 1994).

Despite the growing evidence that surface colonization is a life style that may have encouraged the evolution of structures or polymers that strengthen or stabilize adhesion, there remains the fact that many bacteria never attach to surfaces. Indeed, perpetuation of the species frequently requires mechanisms of dispersal, and a commitment to a sessile mode of living can be a disadvantage when local nutrients are depleted or environmental conditions deteriorate. Even in laboratory studies, often only a small proportion of the suspended organisms attach to available surfaces. It is possible that initial adhesion is in most cases a chance event, caused by the nonspecific adsorption of cell surface polymers with no specific adhesive function. Then once adhesion has occurred, for some cells the surface may induce certain activities, such as EPS production, that foster surface colonization and possibly biofilm development. There may also be mechanisms for release of cells, such as through synthesis of soluble surface polymers or surface-active compounds (Neu, 1996) or degradation of adhesive polymers (see also Chapter 3, section 3.5, this volume). Starvation of a marine *Pseudomonas* species was shown to result in synthesis of a polymer that resulted in cell detachment (Wrangstadh et al., 1990). Production of an alginate lyase by *P. aeruginosa* can result in degradation of alginate and increased cell detachment (Boyd and Chakrabarty, 1994), and similar release mechanisms may occur with aquatic organisms.

1.5. CONCLUSIONS

The interactions between bacteria and solid surfaces in aquatic environments involves a diversity of colonization strategies, bacterial surface compositions, and substratum characteristics. Accordingly, the mechanisms for attachment that have evolved are diverse, and there appear to be a number of adhesive polymers involved at various stages of the attachment process. A fundamental question that remains to be answered is whether there are relatively few types of molecules that can act as bacterial adhesives or whether the suite of macromolecules at the bacterial surface act in concert, through synergistic or antagonistic interactions, to determine the overall adhesiveness of the cell (see also Chapter 8, section 8.4, this volume). There is now evidence that bacteria can "sense" surfaces and that attachment may trigger certain types of gene expression, such as flagella or polysaccharide synthesis, but we have not yet determined whether these are common responses. We also understand very little about the factors that control detachment of bacteria. To what extent can bacteria alter their surface properties or degrade adhesives and under what conditions? By understanding these activities, the conditions under which they occur, and their frequency among bacteria, we will gain important insights into the selection pressures within aquatic environments and the evolution of attachment strategies. Similarly, by developing our understanding of bacterial ecology in oceans and freshwaters, we can better understand the diversity of bacterial adhesion phenomena and their significance for bacterial growth and survival.

ACKNOWLEDGMENTS

Some of the work described in this chapter was supported by the U.S. Department of Energy, Subsurface Science Program, grant DF-FG02-94ER61814. Bacterial strains shown in Table 1.1 were obtained from the U.S. Department of Energy Subsurface Microbial Culture Collection, Florida State University.

REFERENCES

Alldredge AL, Silver MW (1988): Characteristics, dynamics, and significance of marine snow. Prog Oceanogr 20:41–82.

Allison DG, Sutherland IW (1987): The role of exopolysaccharides in adhesion of freshwater bacteria. J Gen Microbiol 133:1319–1327.

Annaka M, Tanaka T (1992): Multiple phases of polymer gels. Nature 355:430–432.

Anwar H, Dasgupta M, Lam K, Costerton JW (1989): Tobramycin resistance of mucoid *Pseudomonas aeruginosa* biofilm grown under iron limitation. J Antimicrob Chemother 24:647–655.

Baier RE, Shafrin EG, Zisman WA (1968): Adhesion: Mechanisms that assist or impede it. Science 162:1360–1368.

Becker K, Wahl M (1991): Influence of substratum surface tension on biofouling of artificial substrata in Kiel Bay (Western Baltic): *In situ* studies. Biofouling 4:275–291.

Belas R, Simon M, Silverman M (1986): Regulation of lateral flagella gene transcription in *Vibrio parahaemolyticus.* J Bacteriol 167:210–218.

Belkin S, Jannasch HW (1989): Microbial mats at deep-sea hydrothermal vents: New observations. In Cohen Y, Rosenberg E (eds): Microbial Mats: Physiological Ecology of Benthic Microbial Communities. Washington, DC: American Society for Microbiology, pp 16–21.

Bellon-Fontaine M-N, Mozes N, van der Mei HC, Sjollema J, Cerf O, Rouxhet PG, Busscher HJ (1990): A comparison of thermodynamic approaches to predict the adhesion of dairy microorganisms to solid substrata. Cell Biophys 17:93–106.

Bidle K, Fletcher M (1995): Comparison of free-living and particle-associated bacterial communities in the Chesapeake Bay by stable low-molecular-weight RNA analysis. Appl Environ Microbiol 61:944–952.

Boyd A, Chakrabarty AM (1994): Role of alginate lyase in cell detachment of *Pseudomonas aeruginosa.* Appl Environ Microbiol 60:2355–2359.

Boyle CD, Reade AE (1983): Characterization of two extracellular polysaccharides from marine bacteria. Appl Environ Microbiol 46:392–399.

Boyle PJ, Mitchell R (1978): Absence of microorganisms in crustacean digestive tracts. Science 200:1157–1159.

Brown MRW, Allison DG, Gilbert P (1988): Resistance of bacterial biofilms to antibiotics: A growth-rate related effect? J Antimicrob Chemother 22:777–783.

Brözel VS, Strydom GM, Cloete TE (1995): A method for the study of *de novo* protein synthesis in *Pseudomonas aeruginosa* after attachment. Biofouling 8:195–201.

Busscher HJ, Sjollema J, van der Mei HC (1990): Relative importance of surface free energy as a measure of hydrophobicity in bacterial adhesion to solid surfaces. In Doyle RJ, Rosenberg M (eds): Microbial Cell Surface Hydrophobicity. Washington, DC: American Society for Microbiology, pp 335–359.

Busscher HJ, Uyen MHWJC, van Pelt AWJ, Weerkamp AH, Arends J (1986a): Kinetics of adhesion of the oral bacterium *Streptococcus sanguis* CH3 to polymers with different surface free energies. Appl Environ Microbiol 51:910–914.

Busscher HJ, Uyen MHWJC, Weerkamp AH, Arends J (1986b): Reversibility of adhesion of oral streptococci to solids. FEMS Microbiol Lett 35:303–306.

Busscher HJ, Weerkamp AH, van der Mei HC, van Pelt AWJ, De Jong HP, Arends J (1984): Measurements of the surface free energy of bacterial cell surfaces and its relevance for adhesion. Appl Environ Microbiol 48:980–983.

Caldwell DE, Germida JJ (1985): Evaluation of difference imagery for visualizing and quantitating microbial growth. Can J Microbiol 31:35–44.

Caldwell DE, Lawrence JR (1986): Growth kinetics of *Pseudomonas fluorescens* microcolonies within the hydrodynamic boundary layers of surface microenvironments. Microb Ecol 12:299–312.

Cohen Y, Rosenberg E (1989): Microbial Mats: Physiological Ecology of Benthic Microbial Communities. Washington, DC: American Society for Microbiology.

Conti E, Flaibani A, O'Regan M, Sutherland IW (1994): Alginate from *Pseudomonas fluorescens* and *P. putida:* Production and properties. Microbiology 140:1125–1132.

Costerton JW, Marrie TJ, Cheng K-J (1985): Phenomena of bacterial adhesion. In Savage DC, Fletcher M (eds): Bacterial Adhesion. New York: Plenum Press, pp 3–43.

Dagostino L, Goodman AE, Marshall KC (1991): Physiological responses induced in bacteria adhering to surfaces. Biofouling 4:113–119.

Dalton HM, Poulsen LK, Halasz P, Angles ML, Goodman AE, Marshall KC (1994): Substratum-induced morphological changes in a marine bacterium and their relevance to biofilm structure. J Bacteriol 176:6900–6906.

Danielsson A, Norkrans B, Björnsson A (1977): On bacterial adhesion—The effect of certain enzymes on adhered cells of a marine *Pseudomonas* sp. Botan Mar 20:13–17.

Davies DG, Chakrabarty AM, Geesey GG (1993): Exopolysaccharide production in biofilms: Substratum activation of alginate gene expression by *Pseudomonas aeruginosa.* Appl Environ Microbiol 59:1181–1186.

Davies DG, Geesey GG (1995): Regulation of the alginate biosynthesis gene *algC* in *Pseudomonas aeruginosa* during biofilm development in continuous culture. Appl Environ Microbiol 61:860–867.

DeFlaun MF, Marshall BM, Kulle E-P, Levy SB (1994): Tn*5* insertion mutants of *Pseudomonas fluorescens* defective in adhesion to soil and seeds. Appl Environ Microbiol 60:2637–2642.

DeFlaun MF, Tanzer AS, McAteer AL, Marshall B, Levy SB (1990): Development of an adhesion assay and characterization of an adhesion-deficient mutant of *Pseudomonas fluorescens.* Appl Environ Microbiol 56:112–119.

DeLong EF, Franks DG, Alldredge AL (1993): Phylogenetic diversity of aggregate-attached vs. free-living marine bacterial assemblages. Limnol Oceanogr 38:924–934.

de Maagd RA, Rao AS, Mulders IHM, Roo LG, van Loosdrecht MCM, Wijffelman CA, Lugtenberg BJJ (1989): Isolation and characterization of mutants of *Rhizobium leguminosarum* bv. *viciae* 248 with altered lipopolysaccharides: Possible role of surface charge or hydrophobicity in bacterial release from the infection thread. J Bacteriol 171:1143–1150.

Dexter SC (1979): Influence of substratum critical surface tension on bacterial adhesion—*in situ* studies. J Colloid Interf Sci 70:346–354.

Dexter SC, Sullivan JD Jr., Williams III J, Watson SW (1975): Influence of substrate wettability on the attachment of marine bacteria to various surfaces. Appl Microbiol 30:298–308.

DiMilla PA, Folkers JP, Biebuyck HA, Haerter R, Lopez GP, Whitesides GM (1994): Wetting and protein adsorption of self-assembled monolayers of alkanethiolates supported on transparent films of gold. J Am Chem Soc 116:2225–2226.

Doyle RJ, Nesbitt WE, Taylor KG (1982): On the mechanism of adherence to *Streptococcus sanguis* to hydroxylapatite. FEMS Microbiol Lett 15:1–5.

Feldner J, Bredt W, Kahane I (1983): Influence of cell shape and surface charge on attachment of *Mycoplasma pneumoniae* to glass surfaces. J Bacteriol 153:1–5.

Fett WF, Osman SF, Dunn MF (1989): Characterization of exopolysaccharides produced by plant-associated fluorescent pseudomonads. Appl Environ Microbiol 55:579–583.

Fett WF, Osman SF, Fishman ML, Siebles III TS (1986): Alginate produced by plant-pathogenic pseudomonads. Appl Environ Microbiol 52:466–473.

Fletcher M, Loeb GI (1979): Influence of substratum characteristics on the attachment of a marine pseudomonad to solid surfaces. Appl Environ Microbiol 37:67–72.

Fletcher M, Marshall KC (1982): Bubble contact angle method for evaluating substratum interfacial characteristics and its relevance to bacterial attachment. Appl Environ Microbiol 44:184–192.

Fletcher RL (1975): Heteroantagonism observed in mixed algal cultures. Nature 253:534–535.

Gannon JT, Manilal VB, Alexander M (1991): Relationship between cell surface properties and transport of bacteria through soil. Appl Environ Microbiol 57:190–193.

Gerhardt DJ, Rittschof D, Mayo SW (1988): Chemical ecology and the search for antifoulants. J Chem Ecol 14:1903–1915.

Gerson DF, Scheer D (1980): Cell surface energy, contact angles and phase partition III. Adhesion of bacterial cells to hydrophobic surfaces. Biochim Biophys Acta 602:506–510.

Goldberg JB, Gorman WL, Flynn JL, Ohman DE (1993): A mutation in *algN* permits *trans* activation of alginate production by *algT* in *Pseudomonas* species. J Bacteriol 175:1303–1308.

Hogt AH, Dankert J, de Vries JA, Feijen J (1983): Adhesion of coagulase-negative staphylococci to biomaterials. J Gen Microbiol 129:2959–2968.

James AM (1991): Charge properties of microbial cell surfaces: In Mozes N, Handley PS, Busscher HJ, Rouxhet PG (eds): Microbial Cell Surface Analysis. New York: VCH, pp 221–262.

Johnson CR, Mann K (1986): The crustose coralline alga, *Phymatolithon foslie,* inhibits the overgrowth of seaweeds without relying on herbivores. J Exp Mar Biol Ecol 96:127–146.

Kastowsky M, Gutberlet T, Bradaczek H (1992): Molecular modeling of the three-dimensional structure and conformational flexibility of bacterial lipopolysaccharide. J Bacteriol 174:4798–4806.

Kawabata N, Hayashi T, Matsumoto T (1983): Removal of bacteria from water by adhesion to cross-linked poly(vinyl-pyridinium) halide. Appl Environ Microbiol 46:203–210.

Keifer P, Rinehart KJ (1986): Renillafoulins, antifouling deterpenes from the sea pansy *Renilla reniformis* (Octocorallia). J Org Chem 51:4450–4454.

Korber DR, Lawrence JR, Sutton B, Caldwell DE (1989): The effect of laminar flow on the kinetics of surface recolonization by mot+ and mot− *Pseudomonas fluorescens.* Microb Ecol 18:1–19.

Lawrence JR, Caldwell DE (1987): Behavior of bacterial stream populations within the hydrodynamic boundary layers of surface microenvironments. Microb Ecol 24:15–27.

Lawrence JR, Delaquis PJ, Korber DR, Caldwell DE (1987): Behavior of *Pseudomonas fluorescens* within the hydrodynamic boundary layers of surface microenvironments. Microb Ecol 14:1–14.

Lawrence JR, Korber DR, Caldwell DE (1992): Behavioral analysis of *Vibrio parahaemolyticus* variants in high- and low-viscosity microenvironments by use of digital image processing. J Bacteriol 174:5732–5739.

Lopez GP, Biebuyck HA, Haerter R, Kumar A, Whitesides GM (1993): Fabrication and imaging of two-dimensional patterns of proteins adsorbed on self-assembled monolayers by scanning electron microscopy. J Am Chem Soc 115:10774–10781.

Ludwicka A, Jansen B, Wadström T, Pulverer G (1984): Attachment of staphylococci to various synthetic polymers. Zbl Bakt Hyg A 256:479–489.

MacRae IC, Evans SK (1983): Factors influencing the adsorption of bacteria to magnetite in water and wastewater. Water Res 17:271–277.

Marshall KC, Stout R, Mitchell R (1971): Mechanisms of the initial events in the sorption of marine bacteria to surfaces. J Gen Microbiol 68:337–348.

Marshall PA, Loeb GI, Cowan MM, Fletcher M (1989): Response of microbial adhesives and biofilm matrix polymers to chemical treatments as determined by interference reflection microscopy and light section microscopy. Appl Environ Microbiol 55:2827–2831.

McEldowney S, Fletcher M (1986): Effect of growth conditions and surface characteristics of aquatic bacteria on their attachment to solid surfaces. J Gen Microbiol 132:513–523.

Molin G, Nilsson I, Stenson-Holst L (1982): Biofilm build-up of *Pseudomonas putida* in a chemostat at different dilution rates. Eur J Appl Microbiol Biotechnol 15:218–222.

Montgomery M, Kirchman DL (1993): Role of chitin-binding proteins in the specific attachment of the marine bacterium *Vibrio harveyi* to chitin. Appl Environ Microbiol 59:373–379.

Montgomery M, Kirchman DL (1994): Induction of chitin-binding proteins during the specific attachment of the marine bacterium *Vibrio harveyi* to chitin. Appl Environ Microbiol 60:4284–4288.

Neu TR (1996): Significance of bacterial surface-active compounds in interaction of bacteria with surfaces. Microbiol Rev 60:151–166.

Nickel JC, Ruseska I, Wright JB, Costerton JW (1985): Tobramycin resistance of *Pseudomonas aeruginosa* cells growing as a biofilm on urinary catheter material. Antimicrob Agents Chemother 27:619–624.

Paul JH, Jeffrey WH (1985): Evidence for separate adhesion mechanisms for hydrophilic and hydrophobic surfaces in *Vibrio proteolytica*. Appl Environ Microbiol 50:431–437.

Paul JH, Loeb GI (1983): Improved microfouling assay employing a DNA-specific fluorochrome and polystyrene as substratum. Appl Environ Microbiol 46:338–343.

Pedersen K, Holmstrom C, Olsson A, Pedersen A (1986): Statistic evaluation of the influence of species variation, culture conditions, surface wettability and fluid shear on attachment and biofilm development of marine bacteria. Arch Mikrobiol 145:1–8.

Pedersen SS (1992): Lung infection with alginate-producing, mucoid *Pseudomonas aeruginosa* in cystic fibrosis. APMIS Suppl 28 100:1–79.

Prime KL, Whitesides GM (1993): Adsorption of proteins onto surfaces containing end-attached oligo(ethylene oxide): A model system using self-assembled monolayers. J Am Chem Soc 115:10714–10721.

Pringle JH, Fletcher M (1983): Influence of substratum wettability on attachment of freshwater bacteria to solid surfaces. Appl Environ Microbiol 45:811–817.

Pringle JH, Fletcher M (1986): The influence of substratum hydration and adsorbed macromolecules on bacterial attachment to surfaces. Appl Environ Microbiol 51:1321–1325.

Pringle JH, Fletcher M, Ellwood DC (1983): Selection of attachment mutants during the continuous culture of *Pseudomonas fluorescens* and relationship between attachment ability and surface composition. J Gen Microbiol 129:2557–2569.

Quintero EJ, Weiner RM (1995): Evidence for the adhesive function of the exopolysaccharide of *Hyphomonas* strain MHS-3 in its attachment to surfaces. Appl Environ Microbiol 61:1897–1903.

Ramphal R, Guay C, Pier GB (1987): *Pseudomonas aeruginosa* mucoid exopolysaccharide in adherence to tracheal cells. Infect Immun 55:600–603.

Read RR, Costerton JW (1987): Purification and characterization of adhesive exopolysaccharides from *Pseudomonas putida* and *Pseudomonas fluorescens*. Can J Microbiol 33:1080–1090.

Rijnaarts HHM, Norde W, Bouwer EJ, Lyklema J, Zehnder AJB (1993): Bacterial adhesion under static and dynamic conditions. Appl Environ Microbiol 59:3255–3265.

Robb ID (1984): Stereo-biochemistry and function of polymers. In Marshall KC (ed): Microbial Adhesion and Aggregation. Berlin: Springer-Verlag, pp 39–49.

Rosenberg M, Gutnick D, Rosenberg E (1980): Adherence of bacteria to hydrocarbons: A simple method for measuring cell-surface hydrophobicity. FEMS Microbiol Lett 9:29–33.

Rosenberg M, Kjelleberg S (1986): Hydrophobic interactions: Role in bacterial adhesion. In Marshall KC (ed): Advances in Aquatic Microbiology. Vol 9. New York: Plenum Press, pp 353–393.

Rosowski JR (1992): Specificity of bacterial attachment sites on the filamentous diatom *Navicula confervacea* (Bacillariophyceae). Can J Microbiol 38:676–686.

Rutter PR, Vincent B (1980): Physicochemical interactions of the substratum, microorganisms, and the fluid phase. In Marshall KC (ed): Microbial Adhesion and Aggregation. Berlin: Springer-Verlag, pp 21–38.

Sengha SS, Anderson AJ, Hacking AJ, Dawes EA (1989): The production of alginate by *Pseudomonas mendocina* in batch and continuous culture. J Gen Microbiol 135:795–804.

Shands JW, Graham JA, Nath K (1967): The morphologic structure of isolated bacterial lipopolysaccharide. J Mol Biol 25:15–21.

Shilo M (1989): The unique characteristics of benthic cyanobacteria. In Cohen Y, Rosenberg E (eds): Microbial Mats: Physiological Ecology of Benthic Microbial Communities. Washington, DC: American Society for Microbiology, pp 207–213.

Sjollema J, van der Mei HC, Uyen HMW, Busscher HJ (1990): The influence of collector and bacterial cell surface properties on the deposition of oral streptococci in a parallel plate flow cell. J Adhesion Sci Tech 4:765–777.

Sledjeski DD, Weiner RM (1991): *Hyphomonas* spp., *Shewanella* spp., and other marine bacteria lack heterogeneous (ladderlike) lipopolysaccharide. Appl Environ Microbiol 57:2094–2096.

Stenström TA (1989): Bacterial hydrophobicity, an overall parameter for the measurement of adhesion potential to soil particles. Appl Environ Microbiol 55:142–147.

Sutherland IW (1983): Microbial exopolysaccharides—Their role in microbial adhesion in aqueous systems. CRC Crit Rev Microbiol 10:173–201.

Todd JS, Zimmerman RC, Crews P, Randall SA (1993): The antifouling activity of natural and synthetic phenolic acid sulfate esters. Phytochemistry 34:401–404.

van der Mei HC, Rosenberg M, Busscher HJ (1991): Assessment of microbial cell surface hydrophobicity. In Mozes N, Handley PS, Busscher HJ, Rouxhet PG (eds): Microbial Cell Surface Analysis. New York: VCH, pp 263–288.

Vandevivere P, Kirchman DL (1993): Attachment stimulates exopolysaccharide synthesis by a bacterium. Appl Environ Microbiol 59:3280–3286.

van Loosdrecht MCM, Lyklema J, Norde W, Schraa G, Zehnder AJB (1987): Electrophoretic mobility and hydrophobicity as a measure to predict the initial steps of bacterial adhesion. Appl Environ Microbiol 53:1898–1901.

van Loosdrecht MCM, Lyklema J, Norde W, Zehnder AJB (1989): Bacterial adhesion: A physicochemical approach. Microb Ecol 17:1–16.

van Pelt AWJ, Weerkamp AH, Uyen MHWJC, Busscher HJ, de Jong HP, Arends J (1985): Adhesion of *Streptococcus sanguis* CH3 to polymers with different surface free energies. Appl Environ Microbiol 49:1270–1275.

Wardell JN, Brown CM, Ellwood DC (1980): A continuous culture study of the attachment of bacteria to surfaces. In Berkeley RCW, Lynch JM, Melling J, Rutter PR, Vincent B (eds): Microbial Adhesion to Surfaces. Chichester, England: Ellis Horwood, pp 221–230.

Weerkamp AH, Uyen HM, Busscher HJ (1988): Effect of zeta potential and surface energy on bacterial adhesion to uncoated and saliva-coated human enamel and dentin. J Dent Res 67:1483–1487.

Wiencek KM (1995): Bacterial Adhesion to Alkanethiol Self-Assembled Monolayers. Ph.D. Thesis, University of Maryland.

Wiencek KM, Fletcher M (1995): Bacterial adhesion to hydroxyl- and methyl-terminated alkanethiol self-assembled monolayers. J Bacteriol 177:1959–1966.

Williams V, Fletcher M (1996): *Pseudomonas fluorescens* adhesion and transport through porous media are affected by lipopolysaccharide composition. Appl Environ Microbiol 62:100–104.

Wrangstadh M, Szewzyk U, Östling J, Kjelleberg S (1990): Starvation-specific formation of a peripheral exopolysaccharide by a marine *Pseudomonas* sp., strain S9. Appl Environ Microbiol 56:2065–2072.

2

BACTERIAL INTERACTIONS WITH SURFACES IN SOILS

AARON L. MILLS
DAVID K. POWELSON

Laboratory of Microbial Ecology, Department of Environmental Sciences, University of Virginia, Charlottesville, Virginia 22903

2.1. INTRODUCTION

Of all the habitats that contain microbes, soil is the most complex. Despite the intensity with which the microbiology of the soil has been studied, it remains the least understood of the microbial habitats. Indeed, it is that complexity that makes the soil–microbe interaction the most wondrous of all. The soil system is composed not only of microbes living in an organic- and inorganic-containing broth, but microbes living in a surface-rich environment in which the surfaces are usually coated with only a thin film of water. It is in this environment, therefore, that the study of microbes attached to or interacting with surfaces is most critical to study—for it is here, more than anywhere, that interaction with surfaces is the life style of the microbial community.

Although wet soils exist in which all or most of the pores are always or nearly always filled with water, unsaturated soils are the dominant form at the earth's surface. All microorganisms are aquatic in that they metabolize and reproduce while immersed in water, even microbes that inhabit unsaturated soils. As is discussed below, the large volumes of air space (and, concomitantly, the small amount of liquid water) in most unsaturated soils dictate that microbes must live on or very near to surfaces. The surfaces of the soil particles are the most obvious habitats, and much discussion of the importance of

Bacterial Adhesion: Molecular and Ecological Diversity, pages 25–57
© *1996 Wiley-Liss, Inc.*

particle–surface attachment has appeared in the literature. A different set of surfaces is also available to microorganisms in unsaturated soils, however, specifically, the gas–water interface (GWI).

Strong evidence has been presented to support the assertion that the GWI is an important surface for microbial attachment. The importance of that interface will vary with the degree of saturation and therefore the amount of GWI surface available. It is likely that the mechanisms that drive attachment to the GWI differ somewhat from those that promote microbial attachment to the solid–water interface (SWI); attachment to the GWI is likely dominated by hydrophobic effects, while sorption to SWI is a combination of hydrophobic and charge effects. Furthermore, the cells attached to the GWI will tend to move with the interface as the soil becomes more or less saturated, whereas attachment of bacteria to particles tends to act as a retarding factor in the transport of microbes in either saturated or unsaturated conditions. This chapter considers both of these surfaces as important habitats for microorganisms, and it attempts whenever possible to point out differences associated with life at each of the interfaces.

2.2. SOIL PARTICLES

Typical medium-textured soils have porosities that range from about 0.25 to 0.60. This means that a volume of soil has no more than 25% to 60% of the total volume available for microbial habitation. In many cases, however, the pore space is formed by the juxtaposition of extremely small particles (clays, 2.0 μm and smaller) in arrangements that result in pores smaller than the size of the bacterial cells. In these cases, bacteria are excluded from some, perhaps sizable, fraction of the pore space. Available pore space, then, may be substantially less than the total pore space that is measured. The remainder of the pore space may be available for diffusion loading of chemicals, bacterial carbon and energy sources, or inorganic nutrients. These "ultramicropores" are also the repositories of liquid water that can be held at high tension due to matric potentials (capillary plus adhesive particle surface–water interactions).

The degree of soil saturation, i.e., the relative proportion of the pore space that is filled with water, is widely variable in both time and space. As the degree of saturation decreases, the amount of air increases and the amount of GWI increases, with the air entering the larger pores first (Fig. 2.1). Given the unsaturated nature of the soil environment, microorganisms must live in close physical proximity to the particle surfaces. Although it is possible that some bacteria exist as free-living cells in the thicker water films scattered throughout the soil, two lines of evidence support the assertion that the majority of microbes in wet or saturated environments such as aquifers live attached to the soil particles. First, direct observation of particle surfaces with electron and confocal laser microscopy shows many of the bacteria in direct contact with the particles, often embedded in an organic matrix produced by the

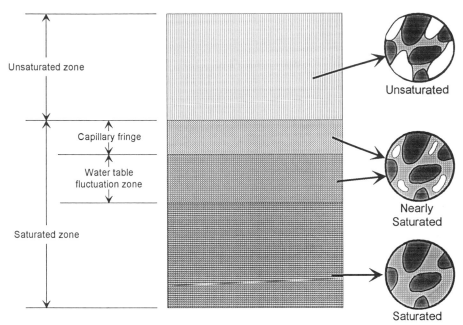

Fig. 2.1. Depiction of the subsurface environment showing the relative degree of saturation of the porous matrix. The soil zone is characteristically unsaturated most of the time, although soils that overlie a shallow water table may be saturated or nearly saturated much of the time. (Redrawn from Hrudey and Pollard, 1993.)

microbes themselves. The second line of evidence is that removal of bacteria from soil samples is never simple. The microbes stick tenaciously to the soil particles and are removed only by extraction with a strong dispersing solution such as sodium pyrophosphate, which displaces the cells from the surface. Given the relatively thin films of water associated with unsaturated soils, it may be that all of the microbes are associated with either the particle surface or the GWI. The easily removed organisms may represent those that exist in association with the GWI and are therefore flushed from the system during extraction. This concept has not been tested experimentally.

2.2.1. Soil Minerals

Soil particles comprise a variety of different minerals, the identities of which depend on the parent material (the original rocks or material transported to the locale) and the weathering processes that alter the minerals in the rocks to generate the soil particles. Different minerals are often associated with particles of different sizes, thus reflecting the degree of weathering of the particles. Furthermore, the different minerals have different charge generation characteristics, i.e., many soils have high affinities for ions, whereas others have only weak attraction for ions. All of these minerals are important for two reasons: First, the ions weathered from the minerals can interact with mineral surfaces to enhance or inhibit the attachment of microbes, and, second, the unweathered portion of the mineral represents the surface that the microbes must interact with in the soil. Some of the most important soil minerals are described in Table 2.1.

2.2.1.1. Primary Silicates

Included in the primary silicates are those minerals that make up the majority of the rocks from which soils are formed. These minerals differ in their weatherability such that the dominant mineralogy (and thus the surface properties of the resultant particles) can change substantially over time. The feldspar group is the most weatherable of the silicates, and young soils are characterized by having substantial amounts of feldspars, along with micas and amphibole. The total mass of quartz tends to change only slowly, so that older soils are dominated by quartz grains and the clay minerals described below. Eventually, even the clays weather away, leaving the quartz and oxides as the dominant mineral matter comprising the soil.

In most soils, the silicate minerals are the primary sources of the dissolved ions in the soil solution. Calcium comes largely from limestone or gypsum

TABLE 2.1. Common Soil Minerals and Representative Chemical Formulas

Mineral group	Formula	Comments
Silicates		
Quartz	SiO_2	Abundant in sand and silt
Feldspar	$(Na,K)AlO_2(SiO_2)_3$	Abundant in soil that is not leached extensively
Amphibole	$(Ca,Na,K)_{2.3}(Mg,Fe,Al)_5(OH)_2[(Si,Al)_4O_{11}]_2$	
Mica	$K_2Al_2O_5(Si_2O_5)_3Al_4(OH)_4$	
Clay minerals		
Kaolinite	$Si_4Al_4O_{10}(OH)_8$	
Smectite ⎫ Vermiculite ⎬ Chlorite ⎭	$M_x(Si,Al)_8(Al,Fe,Mg)_4O_{20}(OH)_4$, where M represents an interlayer cation	
Oxides and hydroxides		
Gibbsite	$Al(OH)_3$	Abundant in leached soils
Goethite	$FeO(OH)$	Most abundant Fe oxide
Hematite	Fe_2O_3	Abundant in warm regions
Ferrihydrite	$Fe_{10}O_{15}\cdot 9H_2O$	Abundant in organic horizons
Birnessite	$(Na,Ca)Mn_7O_{14}\cdot 2.8H_2O$	
Carbonates and sulfates		
Calcite	$CaCO_3$	
Gypsum	$CaSO_4\cdot 2H_2O$	

Reproduced from Sposito (1989), with permission of the publisher.

when present or from feldspars, amphibole, and other minerals (e.g., pyroxene) when limestone is absent. Potassium, sodium, and magnesium come largely from feldspars, micas, amphibole, and pyroxene, with less abundant ions (e.g., iron, zirconium, boron, and titanium) derived from resistant minerals such as epidote, tourmaline, zircon, or rutile.

2.2.1.2. Clay Minerals

The clay minerals (so-called because they tend to dominate the mineralogy of the clay-sized fraction) are weathering products derived from the silicates described above. Clay minerals are characteristically high in charge density, and, overall, the charge associated with these minerals tends to be net negative. Indeed, these minerals are the source of the higher affinity for cations than anions found in most soils. Additionally, the large surface area represented by the clays further amplifies the importance of the clay minerals and their reactivity.

The clay minerals are often referred to as *phyllosilicates,* or layer silicates, because of the laminar structure found in the crystal lattice. These crystals comprise layers of aluminum oxide and silicon oxide arranged in various proportions. Specifically, the most common arrangements are referred to as 1:1 (one aluminum layer + one silicon layer), 2:1 (one aluminum layer sandwiched between two silicon layers), and a group known by a variety of terms (e.g., 2:1:1), but the minerals in this group are simply 2:1 crystals with an additional aluminum-based hydroxide interlayer. A basic soil science textbook should be consulted for details of the structures of the various mineral groups (e.g., Brady, 1984; Foth, 1978; Sposito, 1989).

Clays may have a charge due to isomorphous substitution—commonly Fe^{2+} or Mg^{2+} for Al^{3+} in the octahedral aluminum layer, but some Al^{3+} for Si^{4+} in the tetrahedral silica layer can occur as well. Isomorphous substitution does not involve the dissociation of protons and is therefore not pH dependent. Hence charge arising from such substitution is often referred to as *permanent charge.*

Kaolinite is the dominant mineral in the 1:1 group. This group represents the simplest and most highly weathered of the clay minerals. The repeating silica–alumina sheets are held tightly together by hydrogen bonding, resulting in a situation where the repeating units are at a fixed distance that is too small for bacteria, water, or even ions to intercede. The fixed interlayer spacing and lack of isomorphous substitution means that charge effects are limited to the broken edges of the crystals, where silanol groups (Si–OH) exchange protons (hence the term *pH-dependent charge*) and other cations with the surrounding solution. Kaolinite is low in charge density (kaolinite has a low cation exchange capacity [CEC] of about 10 mEq 100 g^{-1}) and total surface area per unit weight because of the lack of interlayer surfaces.

The 2:1 group is characterized by a less rigid interlayer space. The smectites (typically montmorillonite in soil) and vermiculites (along with other similar minerals) are differentiated on the basis of the substitution of interlayer cations and the degree and identity of substitution of cations in the crystal itself. Of most importance to a discussion of bacterial attachment is the fact that the charge density of 2:1 clay minerals substantially exceeds that of kaolinate: Montmorillonite has a CEC of about 100 mEq 100 g^{-1}. Furthermore, the intercession of ions and water molecules permits the variable interlayer spacing.

The interlayer spaces are an important component of charge generation for ion exchange, but they may also be a potential reservoir of small organic molecules (interlayer spacing for montmorillonite can vary from 1.0 to over 2.0 nm). Given a carbon–carbon bond length of 0.154 nm in propane (lengths of single bonds decrease slightly as a few carbon atoms are added to these smallest organic molecules), a bond length of 0.139 nm in benzene coupled with the covalent radius of 0.067 nm for carbon, and a carbon–hydrogen bond length (in methane) of 0.109 nm, a molecule such as benzene or toluene should not exceed about 1.325 nm. This, of course, is the greatest dimension; molecules such as these are planar (probably not more than 0.1–0.2 nm thick) and should fit into the interlayer spaces as easily as, say, a hydrated calcium ion (ionic radius of 0.099 nm). Indeed, the 2:1 clays are known to sorb large organic cations strongly (e.g., hexadecyltrimethylammonium) in the interlayer spaces (Jaynes and Boyd, 1991), resulting in expansion of smectites to interlayer spacings that exceed 2.2 nm. Inclusion of surfactants like the quaternary ammonium ions or, by analogy, humic or fulvic acids may represent an important modification of mineral surface properties related to bacterial attachment, as discussed below.

2.2.1.3. Oxides and Hydroxides
In many weathered soils, oxides and hydroxides of aluminum, iron, and manganese are important components. In highly weathered soils (oxisols, often called *lateritic soils*), the reactive silicate minerals have disappeared, leaving most of the reactivity due to minerals such as gibbsite, goethite, hematite, ferrihydrite, and birnisite. In many temperate climate soils, the abundance and location of these sesquioxide minerals in the soil profile mark the degree of weathering, with higher amounts of sesquioxide indicative of more extensive weathering. The process of soil formation (podsolization) is a downward translocation of sesquioxides, often accompanied by deposition as a grain or grain coating in a diagnostic subsurface horizon (the spodic horizon). Additionally, these oxidized minerals play a role in soils that may be of direct importance to bacterial attachment. The oxides and hydroxides often accept protons at soil pH values, thereby rendering a positive charge to the surface of the particle. Even if the overall charge is still a net negative, coating of aluminum, iron, or manganese oxide can create centers of positive charge on the surface of the particle. If the reactive particles are grains of the metal oxides, the entire particle may assume a net positive charge. Given the net negative charge usually associated with bacterial cells, charge-related cell sorption should be greatly enhanced in the presence of such oxides.

2.2.1.4. *Carbonates and Sulfates*
The carbonate and sulfate minerals are easily weathered compared to even
the most weatherable of the silicates or oxides. As a result, they are not
generally found in the smaller sand- to clay-sized fractions in humid region
soils, although arid soils may contain large amounts of these minerals due to
the lack of leaching. Calcium and the polyvalent cations help keep clay particles
aggregated in an open structure, thereby facilitating the flow of water and air
and increasing the availability of sites for microbial habitation. Sodium, on
the other hand, tends to disperse clay, resulting in clogging of soil pores. A
primary importance of the carbonates is to supply a substantial amount of
base cation (specifically calcium and some magnesium) to the soil. This has
the obvious effect of delaying acidification of the soil by providing buffer
capacity to the soil water system through cation exchange reactions with
protons from the solution.

2.2.2. Particle Size Distribution

Soil mineral matter is often defined as that fraction passing through a 2 mm
sieve. While there are larger pebbles, cobbles, and boulders that form part
of the macroscopic soil matrix, it is the smaller sizes, the sands, silts, and clays,
that are relevant to microbial habitation. Indeed, when soil texture is described,
it is the fraction passing the 2 mm sieve that is considered.

The texture of a soil is described by the size distribution of the various
mineral grains. Because the mineralogy of the particles often changes from,
for example, quartz and feldspar for the sand- and silt-sized fractions to the
layer silicates as the particle size decreases to become clay, the smaller particles
present not only a larger total surface area, but a more reactive one as well.
The influence of the surfaces of the finer textured materials is, therefore, a
combination of the surface area increase and the specific mineralogy of the
particles. Table 2.2 shows how the surface area increases as the diameter of
soil particles decreases, and Table 2.3 shows how cation exchange capacity
increases with increasing fineness of texture. The range of values for CEC for
each textural category reflects different mineralogy and different amounts of
organic matter present in individual soils. It is important to note that most
soil particles do not present surfaces whose reactivity reflects only the base
mineralogy of the particle. Many, if not all, particles have some portion of
their surface coated with reactive materials, such as iron, aluminum, and
manganese oxides and hydroxides, and organic matter. These coatings can
alter the reactive surfaces of the particles, in some cases changing net negative
surface charges to neutral or positive charges, and they can add reactivity to
otherwise only slightly reactive surfaces. In this way, even quartz sand can
become highly reactive by adsorption of a coat of reactive metal oxide or
organic matter. Obviously the bulk effect of the coatings will still depend on
the surface area of exposed reactive surface.

TABLE 2.2. Soil Particle Size and Resultant Surface Area[a]

Separate	Diameter (mm)[b]	Volume (mm³)	No. of particles per gram	Surface area in 1 g (cm²)
Very coarse sand	2.00–1.00	4.18	90	11
Coarse sand	1.00–0.50	0.524	720	23
Medium sand	0.50–0.25	0.0655	5,700	45
Fine sand	0.25–0.10	0.00818	46,000	91
Very fine sand	0.10–0.05	0.000524	722,000	227
Silt	0.05–0.002	0.0000650	5,776,000	454
Clay	<0.002	0.0000000042	90,260,853,000	8,000,000[c]

[a] Data from Foth (1978).
[b] United States Department of Agriculture classification system.
[c] Surface area of platy montmorillonite determined; all other calculated for spheres of largest size permissible by class.

2.2.3. Organic Particles

As plant material decomposes, the solid mass is broken down into bits of ever-decreasing size. Organic soil horizons (those with >20% organic matter) are classified based on the degree of decomposition and disintegration of the organics present. A great many changes occur in the chemistry of the organics during the course of the decomposition. The initial material is fresh or senescent plant material, but the action of microorganisms soon generates organic matter that bears only slight resemblance to the plants from which it came. As readily metabolizable material is removed, the remnants take on the character both of the refractory components of the plants and of the microbial cells that are generated as a result of the decay process. The remnants have dissolved components as well as solid phase, and the dissolved constituents often become coatings on all of the particulates that make up the soil matrix. The process

TABLE 2.3. Change in Cation Exchange Capacity With Change in Soil Texture[a]

Textural classification	No. of soils	Exchange capacity (mEq 100 g⁻¹) Average	Range
Sand	2	2.8 ± 1.1	2.0– 3.5
Sandy loam	6	6.8 ± 5.8	2.3–17.1
Loam	4	12.2 ± 3.6	7.5–15.9
Silt loam	8	17.8 ± 5.6	9.4–26.3
Clay and clay loam	6	25.3 ± 20.3	4.0–57.5

[a] Data are taken from Brady (1984). Data for averages are expressed as the mean and standard deviation for the soils, and the range represents the minimum and maximum values reported within the textural category.

of conversion of plant material to soil organic matter is often called *humifica-tion* and the organic matter *humus*. This nomenclature reflects the importance given to the class of organics called *humic acids*.

Soil organic matter, whether particulate or dissolved, has properties that include both hydrophobic and electrostatic effects. Fulvic acids are moderately sized, reactive molecules of average molecular weights of 800 to 1,500 Da, whereas humic acids are large molecules (average molecular weights of 1,500 to 4,000 Da [Beckett et al., 1987, 1989], with reported weights up to 200,000 [Thurman, 1985]), also with many reactive sites. These sites are generally dominated by carboxylated phenolic hydroxyl groups that dissociate in water to yield a polyvalent anion, but soil organic matter also contains amines that can carry a positive charge at moderate to low pH. The structures of humic substances can certainly vary, and several structures have been offered; proba-bly all are appropriate for the material that was being studied. An example is given in Figure 2.2. The presence of many carboxyl, carbonyl, and phenol groups makes the electronegative sites on this molecule evident; the peptides and amines are often sites of positive charges. Particles dominated by these types of compounds often exceed even the most reactive clays in their ability to exchange ions. Whereas typical cation exchange capacities for 2:1 clays such as montmorillonite are often about 100 mEq 100 g^{-1}, organic matter frequently has two to three times that capacity.

In addition to electrostatic properties, the large molecules of soil organic matter also tend to be hydrophobic. Molecules that are highly insoluble (e.g., many substituted aromatics such as PCB) rapidly partition into the organic particles and organic mineral grain coatings. Despite the high surface charge associated with humic materials, hydrophobic compounds and, by analogy, hydrophobic bacterial cells will tend to be retained on surfaces that are organic

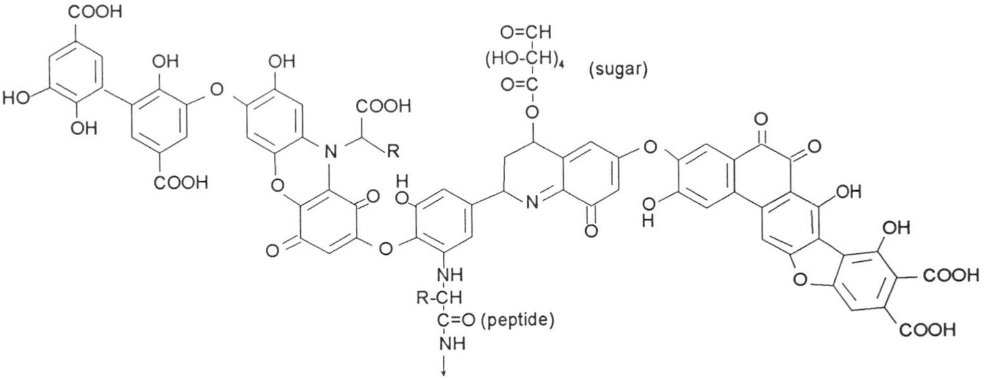

Fig. 2.2. Structure of a sample humic molecule. Note the high proportion of aromatic groups and the opportunities for both positive and negative charge sites on the molecule (Redrawn from Stevenson, 1982).

in nature. Those surfaces may be either organic particles or particles coated with organic matter.

2.2.4. Particle Coatings

Most temperate soils are dominated by silicate minerals such as quartz and the layer silicates. When surface interactions are considered, it is these minerals whose properties are most often considered. The silicates, although abundant, do not occur as pure compounds, nor are their surfaces free from the presence of other materials that may be present as coatings on the mineral grains. There are a number of possible coating types, but those most often considered are the oxides and hydroxides of Fe and Al (and sometimes Mn) and organic matter. Deposition of Fe and Al sesquioxides on the surface of silicate grains is a common process important in the natural weathering of soils. Podsolization of soils involves the downward translocation of Fe, Al, and organic matter. The Fe and Al are dissolved in the surface horizons (often with the assistance of complexation by soil organic matter) and are moved downward in the profile where they are deposited as coatings on sand-, silt-, and clay-sized particles. In some soils, this process so dominates the appearance of the profile that the soils contain a diagnostic horizon called the B_{hs}, or spodic horizon. Spodic horizons are denoted as B horizons, which are characteristically zones of accumulation. Furthermore, the subscript s denotes the accumulation of sesquioxides of Fe and Al. The subscript h refers to the accumulation of humus (organic material). Soils that contain a spodic horizon are termed *spodosols* in the U.S. nomenclature and have classically been called *podzols.*

Mineral grain coatings are not limited to spodic horizons or podzolic soils. Indeed, there is a substantial opinion that nearly all mineral grains have coatings that differ somewhat from the bulk mineralogy of the grains. Given that the surfaces of pure quartz and silicate minerals express a net negative charge, it is reasonable to assume that cations such as $Al(OH)^{2+}$ or $Fe(OH)_2^+$ (products of the hydrolysis reactions of Al^{3+} and Fe^{3+}) would be readily attracted to the silicate surface. Indeed, coatings of amorphous iron and aluminum sesquioxides are common on silicate grains. At typical soil pH values, these layers, as well as the more crystalline coatings such as gibbsite, $Al(OH)_3$, or goethite, $Fe(OH)_3$, can block negative charges from the silicates and in some cases can act as bases and accept protons to impart a positive charge to the mineral surface. Figure 2.3 illustrates how a sesquioxide coating might reduce the electronegativity of a silicate surface or even provide a positive charge site. Note that coatings are probably not continuous over the entire grain surface so that a soil may still retain its net negative charge even though there are abundant sites of less negative or even positive charge distributed over the grains.

Organic molecules (typified by the humics) also have a net negative charge due to the abundant carboxyl groups present. There are sites of positive charge, however, such as the amine groups, and these groups can often associate with

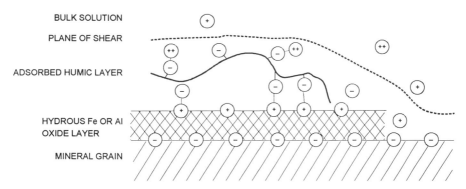

Fig. 2.3. Effect of mineral and organic coatings on the local charge distribution of a mineral grain. While soil particles usually have a net negative charge at most pH values, areas coated with metal sesquioxides or organics may have areas of positive charge at typical soil pH values. The figure indicates the type of surface exposure expected given a discontinuous coat of sesquioxide or humic material over the mineral grain.

the negative charges of the silicate grain. In addition, polyvalent cations can also act as bridges between the negative sites on the organic molecules and those on the silicate surface (this is also true for Fe and Al oxides and bacterial cells). The organic coatings tend to reduce the electronegativity of the silicate surface and often increase the hydrophobicity of the surface (the latter depends on the relative hydrophobicity of the adsorbing compound). Certainly those mineral grains that have a sesquioxide coating will attract the organics to the surface as a secondary coating. Figure 2.3 also shows a partial covering of the sesquioxide surface with a coating of a humic molecule. The nature of these coatings is not well understood, but laboratory investigations suggest that their role in bacterial adsorption to mineral grains is highly important.

The coating of mineral grains with either mineral or organic coatings can have a profound effect on the interaction of the surface with bacteria. Iron sesquioxide coatings tend to provide a surface that is less electronegative than quartz or silicate surfaces and at some pH values may actually confer a net positive charge. Scholl et al. (1990) demonstrated that deposition of iron sesquioxide on quartz and muscovite greatly increased the sorption of bacteria from suspension in batch systems. Similarly, the retention of bacteria in columns of clean quartz sand coated with iron sesquioxide was much greater than for the uncoated analog. Mills et al. (1994) demonstrated that bacterial sorption to clean sand followed a linear isotherm for the range of concentrations of bacterial cells used and that the sorption to iron-coated sand approached an upper limit of 10^8 cells g^{-1}.

At concentrations of up to 10^8, sorption of cells from the bulk solution was nearly complete. Mills et al. (1994) used this observation to generate a conceptual model of sorption for mixtures of iron-coated and clean quartz sand. In that model, bacteria are sorbed from the bulk solution by the iron-coated sand until the maximum of 10^8 cells g^{-1} (of iron-coated sand present) is reached. Because sorption by the iron-coated sand was observed to be irreversible, all bacteria added are removed from the suspension up to the limit. When the iron-coated sand is saturated, sorption is controlled by the linear isotherm observed for the clean quartz sand (note that the intercept for that isotherm is the maximum number of cells adsorbed to the quantity of iron-coated sand in the mixture). This model is significantly different from a standard saturation model (e.g., the Langmuir type) in that the new formulation is concentration independent up to the saturation level of the iron-coated sand. The theory that underlies this simple model could be used to describe any mixture of reversible and irreversible sorption processes.

Organic coatings may make particle surfaces more hydrophobic, thus encouraging the sorption of bacteria with relatively hydrophobic surfaces. Surface charges may also be modified. Richardson (1994) demonstrated that a commercial preparation of humic acid sorbed to iron-coated quartz sand but not to clean quartz sand. This observation was considered to be atypical when compared with the natural environment, but it was thought to reflect the attraction of the relatively electropositive iron coating for the negatively charged organic acid. Bacterial sorption to the iron-coated sand was decreased when a coating of the humic acid was applied. Richardson concluded that while humic acid might increase the attraction of the sand surface for bacterial cells, it provided less attraction to the surface than did the iron coating in the absence of the organic material. The amount of humic acid sorption was not great, even to the iron-coated sand, and competition of the humic acid with bacterial cells for the available sites on the iron-coated surface could not be excluded as a cause for the decrease in sorption of bacteria on application of the organic coating. Figure 2.3 illustrates a conceptual model of how humics might interact with mineral surfaces to generate negatively charged surfaces that is consistent with the observations of Richardson (1994). Indeed, there exists in the literature some evidence that supports the notion that coatings of mineral grains remove all effects due to the properties of the underlying minerals and replaces them with those of the coating material to the extent that different colloids revert to a similar negative surface charge after exposure to dilute solutions of humic substances (Beckett, 1990).

2.3. SOIL WATER

2.3.1. Soil Moisture Tension and Water Content

Most soils exist under unsaturated conditions. As the degree of saturation decreases, the tenacity with which the remaining water is held in the soil pores increases. That tenacity is expressed as soil moisture tension, which represents the amount of suction required to remove the water from the soil. Liquid water that flows easily in response to gravity, is said to be held at 0 tension. Soil moisture tension (SMT) is measured in units of bars or Pascals (1 bar = 760 mm Hg = 0.98692 atmospheres = 10^5 Pa). SMTs arise from adhesive forces of water to solid particles and cohesive forces of water molecules for one another (other forces, such as the attraction of water molecules by solute molecules, are usually, but not always, small with respect to the adhesive and cohesive forces). The amount of water in a soil at any SMT can be determined, and, given some assumptions of uniformity of particles and water films, an "average" film thickness can be computed. Effective film thicknesses have been reported for ranges of SMT in several soils (Figure 2.4). At field capacity (SMT ≈ 0.33 bar) a film thickness of 0.2 to 0.3 μm might be expected. Although this seems too thin to support many microbial cells that may have diameters exceeding 0.5 μm, it is important to remember the assumption of uniform coverage of the grains by the water film. In reality, some pores may be completely filled with water, whereas others may be nearly devoid of liquid

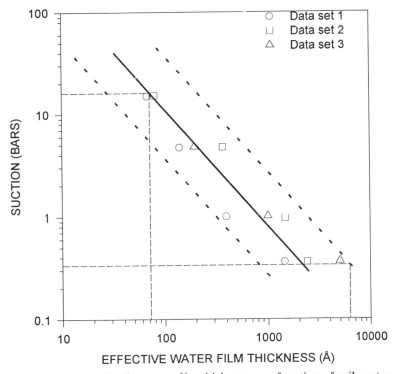

Fig. 2.4. Estimates of effective water film thickness as a function of soil–water pressure. The dashed lines indicate the effective water film thickness that might be expected in moist soil conditions, i.e., between field capacity (about 1/3 bar) and the wilting point (about 15 bar). (Redrawn from Kemper and Rollins [1966]; original data from Gardner [1956], Green et al. [1964], and Holmes et al. [1960].)

water. The principal point is that the thickness of the film declines rapidly as the soil moisture content moves farther from saturation.

Thinning of the water film has several results. First, the amount of GWI increases as the proportion of gas volume increases. For cells that preferentially adhere to the GWI, an increase in the preferred interfacial area would suggest that the relative importance of the GWI would increase as the degree of saturation decreases. A thinning film also forces the cells closer to both the SWI and the GWI, helping to overcome any mass transfer limitations of cells to the interfaces. Additionally, loss of water can have an effect on the composition of the soil solution.

2.3.2. Effect of Unsaturation on Composition of the Soil Solution

As the degree of saturation decreases, the concentration of dissolved and suspended material remaining in the aqueous phase necessarily increases. Some materials may become saturated with respect to some solid phase component and begin to precipitate from the solution. Such an occurrence could change the relative abundances of the various ionic species in the solution, making the final solution more flocculating or dispersing depending on the identity of the ions taken out of solution. In any event, the ionic strength of the solution increases over that found in the saturated state. For those cases in which increasing ionic strength results in increased bacterial attachment to surfaces (e.g., Fontes et al., 1991), attachment should increase. This relationship is presumably due to electrostatic effects brought about by the compression of the double layer as seen in Figure 2.5. At some point the double layer will be compressed to the point that repulsive forces are no longer important. At such a point other effects should dominate sorption to the particle surface, most notably hydrophobic effects.

2.4. BACTERIA IN THE SOIL

2.4.1. Distribution of Bacteria

2.4.1.1. Depth

The highest numbers of bacteria and the most bacterial activity in the soil should be expected where there is copious organic matter, plentiful N, P, S, and so forth, and a balance of saturation such that there is adequate water but that reaeration of the soil with O_2 is not impeded. The ideal conditions most often exist near the soil surface. Indeed, a fundamental concept of soil microbiology (although there are large numbers of exceptions) is that numbers of microorganisms decrease with increasing depth in the profile. This has been accepted as a generality for a long time. While recent studies using current techniques report numbers much larger than those obtained in the classic experiments (Table 2.4), they do not suggest that another paradigm should

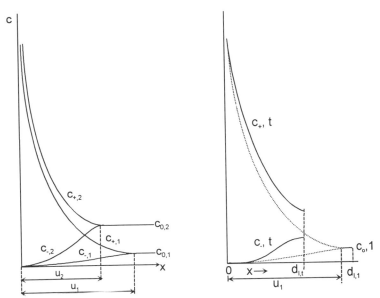

Fig. 2.5. The effect of ionic strength and thining of the water layer (x) on the concentration of ions in the diffuse double layer and the thickness of the double layer. In each panel, the y-axis represents the concentration of ions and the x-axis represents the relative distance from the charged surface. The left panel shows the concentration of counterions (in this case cations; $c_{+,1}$, $c_{+,2}$)and co-ions (in this case anions; $c_{-,1}$, $c_{-,2}$) near a negatively charged surface under saturated conditions or with a thick water film. Two equilibrium (bulk solution) concentrations are depicted ($c_{0,1}$ and $c_{0,2}$). Note that the higher equilibium ion concentration (i.e., higher ionic strength) results in a thinner double layer (u_1 vs. u_2). The thinner double layer would allow closer approach by a negatively charged particle, such as a bacterial cell.

In the right panel, the thickness of the water film has been decreased from distance $d_{1,1}$ to $d_{1,t}$. Because the water evaporates, leaving the salt behind, the concentration increases dramatically ($c_{+,t}$; $c_{-,t}$). The double layer becomes truncated, however, because the thinness of the water layer forces the ions closer to the surface than they would normally be at equilibrium (i.e., if the film were thicker than $d_{1,1}$). (Reproduced from Bolt and Bruggenwert, 1976, with permission of the publisher.)

be adopted. Given the distribution of other soil components, most notably organic matter, it could be inferred that the role of organic surfaces in the attachment of microbes to interfaces may decrease with increasing depth, such that direct interaction with mineral grains becomes more important. This assertion has not been tested directly, however.

2.4.1.2. Soil Solution Vs. Soil Particles

It is observed that whenever microorganisms have access to surfaces and interfaces, those habitats are rapidly colonized. A basic ecological principle dictates that when organisms inhabit a site, there is an advantage to the

TABLE 2.4. Distribution of Some Microorganisms in Various Horizons of the Soil Profile[a]

Depth (cm)	Organisms g^{-1} soil × 10^3		
	Aerobic bacteria	Anaerobic bacteria	Actinomycetes
3–8	7,800	1,950	2,080
20–25	1,800	379	245
35–40	472	98	49
65–75	10	1	5
135–145	1	0.4	—

[a] Data from Alexander (1977).

organism to do so. The high frequency of microbial colonization of surfaces and interfaces, therefore, suggests that there is an advantage to the microbes to associate with the interfaces. What that advantage is, however, is not completely understood. A common argument for the adhesion of microbes to surfaces is that to do so allows the organisms to utilize a higher concentration of nutrients, often energy sources, that are also found associated with the interface (see Chapter 3, this volume). A number of studies have shown that advantage. In a study using starved bacteria, Power and Marshall (1988) demonstrated that growth occurred on a surface where the only source of available energy was the stearic acid bound to that surface. Similarly, Kjelleberg et al. (1982) observed cell growth to occur on a membrane surface when the energy substrate was too dilute in the bulk medium to support growth. Bacteria have been shown to be able to scavenge fatty acids from glass surfaces (Kefford et al., 1982; Hermansson and Marshall, 1985). Marshall (1990) argued that such an advantage might not be of great importance in static water systems because the amount of organic matter associated with surfaces may be very limited there. He suggested that such an advantageous situation would be maximized in flowing systems where continued replenishment of energy source to the surface might provide an adequate supply for growth.

How this situation relates to soils is not clear. The thin films of water present near soil particles under unsaturated conditions suggests that the organic concentration associated with the grains should exceed that envisioned by Marshall (1990). Furthermore, soil-solution systems are rarely, if ever, in the "flowing" situation alluded to by Marshall. Clearly, more work is necessary to understand the situations under which surfaces and interfaces permit or enhance, or in some cases inhibit, microbial growth activity.

For some compounds, particularly many anthropogenic contaminants, association of the compound with surfaces can decrease the rate of biodegradation. Often, otherwise metabolizable compounds are rendered nondegradable when associated with particle surfaces. Mills and Eaton (1984) noted nearly complete inhibition of degradation of bromobenzene when sand was added to the reaction. Guerin and Boyd (1992) observed reduced degradation of naphthalene in some of the soils tested to determine the effect of particle sorption on bioavailability of the compound, but in no case was degradation enhanced when the soil particles were present. Other work examining different contaminant compounds showed reduced or completely inhibited biodegradative activity in the presence of sorptive particles (e.g., Gordon and Millero, 1985; Ogram et al., 1985).

Difference in degradability is certainly associated with availability of compound. For example, bovine serum albumin was degraded by attached bacteria but not by unattached bacteria, whereas suspended bacteria metabolized a readily desorbable dipeptide that was unavailable to the attached bacteria (Griffith and Fletcher, 1991). In the cases where availability to organisms is decreased by sorption to particles, there appears to be a competition between the microbes and surface for the compound. If the microbes are able to extract the compound from the surface, there may be no change or a possible enhancement of degradation. If the surface attraction is stronger, then the compound will be less available. It is also possible that sorption to the interlayer spaces of expanding lattice clays may further affect the situation. For example, Weber and Coble (1968) observed that diquat could be degraded when it was sorbed to the external exchange sites of kaolinite (a nonexpanding clay), but it was not available when bound in the interlayers of montmorillonite (an expanding clay mineral).

The extent of particle coverage varies from spot to spot, but an oversimplified example calculation suggests that, overall, the coverage is quite low. If it is assumed that the soil is pure silt composed of spherical grains 0.05 mm in diameter, and if it is further assumed that the soil bacteria are of a uniform shape and size (a 1 μm long by 0.5 μm wide rod with spherical ends), then 10^{10} bacteria per gram of soil would yield an average surface coverage of about 10% of the total surface area. While these assumptions are clearly hypothetical, they can be used to generate a figure that likely overestimates the actual coverage of soil particles. The particle size used is that of the largest silt size listed in Table 2.2. Presence of smaller particles would increase the surface area geometrically as the diameter decreases. The area of available surface is certainly much greater than the area actually utilized as bacterial habitat.

Adsorption in porous media has generally been assumed to occur exclusively on solid surfaces. Groundwater bacteria may also interact with air interfaces in the vadose zone above the water table or below the water table at the surface of gas bubbles (Fig. 2.6). Sorption to the GWI has been observed in aquatic systems. Given the large amount of gas, and therefore GWI, present in soil, this surface is likely of great importance to the soil microbes.

2.4.2. Sorption of Bacteria to the GWI

2.4.2.1 The GWI in Oceans and Lakes
Although microbial adsorption to GWIs has largely been ignored in soil science and hydrology, with a few exceptions (e.g., Powelson et al., 1990; Wan

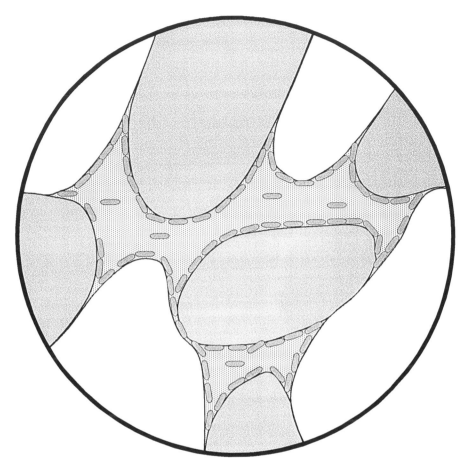

Fig. 2.6.   Attachment of bacteria to soil surfaces (the SWI) and to the GWI.

et al., 1994) this phenomenon has been carefully examined in marine and freshwater biology. Information collected from aquatic systems can inform studies of the GWI in soils. Beneath the air surface of oceans and lakes is a "microlayer," approximately 0.1 mm thick and consisting of partly hydrophobic organic molecules, that behaves like a polysaccharide–protein complex (Sieburth, 1976; Baier, 1975). Associated with the surface microlayer, organic material, particles, and microbes are enriched relative to bulk water (MacIntyre, 1974; Harvey, 1966; Hatcher and Parker, 1974; Norkrans, 1980; Plusquellec et al., 1991; Sodergren, 1993). Inhabitants of the surface film are important enough for marine biologists to coin the term *neuston* for these organisms. Given the tendency for microbial enrichment in quiescent surface waters, a similar enrichment would be expected at the GWIs in porous media.

Increased concentrations of particles and microorganisms have been observed at the surface of many different bodies of water. Guerin (1989) found that total heterotrophic bacteria and humic materials concentrated in an estuary microlayer. Particulate amino acids were found to adsorb to the sea surface microlayer (Hendricks and Williams, 1985). In lakes, organic N, P, and C in the microlayer are enriched by factors of 1.6 to 45 (concentration in the microlayer divided by concentration in the subsurface), and bacteria by factors of 6.4 to 10.7, compared with subsurface water (Sodergren, 1993). In the ocean near sewage outfalls, Plusquellec et al. (1991) found bacterial surface enrichment factors of 32 to 341. Many of the surface bacteria may be attached to particles, which in turn are concentrated in the surface microlayer. Harvey and Young (1980) found that the degrees by which bacteria were concentrated into the surface microlayer were linearly dependent on surface concentration of particulate material. Hardy and Apts (1984) found not only that the sea surface microlayer is enriched in total microalgae or "phytoneuston" but also that it contains distinct types of algae compared with bulk water.

2.4.2.2. *Microbial Adsorption to the GWI in Porous Media*
Microlayer enrichments of bacteria in soil may be even greater than in the ocean since the interfaces in soil would not have the bacterial stress factors of ultraviolet light, air-borne pollutants, and rapid temperature and salinity changes that are found in oceans (Lion and Leckie, 1981). Gas phases in porous media are frequently anchored to solid particles and consequently may act as an additional immobile adsorptive surface.

Recent work by Wan and Wilson (1994) and Wan et al. (1994) indicates that bacterial adsorption to the GWI may be an important and often overlooked factor affecting transport of bacteria in porous media. Wan and Wilson (1994) used glass micromodels of porous media and flowing water to observe qualitatively colloidal polystyrene beads, clay particles, and bacteria concentrating at air–water interfaces. Sorption appeared to increase with particle hydrophobicity, solution ionic strength, and positive electric charge of the particles. They suggested that initial adsorption was due to van der Waals and electrostatic interactions, followed by essentially irreversible adsorption due to capillary force. Wan and Wilson (1994) predict that "for a relatively hydrophobic strain of bacteria, even a small amount of residual gas can dramatically reduce . . . transport." Also, under microscopic examination, moving bubbles were observed to sweep glass and polymer surfaces clean of adsorbed bacteria (Pitt et al., 1993).

Wan et al. (1994) extended the work of Wan and Wilson (1994) by measuring breakthrough curves of two bacterial strains in sand columns with a range of air contents. They found increasing the air volume in the columns decreased the bacterial breakthrough concentrations for both the hydrophilic and relatively hydrophobic strains of bacteria.

Adsorption to gas surfaces in unsaturated porous media has usually been found to reduce transport of microbes. Powelson et al. (1990) found that MS2 bacteriophage was not removed during passage through 1 m of saturated soil, but was 95% removed by unsaturated soil. The authors suggested that the partially hydrophobic virus particles adsorbed to air–water interfaces in the unsaturated soil and were degraded by physical disruption of viral structure. Poletika et al. (1995) attempted to predict retardation of MS2 bacteriophage transport in unsaturated soil from equilibrium adsorption to batches of soil slurry. The batch adsorption indicated that the virus did not adsorb to the soil solids. Modeling of the transport breakthrough, however, yielded a retardation factor of 254, indicating strong adsorption. Poletika et al. (1995) suggested that the difference in slurry and unsaturated soil adsorption may be due to the interaction of viral particles with air–water interfacial forces in unsaturated soil. Tan et al. (1992), however, were able to predict transport of bacteria through unsaturated sand columns from batch adsorption to the solids without considering additional adsorption due to air interfaces. This may have been due to use of a weakly adsorbing bacterial strain (*Pseudomonas fluorescens* strain 2-79), which was readily desorbed from the sand simply by vortex mixing for 30 s with distilled water. Huysman and Verstraete (1993a,b) attributed greater adsorption of bacteria in drier soil conditions to slower water flow rates rather than to interaction with air interfaces.

Microbes may die at faster rates when exposed to GWIs. Adams (1948) found that shaking virus suspensions reduced virus concentrations and that addition of 1 mg/L of gelatin protected the viruses from inactivation, presumably by reducing virus adsorption to the air–water interface. Kibbey et al. (1978) found that the time for 95% reduction of *Streptococcus faecalis* in soil declined from 53 days under saturated conditions to 38 days at −0.3 bar matric potential and to 22 days −7.5 bar matric potential. Decreased moisture content was also found to reduce survival of *Escherichia coli* in soil (Boyd et al., 1969).

All of the concepts discussed for microbial interactions with soil apply to microbes in ground water. The operational separation between the two regions—groundwater will fill a well (matric pressure greater than atmospheric), whereas soil water may require suction extraction—does not reflect a separation in other physical, chemical, or biological characteristics. Generally, ground water is obtained from coarse formations where there is less clay and organic matter. Groundwater is also exposed to fewer GWIs, yet the GWI may affect bacterial transport below the water table at the capillary fringe and where bubbles form due to biogenic gas production of gas entrapment.

2.5. ATTACHMENT MECHANISMS

2.5.1. Mechanisms of Attachment to Solids

Attachment of bacteria to solids involves surface characteristics of the solid and the bacteria. Harvey (1991) reviewed two approaches to modeling this interaction: colloid stability theory and surface free energy. Most researchers who work with soil or subsurface environments use colloid stability theory, which will be covered in this section. As discussed in preceding sections, soil solids generally have a negative charge, although oxide coatings may produce regions of positive charge. Organic solids such as humic acids have pH-sensitive charge and may also have sufficient noncharged area to provide hydrophobic adsorption sites. This section will focus on characteristics of bacteria that may affect adsorption, the interaction of bacterial and solid surfaces, and models of adsorption and desorption.

2.5.2. The Bacterial Surface

There is considerable diversity in the cell surface chemistries of soil and subsurface bacteria (see also Chapter 1, this volume). Surfaces of bacteria are dominated by teichoic acid (gram-positive strains) or polysaccharides (gram-negative strains) (Brock and Madigan, 1991). Some bacteria also have protein-aceous appendages (fimbriae, pili, or flagella), which may affect surface reactions. In general these compounds are configured so that hydrophobic sections are directed inward toward the hydrocarbon-rich interior of the cell, and hydrophilic sections are facing the water environment. The degree of hydrophilicity is relative, however, which allows for the possibility of hydrophobic interactions.

The water interface of teichoic acid, polysaccharides, proteins, and other organic compounds is generally dominated by oxygen and nitrogen. These atoms are strongly electronegative and tend to attract polar water molecules. Water may contribute one or more protons to O or N, depending on the pH, resulting in a variably charged surface. The most common surface group is hydroxyl ($-OH$), which is found in all the bacterial surface compounds mentioned above. The amino acids of proteins and the alanine segment of teichoic acid also contains the amine group ($-NH-$), which may be involved in surface reactions. Figure 2.7 illustrates how these groups may gain or lose a proton, depending on the pH of the suspension, and thereby alter the net charge on the bacteria or solid.

2.5.2.1. Electrostatic and Electrodynamic Interactions

The pH at which electrical charges on a particle balance is the isoelectric pH (pH_{iep}). Harden and Harris (1953) found that bacterial pH_{iep} varied from 1.75 to 4.15 for 18 gram-positive species and from 2.07 to 3.65 for 13 gram-negative species. Typical soil solid pH_{iep} values are 2.0 for quartz and 4.6 for kaolinite (Stumm and Morgan, 1981). Both bacteria and common soil components have pH_{iep} values lower than the typical pH of soil solution, pH 5 to 8. Consequently, bacteria and solids generally will have net negative charges and electrostatically repel each other. It is important to keep in mind that some soil components may be positively charged in near-neutral pH conditions, e.g., amorphous $Fe(OH)_3$, which has a pH_{iep} of 8.5 (Stumm and Morgan, 1981). Scholl et al. (1990) found that attachment of negatively charged bacteria was much greater to positively charged surfaces of limestone, $Fe(OH)_3$-coated quartz, and $Fe(OH)_3$-coated muscovite than to uncoated quartz and muscovite. Mills and

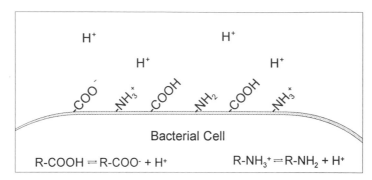

Fig. 2.7. Generation of charges on the surface of the bacterial cell. The degree of positive and negative charge on any cell will reflect the relative proportion of acids and bases exposed to the solution, the pK_a for any proton donating-accepting group, and the pH of the surrounding solution. Other groups may be involved with this type of reaction; carboxyl groups and amines are common groups and are shown for the sake of simplicity.

Maubrey (1981), however, found that quartz was colonized more rapidly than calcite during immersion in ponds.

The negative charge on bacteria and solids is neutralized by a swarm of cations that becomes progressively dense away from the surface, resulting in a diffuse double layer of charge. The approximate thickness of the diffuse layer is κ^{-1}. At 20°C,

$$\kappa \approx \frac{I^{0.5}}{0.28 \text{ nm}} \tag{1}$$

where I is the ionic strength expressed in units of molality (m; note that many authors use molarity, M, to quantify ionic strength; although not strictly correct, molarity is a good approximation for dilute solutions.) (Stumm and Morgan, 1981). For example, when $I = 0.001\ m$, $\kappa^{-1} = 8.9$ nm.

If a bacterial suspension with pH $>$ pH$_{\text{iep}}$ is placed in an electric field, the cells will be drawn toward the positive pole, and the cations farthest away from the surface of a cell will be sheared off. The resulting potential at the outside of the diffuse layer, determined from the bacterial velocity, is called the *zeta potential*. This potential is dependent on the ionic strength and pH of the suspension as well as density of charge on the bacteria. Gannon et al. (1991a) measured zeta potentials for 19 bacterial strains suspended in deionized water that ranged from -8 to -36 mV.

Due to the diffuse layer, the electrostatic potential that repels ions of like charge (negatively charged particles, for the case considered here) increases as the particle approaches the solid. For the case of a negatively charged, 1 μm bacterium approaching a soil particle, the two surfaces may be approximated as spheres. The repulsive electrostatic potential (V_R) is

$$V_R \approx \frac{cr^2\,e^{-\kappa x}}{(2r + x)\kappa^2} \qquad (2)$$

where c is a constant for fixed temperature and electrochemical conditions (details are discussed by Stumm and Morgan, 1981), r is the particle radius, and x is separation distance. As ionic strength increases, κ increases, which reduces the overall repulsive force and decreases the separation distance for a given level of repulsion.

If electrostatic repulsion is not strong enough to keep bacteria away from a solid, close approach may result in strong adsorption due to van der Waals force. This force is sometimes referred to as *electrodynamic* or *fluctuating dipole-induced dipole,* since it results from the motion of electrons in their orbitals (Weber et al., 1991). This potential (V_A) is always attractive; for spheres,

$$V_A \approx \frac{Ar}{12x} \qquad (3)$$

where A is the Hamaker constant (about 100 g nm^2 s^{-2}) (Stumm and Morgan, 1981).

The total interaction potential is $V_R + V_A$. This is the basis of what is called the DLVO theory (for the researchers Derjaguin and Landau [1941] and Verwey and Overbeek [1948]; see also Chapter 1, section 1.2.2 and Chapter 4, section 4.3.3, this volume). According to the theory, for most groundwater ($I < 0.01\ m$), suspended bacteria are electrostatically repelled from solids with increasing force as they approach the solid until a separation distance of about 1 nm is reached. Brownian motion, convection, or motility may push the bacteria over the repulsive barrier. If the separation distance becomes less than 1 nm, the potential rapidly decreases to a strongly adsorptive van der Waals minimum. At very close range, other mechanisms such as hydrogen bonding may also contribute to adsorption. At very close range, other mechanisms such as hydrogen bonding may also contribute to adsorption. Increasing ionic strength decreases the magnitude of the barrier (Eq. 2). Mills et al. (1993) found that increasing the ionic strength from 0.001 to 0.01 reduced recovery of bacteria from sand columns by as much as an order of magnitude. Similarly, Gross and Logan (1995) found that increasing ionic strength from 0.01 to 1 M increased adsorption of bacteria to borosilicate beads by a factor of 9. For high ionic strengths (generally $I > 0.1$ M) the diffuse-layer potential is so weak that the total potential is always negative (attractive), and there will be no net energy barrier to contact. Murray and Parks (1980) discuss details of the DLVO equations for the interaction of a virus particle and an oxide surface. They emphasize the importance of differences in van der Waals potentials for different materials, with metals having strong potential and organics weak potential.

In addition to the primary van der Waals minimum described above, a "secondary minimum" may be important for reversible adsorption of bacteria. The total interaction potential of a sphere and a plate (using representative values for the Hamaker constant and electrostatic potential and for ionic strengths around 0.01 M) is negative at about 5 to 20 nm separation, as well as at <1 nm (van Loosdrecht et al., 1989; McDowell-Boyer et al., 1986). Support for this hypothesis is found in the attachment energy calculated from adsorption isotherms. In many cases bacteria reversibly adsorb to solids such that there is a linear relationship between suspension concentration and adsorbed cells per gram of solid (the adsorption isotherm). The slope of this relationship (K_d) and the number of adsorption sites per gram (M) may be used to calculate the energy of bacterial adhesion with the van't Hoff equation

$$\ln \left(\frac{K_d}{M} \right) = -\frac{\Delta H^0}{RT} + C, \tag{4}$$

where ΔH^0 is standard state enthalpy, R is the gas constant, T is absolute temperature, and C is a constant (Hendricks et al., 1979). van Loosdrecht et al. (1989) found that the adhesion energy calculated from adsorption isotherms is about the same as the predicted potential at the secondary minimum using the DLVO equations and suggested that initial bacterial adhesion takes place at the secondary minimum.

Electrostatic repulsion may result in enhanced transport of microbes relative to the average mass of water by keeping them exposed to greater velocities in the central "fast lane" of flowing soil pore water. Water adheres to pore walls and slows adjacent flow according to Newton's law of viscosity. Consequently, water and entrained substances in the center of a pore move the most rapidly. At the molecular level this entire process is called *anion exclusion*.

2.5.2.2. Hydrophobic Interactions

Microbes have hydrophobic regions on their surfaces where there are high densities of hydrocarbon (C–H) groups. Water molecules "prefer" (have lower free energy) to associate with other water molecules or hydrophilic surfaces. Water molecules in contact with hydrophobic surfaces exhibit a surface tension that tends to minimize water contact with hydrophobic regions. Surface tension may squeeze hydrophobic particles out of the water environment and into other hydropobic regions. Murray and Parks (1980) defined hydrophobic bonding as "the aggregation of nonpolar surfaces resulting from the minimization of reoriented and thus higher (than bulk) free energy water structure adjacent to the nonpolar surfaces." For a bacterium attaching to a solid surface, that change in free energy is

$$\Delta G_{adh} = \gamma_{BS} - \gamma_{BL} - \gamma_{SL} \tag{5}$$

where γ_{BS} is interfacial tension of the bacterium–solid surface γ_{BL} is that of the bacterium–liquid, and γ_{SL} is that of the solid surface–liquid. Values of the interfacial tensions are calculated from Young's equation by the relationship

$$\gamma_{SV} - \gamma_{SL} = \gamma_{LV} \cos\theta \tag{6}$$

where γ_{SV}, γ_{SL}, and γ_{LV} are the solid–air, solid–water, and air–water interfacial tensions and θ is the contact angle of the water on the solid (Marshall, 1990; Neuman et al., 1980).

Bacterial hydrophobicity is, therefore, most appropriately evaluated by contact angle measurement, but indexes of cell surface hydrophobicity have also been obtained by measuring bacterial adherence to hydrocarbons (BATH) and by hydrophobic interaction chromatography (HIC). The bacterial contact angle is the angle formed when a drop of water contacts a surface such as a lawn of bacterial cells. More hydrophobic bacterial surfaces cause the water drop to "ball up," resulting in larger contact angles. Wan et al. (1994) concentrated three strains of bacteria on filters and used a goniometer eyepiece to observe the angle formed by a drop of 1 mM $NaNO_3$ on the bacteria. The contact angles ranged from 77.1° (relatively hydrophobic bacteria) to 24.7° (relatively hydrophilic).

Huysman and Verstraete (1993a) used the BATH method with octane to measure hydrophobicity of *E. coli, S. faecalis,* and seven strains of *Lactobacillus.* Hydrophobicity, measured as percent removal by octane, ranged from 2% for *E. coli* to 94% for *Lactobacillus* strain Lc4. They then tested bacterial adsorption to sand and transport through sand columns and found significant correlations between hydrophobicity and adhesion to sand. The hydrophobic characteristics of poliovirus (which has a protein surface) were evaluated by Murray and Parks (1980) by testing its sorption to $C_2Cl_3F_3$. There was no significant difference in virus concentration between a viral suspension mixed with this hydrophobic liquid and a control suspension similarly mixed without it. The authors concluded that the protein surface was very hydrophilic. They neglected to account for sorption of virus to the air–water interface, however, which could explain an equal loss in virus concentration for the control and $C_2Cl_3F_3$ treatments.

HIC measures the amount of bacteria retained by a hydrophobic gel. Gannon et al. (1991) used this method as well as the BATH method to estimate hydrophobicity of 19 bacterial strains. The HIC assay retained from 7% to 91% of the cells, and the BATH assay retained from 5% to 85% of the cells, although the correlation between the HIC and BATH assays was poor. Furthermore, there was no significant relationship between the HIC or BATH results and transport of cells through loam columns.

Hydrophobic interactions are not simple relationships since humic material (generally the most important hydrophobic component of soil solids) occurs in solution as well as associated with solids. Dissolved organic molecules may compete with microbes for adsorption sites or modify interfacial tensions and thereby interfere with hydrophobic adsorption (Powelson et al., 1991).

2.5.2.3. Attachment Polymers

After initial adhesion, bacteria may become more permanently moored with polymers that extend beyond the bacterial and solid surfaces. Because of their small diameter, these fibers may be able to overcome electrostatic repulsion to link the bacterium with the solid by a process called *bridging* (van Loos-drecht et al., 1989). Many bacterial cells secrete a polysaccharide layer that may effectively cement them to solids and to each other. Some bacteria may use stalks, pili, or other fibrils to anchor them to solids. Gannon et al. (1991), however, found that the presence of polysaccharide capsules or the presence of flagella was not correlated with soil transport at a flow rate of 2.5 cm^{-1}. It may be that polysaccharides and appendages are more important at slower flow rates.

2.5.3. Mechanisms of Adsorption at the GWI

It is not known if GWIs and SWIs react with bacteria in a similar manner. The two interfaces are likely to react differently in several ways. The SWI generally has an electrostatic component that may dominate adsorption, unlike the essentially noncharged GWI. Although not conclusively demonstrated, it is likely that bacterial association with the GWI is dominated by hydrophobic effects. If so, surfactants may have particularly strong effects on adsorption at the GWI.

Polystyrene beads have been used to elucidate some of the basic mecha-nisms of particle interaction with the GWI. Butt (1994) directly measured the force between hydrophobic particles in water and air bubbles. The particles "snapped" into the air bubble, a process that is thermodynamically irrevers-ible. Although Butt (1994) did not elaborate on this irreversibility, examination of the figures presented indicates that the force necessary to pull a 20 μm hydrophobic particle (contact angle 110°–120°) away from the air–water inter-face into water was about 1,000 nN, while suspension agitation (probably Brownian motion) appeared to provide less than 2 nN. Williams and Berg (1992) observed polystyrene beads accumulating and aggregating at the surface of a drop of water. They found that the rate of arrival of beads at the microlayer was constant over 1 hour and that the rate increased with salt content. Bead aggregates formed at the surface in salt concentrations only 1% of that required for aggregation in bulk suspension.

Surfactants and other organic matter appear to have important effects on microbial adsorption at the microlayer. After reviewing the research in bacte-rial adsorption to liquid–liquid and gas–liquid interfaces, Marshall (1976) suggested that hydrophobic bacteria should be more attracted to the gas phase and that surfactants should reduce adsorption. Powelson et al. (1991) found that dissolved organic matter increased by a factor of nine the transport of MS2 virus in unsaturated soil columns. There are at least three possible explanations for this effect. First, organic matter may have lowered surface tension and thereby reduced strength of attachment of surface-adsorbed virus proteins. Almost all organic substances found in natural waters reduce the interfacial tension (Lion and Leckie, 1981). Hunter and Liss (1981) report that the adsorption of surface-active species on lake water may reduce the surface tension by 40%. MacRitchie and Alexander (1963) found that the logarithm of adsorption rate of proteins at air interfaces declines linearly with surface tension. Second, the increase in virus transport with dissolved organic matter may be due to the effect of surface-active organic molecules on the structure of water. In pure water, without, an organic microlayer, layers thicker than 1 μm can be considered to have bulk liquid properties (Clifford, 1975). The presence of surface-active organic films, however, may cause water molecules next to the interface to orient into "ice-like," clathrate (lattice-like) structures, resulting in additional stabilization of the boundary layer to a depth of 50 μm (Hardy et al., 1987). It is possible that this stabilization reduces access of the partly hydropobic virus to the air interface. Finally, the presence of organic matter may facilitate virus transport because organic matter may compete with virus particles for air–surface adsorption sites (Trouwborst et al., 1974).

Surface microlayers may be altered by physical, chemical, or biological factors. A bubble selectively adsorbs inorganic and organic matter and microorganisms during its upward path through water, resulting in selective concentration of these substances in thin films at the air–water interface (Blanchard and Parker, 1977). A similar bubble-transport process may occur in the subsurface. Baier's observation (1972) of a coherent surface matrix formed by bacteria at the interface between lubricating oil and water may also apply to the air–water interface.

Powelson and Gerba (1995) reviewed the studies that have measured microbial concentrations in porous media after exposure to water-saturated and -unsaturated conditions and found that, in every case, recovery of microbes was less in unsaturated conditions. Some of the authors mentioned (Boyd et al., 1969; Kibbey et al., 1978) found a correlation of loss of microbes with degree of unsaturation that Powelson and Gerba deemed consistent with the hypothesis of strong sorption to the GWI. Later work (Powelson et al., 1990, Powelson et al., 1993; Powelson and Gerba, 1995; Wan et al., 1994) directly attributed the loss of the microbes to adsorption to the GWI. From a practical point of view, bacterial adsorption to a GWI in soil may be detrimental by slowing transport of organisms intended to degrade a pollutant or may be beneficial by slowing transport of pathogens to drinking water aquifers. In the vadose zone it may be possible to "chase" pollutants spilled on the ground surface with bacteria capable of degrading the chemicals by minimizing bacterial adsorption to GWIs. Near the water table, air may interfinger with water in the capillary fringe, or a rising water table may trap air bubbles. This region is important for bioremediation because light nonaqueous-phase liquids concentrate here. Deeper in an aquifer, gas bubbles may develop in regions of biogenic gas production. Manipulation of bacterial interaction with the GWI may help to remove hazardous colloids such as pathogenic bacteria and viruses and toxic compounds adsorbed to mobile colloids from percolating water.

Below the water table it may be possible to utilize bacterial adsorption to gas bubbles to "raft" bacteria to polluted zones along with groundwater flow. This process would be a variation of the air sparging techniques currently being used to increase oxygen supplies for biodegradation below the water table. By generating tiny bubbles and optimizing bacterial adsorption to them, it may be possible to entrain bacteria in the gas dispersion while simultaneously supplying oxygen for biodegradation. Anaerobic degradation may be promoted by using N_2 or CH_4 sparging to carry anaerobic bacteria.

2.6. CONCLUSIONS

Microbial interactions with surfaces in soil are complex and varied. The surface properties of soil and subsurface bacteria have not been studied in detail, but they are likely to demonstrate the diversity and variations in characteristics exhibited by bacteria in other habitats. Moreover, the soil environment is extremely complex—both chemically and structurally. The several processes that act to hold cells to surfaces may be in effect at different locations; they may act separately or together in various combinations. Cell-associated phenomena may behave differently when faced with different physical–chemical situations. The apparent importance of the gas–liquid interface adds an entire new facet to the overall phenomenon to be studied. While rapid progress is being made to develop a more solid understanding of the mechanisms and controls on the interactive processes, it is clear that much more work is neccessary to approach a comprehensive understanding of the interactions in soils under field conditions.

REFERENCES

Adams MH (1948): Surface inactivation of bacterial viruses and of proteins. J Gen Physiol 31:417–431.

Alexander M (1977): Introduction to Soil Microbiology. New York: John Wiley & Sons.

Baier RE (1972): Organic films on natural waters; their retrieval, identification and modes of elimination. J Geophys Res 77:5062.

Baier RE (1975): Applied chemistry at protein interfaces. Adv Chem Ser 145:1–25.

Beckett R (1990): The surface chemistry of humic substances in aquatic systems. In Beckett R (ed): Surface and Colloid Chemistry in Natural Waters and Water Treatment. New York: Plenum Press, pp 3–20.

Beckett R, Bigelow JC, Zhang J, Giddings JC (1989): Analysis of humic substances using flow field-flow fractionation. In MacCarthy P, Suflett IH (eds): The Influence of Aquatic Humic Substances on the Fate and Treatment of Pollutants. ACS Advances in Chemistry Series No. 219. Washington, DC: American Chemical Society, pp 65–80.

Beckett R, Zhang J, Giddings JC (1987): Determinations of molecular weight distributions of fulvic and humic acids using flow field-flow fractionation. Environ Sci Technol 21:289–295.

Blanchard DC, Parker BC (1977): The freshwater to air transfer of microorganisms and organic matter. In Cairns JJ (ed): Aquatic Microbial Communities. New York: Garland, pp 627–658.

Bolt GH, Bruggenwert MGM (1976): Soil Chemistry, A. Basic Elements. Amsterdam: Elsevier Scientific.

Boyd JW, Yoshida T, Vereen LE, Cada RL, Morrison SM (1969): Bacterial Response to the Soil Environment. Sanitary Engineering Papers, No. 5. Fort Collins: Colorado State University.

Brady NC (1984): The Nature and Properties of Soils. New York: MacMillin.

Brock TD, Madigan MT (1991): Biology of Microorganisms. Englewood Cliffs, NJ: Prentice-Hall.

Butt H-J (1994): A technique for measuring the force between a colloidal particle in water and a bubble. J Colloid Interface Sci 166:109–117.

Clifford J (1975): Properties of water in capillaries and thin films. In Franks F (ed): Water: A Comprehensive Treatise. New York: Plenum Press.

Corapcioglu MY, Haridas A (1985): Microbial transport in soils and groundwater: A numerical model. Adv Water Resources 8:188–200.

Derjaguin BV, Landau L (1941): Theory of stability of strongly charged lyophobic sols and of the adhesion of strongly charged particles in solutions of electrolytes. Acta Physicochim URSS 14:633–662.

Fontes DE, Mills AL, Hornberger GM, Herman JS (1991): Physical and chemical factors influencing transport of microorganisms through porous media. Appl Environ Microbiol 57:2473–2481.

Foth HD (1978): Fundamentals of Soil Science. Ed 6. New York: John Wiley & Sons.

Gannon JT, Manilal VB, Alexander M (1991a): Relationship between cell surface properties and transport of bacteria through soil. Appl Environ Microbiol 57:190–193.

Gardner WR (1956): Calculation of capillary conductivity from pressure plate outflow data. Soil Sci Soc Am Proc 20:317–320.

Gordon AS, Millero FJ (1985): Adsorption mediated decrease in the biodegradation rate of organic compounds. Microb Ecol 11:289–298.

Green RE, Hanks RJ, Larson WE (1964): Estimates of field infiltration by numerical solution of the moisture flow equation. Soil Sci Soc Am Proc 28:15–19.

Griffith PC, Fletcher M (1991): Hydrolysis of protein and model dipeptide substrates by attached and nonattached marine *Pseudomonas* sp. strain NCIBM 2021. Appl Environ Microbiol 57:2186–2191.

Gross MJ, Logan BE (1995): Influence of different chemical treatments on transport of *Alcaligenes paradoxus* in porous media. Appl Environ Microbiol 61:1750–1756.

Guerin WF (1989): Phenanthrene degradation by estuarine surface microlayer and bulk water microbial populations. Microb: Ecol 17:89–104.

Guerin WF, Boyd SA (1992): Differential bioavailability of soil-sorbed napthalene to two bacterial species. Appl Environ Microbiol 58:1142–1152.

Harden VP, Harris JO (1953): The isoelectric point of bacterial cells. J Bacteriol 65:198–202.

Hardy JT, Apts CW (1984): The sea-surface microlayer: Phytoneuston productivity and effects of atmospheric particulate matter. Mar Biol 82:293–300.

Hardy JT, Crecelius EA, Antrim LD, Broadhurst VL, Apts CW, Gurtisen JM, Fortman TJ (1987): The sea-surface microlayer of Puget Sound: Part II, concentrations of contaminants and relation to toxicity. Mar Environ Resources 23:251–271.

Harvey GW (1966): Microlayer collection from the sea surface: A new method and initial results. Limnol Oceanog 11:608–613.

Harvey RW (1991): Parameters involved in modeling movement of bacteria in groundwater. pp. 89–114 In Hurst CJ (ed): Modeling the Environmental Fate of Microorganisms. Washington, DC: American Society of Microbiology.

Harvey RW, Young LY (1980): Enrichment and association of bacteria and particulates in salt marsh surface water. Appl Environ Microbiol 39:894–899.

Hatcher RF, Parker BC (1974): Microbiological and chemical enrichment of freshwater-surface microlayers relative to the bulk-subsurface water. Can J Microbiol 20:1051–1057.

Hendricks DW, Post FJ, Khairnar DR (1979): Adsorption of bacteria on soils: Experiments, thermodynamic rationale, and application. Water Air Soil Pollut 12:219–232.

Hendricks SM, Williams PM (1985): Dissolved and particulate amino acids and carbohydrates in the sea surface microlayer. Mar Chem 17:141–163.

Hermansson M, Marshall KC (1985): Utilization of surface localized substrate by nonadhesive marine bacteria. Microb Ecol 12:91.

Holmes JW, Greacen EL, Gurr CG (1960): Transactions of the 7th International Congress of Soil Science 1:188–194.

Hrudey SE, Pollard SJ, (1993): The challenge of contaminated sites: Remediation approaches in North America. Environ Rev 1:55–72.

Hunter KA Liss PS (1981): Organic sea surface films. In Duursma EK, Dawson R (ed): Marine Organic Chemistry. Amsterdam: Elsevier, pp. 259–298.

Huysman, F, Verstraete W (1993a): Water-facilitated transport of bacteria in unsaturated soil columns: Influence of cell surface hydrophobicity and soil properties. Soil Biol Biochem 25:83–90.

Huysman F, Verstraete W (1993b): Water-facilitated transport of bacteria in unsaturated soil columns: Influence of inoculation and irrigation methods. Soil Bio Biochem 25:91–97.

Jaynes WF, Boyd SA (1991): Clay mineral type and organic compound sorption by headecyltrimethylammonium-exchanged clays. Soil Sci Soc Am J 55:43–48.

Kefford B, Kjelleberg S, Marshall KC (1982): Bacterial scavenging: Utilization of fatty acids localized at a solid–liquid interface. Arch Microbiol 133:257.

Kemper WD, Rollins JB (1966): Osmotic efficiency coefficients across compacted clays. Soil Sci Soc Am Proc 30:529–534.

Kibbey HJ, Hagedorn C, McCoy EL (1978): Use of fecal streptococci as indicators of pollution in soil. Appl Environ Microbiol 35:711–717.

Kjelleberg S, Humphrey BA, Marshall KC (1982): The effect of interfaces on small starved marine bacteria. Appl Environ Microbiol 45:1106.

Lindqvist R, Cho JS, Enfield CG (1994): A kinetic model for cell density dependent bacterial transport in porous media. Water Resources Res 30:3291–3299.

Lindqvist R, Enfield CG (1992): Cell density and non-equilibrium sorption effects on bacterial dispersal in groundwater microcosms. Microb Ecol 23:25–41.

Lion LW, Leckie JO (1981): The biogeochemistry of the air–sea interface. Annu Rev Earth Planet Sci 9:449–486.

MacIntyre F (1974): The top millimeter of the ocean. Sci Am 230:62–77.

MacRitchie F, Alexander AE (1963): Kinetics of adsorption of proteins at interfaces. J Colloid Sci 18:453–469.

Marshall KC (1976): Interfaces in Microbial Ecology. Cambridge, MA: Harvard University Press.

Marshall KC (1990): Microbial processes occurring at surfaces. In Beckett R (ed): Surface and Colloid Chemistry in Natural Waters and Water Treatment. New York: Plenum Press, pp 21–26.

McDowell-Boyer LM, Hunt JR, Sitar N (1986): Particle transport through porous media. Water Resources Res 22:1901–1921.

Mills AL, Eaton WD (1984): Biodegradation of bromobenzene in simulated groundwater conditions. In Llewellyn GC, O'Rear CE (ed): Biodegradation 6. Slough, England: CAB International, pp 9–13.

Mills AL, Fontes DE, Hornberger GM, Herman JS (1993): Physical and chemical controls on the advective transport of bacteria. McCarthy JF, Wobber FJ (eds): Manipulation of Groundwater Colloids for Environmental Restoration Boca Raton, FL: Lewis, pp. 75–80.

Mills AL, Herman JS, Hornberger GM, deJesus TH (1994): Effect of solution ionic strength on mineral grains on the sorption of bacterial cells to quartz sand. Appl Environ Microbiol 60:3600–3606.

Mills AL, Maubrey R (1981): Effect of mineral composition on bacterial attachment to submerged rock surfaces. Microb Ecol 7:315–322.

Murray JP, Parks GA (1980): Poliovirus adsorption on oxide surfaces. Kavenaugh MC, Leckie JO (eds): Particulates in Water. Advances in Chemistry Series, Vol 189: Washington, DC: American Chemical Society, pp 97–133.

Neumann AW, Hum OS, Francis DW, Zingg W, van Oss CJ (1980): Kinetic and thermodynamic aspects of platelet adhesion from suspension to various substrates. J Biomed Mater Res 14:499.

Norkrans B (1980): Surface microlayers in aquatic environments. Adv Microb Ecol 4:51–81.

Ogram AV, Jessup RE, Lou LT, Rao PSC (1985): Effects of sorption on biological degradation rates of (2,4-dichlorophenoxy)acetic acid in soils. Appl Environ Microbiol 49:582–587.

Pitt WG, McBride MO, Barton AJ, Sagers RD (1993): Air–water interface displaces adsorbed bacteria. Biomaterials 14:605–608.

Plusquellec A, Beucher M, LeLay C, LeGal Y, Cleret JJ (1991): Quantitative and qualitative bacteriology of the marine water surface microlayer in a sewage-polluted area. Mar Environ Res 31:227–239.

Poletika NN, Jury WA, Yates MV (1995): Transport of bromide, simazine, and MS-2 coliphage in a lysimeter containing undisturbed, unsaturated soil. Water Resources Res 31:801–810.

Powelson DK, Gerba CP (1995): Fate and transport of microorganisms in the vadose zone. In Wilson LG et al (eds): Handbook of Vadose Zone Characterization and Monitoring. Boca Raton, FL: Lewis, pp. 123–135.

Powelson DK, Gerba CP, Yahya MT (1993): Virus transport and removal in wastewater during aquifer recharge. Water Res 27:583–590.

Powelson DK, Simpson JR, Gerba CP (1990): Virus transport and survival in saturated and unsaturated flow through soil columns. J Environ Q 19:396–401.

Powelson DK, Simpson JR, Gerba CP (1991): Effects of organic matter on virus transport in unsaturated flow. Appl Environ Microbiol 57:2192–2196.

Power K, Marshall KC (1988): Cellular growth and reproduction of marine bacteria on surface-bound substrate. Biofouling 1:163.

Richardson RR (1994): Interactive effects of bacterial cell surface properties, mineral grain coatings, and organic matter on processes affecting transport of bacteria in porous media. M.S. Thesis, University of Virginia, Charlottesville, VA.

Scholl MA, Mills AL, Herman JS, Hornberger GM (1990): The influence of mineralogy and solution chemistry on the attachment of bacteria to representative aquifer materials. J Contam Hydrol 6:321–336.

Sieburth JM (1976): Dissolved organic matter and heterotrophic microneuston in the surface microlayers of the North Atlantic. Science 194:1415–1418.

Sodergren A (1993): Role of aquatic surface microlayer in the dynamics of nutrients and organic compounds in lakes, with implications for their ecotones. Hydrobiologia 251:217–225.

Sposito G (1989): The Chemistry of Soils. New York: Oxford University Press.

Stevenson FJ (1982): Humus Chemistry. New York: John Wiley.

Stumm W, Morgan JJ (1981): Aquatic Chemistry. New York: John Wiley.

Tan Y, Bond WJ, Griffin DM (1992): Transport of bacteria during unsteady unsaturated soil water flow. Soil Sci Soc Am J 56:1331–1340.

Thurman EM (1985): Organic Geochemistry of Natural Waters. Dordrecht, Netherlands: Nijhoff/Junk.

Trouwborst T, Kuyper S, DeJong JC, Plantinga AD (1974): Inactivation of some bacterial and animal viruses by exposure to liquid–air interfaces. J Gen Virol 24:155–165.

van Loosdrecht MCM, Lyklema J, Norde W, Zehnder AJB (1989): Bacterial adhesion: A physiochemical approach. Microb Ecol 17:1–15.

Verwey EJW, Overbeek JTG (1948): Theory of the Solubility of Lyophobic Colloids. Amsterdam: Elsevier.

Wan J, Wilson JL (1994): Visualization of the role of the gas–water interface on the fate and transport of colloids in porous media. Water Resources Res 30:11–23.

Wan J, Wilson JL, Kieft TL (1994): Influence of the gas–water interface on transport of microorganisms through unsaturated porous media. Appl Environ Microbiol 60:509–516.

Weber JB, Coble HD (1968): Microbial decomposition of diquat adsorbed on montmorillonite and kaolinite clays. J Agric Food Chem 16:475–478.

Weber WJ, McGinley PM, Katz LE (1991): Sorption phenomena in subsurface systems: Concepts, models and effects on contaminant fate and transport. Water Res 25:499–528.

Williams DF, Berg JC (1992): The aggregation of colloidal particles at the air–water interface. J Colloid Interface Sci 152:218–229.

Yao, K-M, Habibian MT, O'Melia CR (1971): Water and waste water filtration: Concepts and applications. Environ Sci Technol 5:1105–1112.

3

ADHESION AS A STRATEGY
FOR ACCESS TO NUTRIENTS

KEVIN C. MARSHALL

School of Microbiology and Immunology, The University of New South Wales, Sydney, New South Wales 2052, Australia

3.1. INTRODUCTION

Early investigators suggested that the *raison d'étre* for the almost universal association of microorganisms, and of heterotrophic bacteria in particular, with surfaces was the easy access to nutrients at such surfaces. This idea evolved from observations of the so-called bottle effect, whereby samples taken from aquatic environments in bottles and stored for short periods exhibited increases in bacterial numbers several orders of magnitude greater than those in the original samples (Waksman and Carey, 1935). This effect was believed to be related to the substantial increase in available surface area in the sample bottles, an assumption confirmed by ZoBell and Anderson (1936), who demonstrated that increases in bacterial growth in stored seawater samples were proportional to the surface-to-volume ratio of the containers employed. Subsequently, Stark et al. (1938) reported a significant accumulation of organic matter on clean slides immersed in sterile freshwater samples. Further investigations revealed that increased surface area, achieved by the addition of glass beads or various particulate materials to liquid cultures, stimulated bacterial growth compared with control liquid cultures at low nutrient concentrations, but not at high nutrient concentrations (Heukelekian and Heller, 1940; ZoBell, 1943; Jannasch, 1973). These results emphasized the perception that surfaces in oligotrophic conditions served as sites of nutrient concentration and that surface-associated bacteria utilized these nutrients for growth even though they were unable to multiply in the bulk aqueous phase.

Bacterial Adhesion: Molecular and Ecological Diversity, pages 59–87
© *1996 Wiley-Liss, Inc.*

ZoBell (1943) went a step further by differentiating between the types of organic nutrients supplied in these experiments. He observed beneficial effects of surfaces on bacterial growth in solutions enriched by 5 mg 1^{-1} of sodium caseinate, lignoprotein, or an emulsified chitin preparation, but not with glycerol, glucose, or lactate. This led ZoBell to propose that bacterial growth at surfaces is stimulated only when microbially produced extracellular enzymes are involved, such as in the metabolism of macromolecular or particulate substrates. ZoBell hypothesized that contact between enzyme and substrate in the aqueous phase would be rare, and products of enzymatic hydrolysis would diffuse away and be unavailable to the bacterium. At a surface, on the other hand, the enzyme and substrate would be in close juxtaposition, and, provided the enzyme remained active in the adsorbed state, the bacterium should benefit from the end products of the reaction. We will see below (section 3.4.1) that this idea was overly simplistic, yet it provided a basis for further studies on the reasons for rapid bacterial growth at surfaces in aquatic environments.

It is of historical interest to note that Heukelekian and Heller (1940) found that the *Escherichia coli* strain they employed did not adhere to surfaces, its association with surface being only superficial, and yet its growth appeared to be stimulated by surface-found substrates. More detailed studies many years later provided concrete evidence for the utilization of surface-bound substrates by poorly adhering bacteria (see sections 3.4.1 and 3.4.3).

The studies reported above presented circumstantial evidence for the role of surfaces in providing a concentrated source of nutrient for bacteria in aquatic environments, particularly those of an oligotrophic nature. Recent evidence, based on studies on the adsorption of organics to surfaces and the bacterial metabolism of and growth on these surface-bound substrates, has confirmed and refined the ideas propounded in the 1930s and 1940s.

3.2. SURFACE CONDITIONING FILMS

3.2.1. Adsorption of Organic Molecules at Surfaces

Immersion of a clean surface into a natural aqueous environment results in the instantaneous formation of a layer of adsorbed organic molecules, termed a *conditioning film,* (CF), that alters the physicochemical properties of the surface (section 3.2.2; see also Chapter 4, section 4.3.2 , this volume). The nature and amount of the molecules adsorbed and their configuration at a surface are dictated, to a large extent, by the original properties of the substratum in question and by the molecular species present in the aqueous phase (section 3.2.3). These organic molecules originate by excretion from or lysis of micro- and macroorganisms or by leaching from terrestrial environments.

Molecular adsorption can occur in two ways, and, in natural environments, both processes probably occur simultaneously. *Physical adsorption* is characterized by a rapid establishment of an adsorption equilibrium, reversibility, and low heats of adsorption (ΔH_{ads} = <50 kJ mole^{-1}).

The adsorbed layer is often more than one molecule thick, and the attractive forces involved are of the nonspecific London–van der Waals type. *Chemical adsorption* involves specific chemical bonds (electrostatic, covalent, and hydrogen bonds), dipole interactions, and hydrophobic interactions and is characterized by the formation of monomolecular layers and high heats of adsorption (ΔH_{ads} = 80 to 420 kJ mole^{-1}).

3.2.2. Methods for Determining the Existence and Properties of CF on Surfaces

The adsorption of organic molecules to a previously clean surface results in an alteration to the physicochemical properties of the surface, and their presence can be detected by the application of a variety of physical and chemical techniques.

3.2.2.1. *Microelectrophoresis*

Because of the imbalance of charges existing at a sold–liquid interface, all surfaces possess a characteristic surface charge. The charge is independent of size of the solid and can be determined as mean velocities of small particles in a microelectrophoresis cell (Abramson et al., 1942). The zeta potential (the potential at the plane of shear of the moving particle) and surface charge can be calculated from the electrophoretic mobility (Abramson et al., 1942), but, because of errors resulting from particle surface conductance (Gittens and James, 1963; Einolf and Carstensen, 1967), most authors express their results as electrophoretic mobilities.

Neihof and Leob (1972) found considerable differences between the electrophoretic mobilities of a range of particles suspended in an artificial seawater lacking organics, whereas the same particles in natural seawater, where organcis were present, all tended toward a low negative charge (Table 3.1.). Subsequent studies (Neihof and Leob, 1974) revealed that photo-oxidation of the

TABLE 3.1. Electrophoretic Mobilities of Model Particles in Organic-Free Artificial Seawater and in Natural Seawater Containing Organic Molecules[a]

Model particle	Electrophoretic mobility (μm s^{-1} V^{-1} cm^{-1})	
	Artificial seawater	Natural seawater
Glass	−0.75	−0.87
Anion exchange resin	+1.19	−0.38
Bentonite	−1.49	−1.11
Calcium carbonate	−1.19	−0.05
Chlorinated wax	−1.13	−0.68
Sephadex	0.0	−0.15

[a] Adapted from Neihof and Loeb (1972), with permission of the publisher.

organic molecules in natural seawater using ultraviolet irradation resulted in electrophoretic mobilities resembling those found in artificial seawater. These results demonstrated that absorbed organic molecules modify the apparent surface charge of substrata immersed in natural aquatic environments. Similar results have been reported by Hunter and Liss (1982).

3.2.2.2. Wettability of Surfaces

Modifications to the wettability of solid surfaces following the adsorption of macromolecules have been determined using contact angle measurements. A drop of liquid may spread on a surface (completely wettable) or remain as a discrete drop making a finite angle of contact (θ) with the solid surface. As θ increases, the degree of wettability decreases.

Changes in wettability resulting from molecular adsorption can be determined as the as the critical surface tension (γ_c), a measure of the highest surface tension a liquid can have and still spread over a given surface (Baier et al., 1968). The γ_c for wetting by an homologous series of liquids is defined on a "Zisman plot" as the intercept of the horizontal axis at $\cos\theta = 1$ with the extrapolated straight line plot of $\cos\theta$ against the surface tension of the various liquids employed. Using this technique, Baier et al. (1968) demonstrated that monolayers of stearic acid, primary octadecyl amine, or octadecyl alcohol reduced the γ_c of a variety of surfaces to approximately the same value ($20 - 24 \, \text{mNm}^{-1}$), indicating a preponderance of methyl groups exposed at the surfaces. Substratum wettability has also been expressed in terms of the water contact angle (Fletcher and Loeb, 1979), bubble contact angles (Fletcher and Marshall, 1982), the work of adhesion (W_A) between water and the substratum (Pringle and Flecther, 1983), and the surface free energy (γ_s) engery determined from a single contact angle measurement using the equation-of-state approach (Neumann et al., 1974). Another approach is the use of gradient wettability plates to study protein adsorption over a range of wettabilies (Elwing and Gölander, 1990).

3.2.2.3. Chemical Techniques

Evidence for the adsorption of macromolecules to solid surfaces (Fig. 3.1) has been obtained with the use of infrared/internal reflectance spectrometry (Baier, 1980), but the technique is limited because of the need to dry the prisms used as substrata prior to insertion of the spectrometer. This problem has been overcome by the application of Fourier transform infrared (FTIR) spectrometry (Nichols et at. 1985), which allows the water signal to be subtracted to reveal signals from adsorbed organic molecules. When combined with internal reflectance in a circle cell assembly (Bartick and Messerschmidt, 1984), the aqueous phase can flow through continuously and a time-course study of macromolecular adsorption can be undertaken (Nichols et al., 1985). Other techniques include Auger and x-ray photoelectron spectroscopy (XPS or ESCA) (Kristoffersen et al., 1982) and pyrolysis–chemical ionization combined with mass spectrometry (Zsolnay and Little, 1983). Some of these tech-

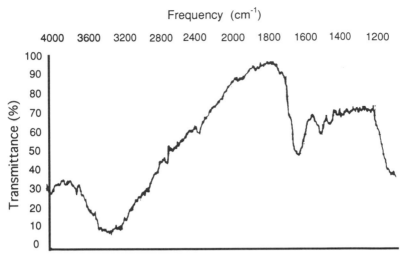

Fig. 3.1. Multiple attenuated internal reflection infrared spectrum of a germanium prism after 2 h exposure to a bacterial culture showing the adsorption of possibly protein (1,650 and 1,560 cm^{-1} absorption bands) and polysaccharide (3,250 cm^{-1} band). (Adapted from Baier, 1980, with permission of the publisher.)

niques have the disadvantage of being necessarily destructive and only suitable for fragment analysis.

3.2.3. Some Generalizations Regarding Adsorbed Organics

It is necessary to distinguish between macromolecules and low-molecular-weight organics in any discussion of CFs (Marshall, 1988, 1992b) because of differences in their degrees and modes of adsorption. Many small, uncharged organics (e.g., glucose) do not adsorb to surfaces and, in any case, are probably rapidly utilized by bacteria suspended in the aqueous phase. Charged, low-molecular-weight molecules are also susceptible to rapid utilization in the aqueous phase but, if adsorbed, may be of limited availability to bacteria attached to surfaces (Gordon and Millero, 1985). Small hydrophobic molecules (fatty acids, lipids), on the other hand, rapidly partition at surfaces and yet may still be available for bacterial utilization (Kefford et al. 1982) (see also section 3.4.1).

Macromolecules, whether in solution or adsorbed to surfaces, require digestion by extracellular enzymes prior to utilization by the microorganisms. In the adsorbed state, potential utilization of such molecules by microorganisms is complicated by their attachment to the substratum at multiple sites along their length, leading to varying configurations and amounts occurring on different substrata and, thus, different degrees of availability to various microorganisms. Robb (1984) has defined four important interactions involved in the adsorption of a polymer at a surface, namely, (1) polymer segment–surface, (2) polymer segment–solvent, (3) polymer segment–other solvent components, and (4) solvent–surface interactions (Fig. 3.2). Polymers may have 10^5 segments per chain, with a random coil polymer having up to 10^4 segments in contact, by means of chemical bonds, dipole interactions, and hydropobic interactions, with the substratum (Robb, 1984). Once adsorbed, it is difficult to desorb such a polymer as all the bonds need to be broken simultaneously. For a single polymer type, the combination of bonds involved will vary with the chemistry of the substratum and will result in differences in molecular configuration. The varying amounts of particular macromolecules adsorbed to surfaces with differing wettabilities is well illustrated by the surface-gradient technique of Elwing and Gölander (1990). Proteins such as fibrinogen and γ-globulin adsorbed at significantly higher levels to the hydrophobic than to the hydrophilic portion of the gradient, whereas lysozyme adsorbed only slightly more to the hydrophobic portion. Electron microscopy revealed that fibrinogen adsorbed to a hydrophobic surface as a bi- or trinodular structure, whereas it had a filamentous appearance on a hydrophilic surface (quoted by Elwing and Gölander, 1990). In view of these results, it is not surprising that bovine glycoprotein exhibits differential susceptibility to enzymatic degradation when adsorbed at a hydrophobic compared with a hydrophilic surface (Fletcher and Marshall, 1982).

Mixed molecular species, rather than a single molecular species, are the norm in natural environments, and, as a result, it is necessary to consider interactions between various molecules at or near a surface. Some of these will include interactions between polymers and smaller molecules present in the solvent phase (water), as indicated in Figure 3.2, whereby the interaction alters the configuration of the native polymer. The well-known interaction between lipids and bovine serum albumin (BSA) illustrates this point, with

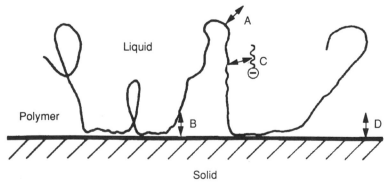

Fig. 3.2. Diagrammatic representation of a polymer adsorbed at a solid–liquid interface, showing the following interactions: **(A)** polymer segment–solvent; **(B)** polymer segment—surface; **(C)** polymer segment–other solvent components; and **(D)** surface–solvent. (Adapted from Robb, 1984, with permission of the publisher.)

the lipid-free BSA having different properties from the lipid-complexed form. Furthermore, polymers compete for adsorption sites on surfaces. Differences in the diffusion constants of various polymers result in smaller molecules arriving at the surface first, but these tend to exchange with larger molecules having higher surface affinities, an exchange often termed the *Vroman effect*. In a study of interactions between fibrinogen and high-molecular-weight kinogen (HMWK) in blood plasma using a wettability gradient surface, Elwing and Gölander (1990) found three distinct interaction zones: (1) on the hydrophilic portion, fibrinogen is adsorbed first, then desorbed, and replaced by HMWK; (2) at the intermediate wettability region, the fibrinogen is adsorbed, but little is desorbed after prolonged incubation in plasma; and (3) on the hydrophobic portion, fibrinogen is adsorbed from plasma, but not desorbed with time. The fibrinogen molecules adsorbed at the hydrophobic side showed a loss of antibody-binding capacity as a result of conformational changes.

3.3. MODIFICATIONS TO BACTERIAL ADHESION BY CF

Although the presence of CF has long been recognized as a modifying factor in adhesion processes (Baier et al., 1968), very few studies on bacterial adhesion have taken this factor into account. Fletcher (1976) reported that adsorption of some proteins, e.g., BSA, inhibited adhesion of a marine pseudomonad to polystyrene surfaces, whereas other proteins, e.g., protamine sulfate, had little or no effect on adhesion. Similar results were obtained by Ørstavik (1977) in a study of the adhesion of oral streptococci to glass.

Baier (1980) compared the physical properties of clean, high-energy germanium surfaces with low-energy polydimethylsiloxane-modified germanium surfaces both before and after exposure to a bacterial culture for varying periods. He concluded that the original substratum surface properties do influence the nature, rate, and colonization success of bacteria despite the instantaneous adsorption of CF (mostly proteins and polysaccharides; see Fig. 3.1) to different surfaces. Using bubble contact angles to monitor changes in surface characteristics following protein adsorption, Fletcher and Marshall (1982) found that BSA and bovine glycoprotein (BGP), but not protamine sulfate, rendered both hydrophobic and partly hydrophilic surfaces more hydrophilic. Following adsorption of proteins to the substrata, reaction with the proteolytic enzyme pronase increased the contact angle for both BSA and BGP on the hydrophilic substrata and for BSA on the hydrophobic substratum, but did not alter the contact angle for BGP on the hydrophobic substratum. An increase in contact angle represents a decrease in wettability, suggesting digestion of the proteins by pronase. The lack of alteration in the contact angle for BGP on the hydrophobic substratum suggests a configurational alteration in the BGP molecule on the surface. Fletcher and Marshall (1982) reported that, in general, the effects of adsorbed proteins on the adhesion of *Pseudomonas* NCIMB 2021 to both substrata were consistent with changes in bubble contact angles, as were alterations in protein-induced inhibition of bacterial adhesion to hydro-phobic substrata by treatment with pronase.

CF from different sources (marine and freshwater samples, as well as BSA, β-lactoglobulin, myoglobin, and humic acid) modified the retention of a marine bacterium, SW8, under laminar flow conditions (Schneider and Marshall, 1994) in the following ways: (1) some CF appeared to control bacterial retention irrespective of substratum type, (2) some CF either decreased or increased retention on all substrata without altering the relative order of retention on the substrata, and (3) some CF affected retention in an apparently substratum-specific manner. In general, retention was controlled by complex interactions between the substratum properties, the physiological state of the bacterium, and the source of the CF.

Attempts to control the nature of CF in order to manipulate bacterial adhesion to surfaces appears an attractive proposition. Chet et al. (1975) reported that adsorption at surfaces of tannins secreted by algae produced a negative chemotaxis in bacteria and reduced the incidence of bacterial colonization of the surfaces. The treatment of hydrophobic surfaces with a tri-block copolymer of polyethylene oxide and polypropylene oxide ("Synper-onic" F108) prevented bacterial adhesion (Owens et al., 1987) in short-term experiments. The copolymer apparently adsorbs to surfaces via its hydropho-bic segment, leaving the hydrophilic segments to extend into the aqueous phase and, presumably, inhibit bacterial adhesion by steric repulsion of the extracellular adhesive polymers of the bacteria. Short-term laboratory trials with a range of marine bacteria confirmed this result, but tests in natural marine environments showed an initial inhibition of bacterial adhesion followed by high levels of bacterial colonization of surfaces, equivalent to those of un-treated controls, after 5 days of exposure (Marshall and Blainey, 1991; Blainey and Marshall, 1991). The poor long-term effectiveness of the copolymer was attributed to one or more of the following: (1) leaching of the copolymer from the treated surface, (2) removal by abrasion, (3) interference with the copolymer by adsorbed CF, (4) breaching of the copolymer barrier by bacteria with unique adhesion mechanisms, and (5) biodegradation of the copolymer by specific bacterial populations in the seawater. Another possible approach is the application to surfaces of biological products known to inhibit bacterial and/or eukaryote growth (Maki et al., 1992; Holmström and Kjelleberg, 1994).

3.4. CF AS AN ORGANIC NUTRIENT SOURCE FOR ADHERING BACTERIA

As stated in section 3.1, a great deal of circumstantial evidence is available to suggest that bacteria benefit from the organic substrates adsorbed to surfaces in natural environments. Modern techniques, such as the use of radiolabeled substrates and phase-contrast microscopy combined with image analysis and time-lapse methods, have provided direct evidence for the concept of active microbial growth at surfaces.

3.4.1. Diversity in the Utilization of Small, Hydrophobic Molecules

ZoBell's simplistic suggestion (1943) (see section 3.1) that macromolecules were responsible for the growth response by bacteria at surfaces ignored the role played by smaller hydrophobic organics that partition strongly at surfaces. Various bacteria are able to scavenge, and subsequently metabolize, surface-localized ^{14}C-labeled stearic as a sole carbon source. Organisms examined by this technique include *Leptospira biflexa patoc* 1, the pigmented *Serratia marcescens* EF190 and a nonpigmented mutant of EF190 (Kefford et al., 1982), *Vibrio* DW1 and *Pseudomonas* S9 (Kjelleberg et al., 1983), and *Vibrio* MH3 and other marine isolates (Hermansson and Marshall, 1985). These bacteria exhibited different patterns of behavior depending on their adhesive and cell surface properties.

The leptospire, which is highly motile and adheres to surfaces reversibly, scavenged most of the ^{14}C-stearic acid within 24 h (Fig. 3.3A), with a significant amount of the ^{14}C label detected in bacteria returning to the aqueous phase (Fig. 3.3B) and in the culture supernatant (Fig. 3.3C). The efficient scavenging of the surface-bound substrate was attributed by Kefford et al. (1982) to the ability of leptospires to "crawl" across surfaces (Cox and Twigg, 1974). *Vibrio* MH3, which was isolated from seawater by deliberate selection for a nonadhesive bacterium, also scavenged ^{14}C-stearic acid from surfaces, and large numbers of labeled bacteria were found in the aqueous phase, as expected when an organism is able to alternate between the semisessile and planktonic states (Hermansson and Marshall, 1985).

The hydrophobic, irreversibly adhering *S. marcescens* EF190 initially showed a very fast rate of substrate scavenging, but the rate then slowed considerably (Fig. 3.3A). Little of the label was associated with bacteria in the aqueous phase, indicative of the firm adhesion of EF190 to the surface, and a moderate amount was found in the supernatant (Figs. 3.3B,C). The slow rate of substrate scavenging by the nonpigmented mutant of EF190 (Fig. 3.3A) was attributed by Kefford et al. (1982) partly to the more hydrophilic surface of the mutant leading to less efficient adhesion to the test surface. Similarly, early phase (2 h) starved cells of the hydrophobic *Pseudomonas* S9 scavenged more surface-bound stearic acid than did cells of a similar age of the hydrophilic *Vibrio* DW1 (Kjelleberg et al., 1983). Longer term (18 h) starvation prior to testing resulted in much lower utilization of the substrate, with no differences observed between the scavenging rates of the two starved bacteria.

3.4.2. Utilization of Macromolecules

3.4.2.1. Adsorbed Macromolecules
Numerous studies have been made on the adsorption of proteins, DNA, and

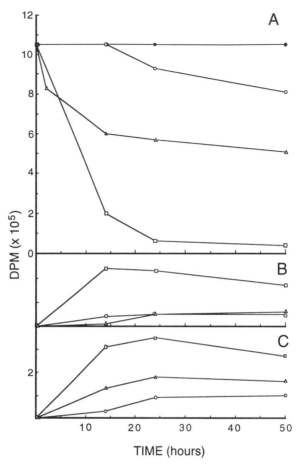

Fig. 3.3. Utilization of surface–bound stearic acid by bacteria: **(A)** ^{14}C remaining on surface after exposure to bacteria for varying times and **(B)** ^{14}C associated with bacteria in the aqueous phase; **(C)** ^{14}C in bacterial free supernatant. △, *Serratia marcescens* EF190; ○, *S. marcescens* (nonpigmented); □, *Leptospira biflexa patoc* 1; ●, Control. (Adapted from Kefford et al., 1982, with permission of the publisher.)

other macromolecules to soil and sediment components, such as sands and clays (Marshall, 1971; Marshman and Marshall, 1981a,b; Lorenz and Wackernagel, 1987; and the excellent, extensive review by Stotzky, 1986) (see also Chapter 2, section 2.1.1.4, this volume). Because of the almost infinite variability of soil components and, to a lesser extent, those of sediments, macromolecule–particle interactions range from inhibitory, to neutral, to stimulatory effects on extracellular enzyme activity, on microbial growth, and on microbial interactions. In a simple two-organism system with starch as substrate, there was a marked increase in the complexity of flow of nutrients to the organisms in the presence of montmorillonitic clay (Marshman and Marshall, 1981b). The clay probably interacted with all components of the system—the microorganisms, the substrates, the enzymes, and the growth factors. Taking the immense complexity of natural soils into consideration, even with some knowledge of the microfabric of the soil (Marshall, 1971), it is almost impossible to predict the nature and extent of interactions that may occur between the biological and nonbiological components of the system (see Chapter 2, this volume).

Samuelsson and Kirchman (1991), in a study the degradation of adsorbed ^{3}H-labeled ribulose-1,5-bisphosphate carboxylase (RuBPCase) by the marine bacterium *Pseudomonas* S9, found that the percent degradation decreased with increasing hydrophobicity of the substratum. Growth rates of attached bacteria were initially higher on hydrophilic glass than on hydrophobic polyethylene, but, on prolonged incubation (6 h), growth rates increased with substratum hydrophobicity because of the availability of increasing amounts of adsorbed RuBPCase on such surfaces (Kirchman et al., 1989). Griffith and Fletcher (1991) demonstrated that the dipeptide analogue methyl-coumarinyl-amide-leucine (MCA-leucine) adsorbed poorly to particles derived from diatoms and was hydrolyzed faster by nonattached than attached cells of the marine strain *Pseudomonas* NCIMB 2021. In contrast, almost 100% of ^{14}C-labeled bovine serum albumin adsorbed to the particles, was readily hydrolyzed by attached bacteria, and was poorly available to nonattached bacteria.

3.4.2.2.   *Substrata as Substrates*

Many of the particles encountered by microorganisms in natural environments are derived from comminution of living debris and, as such, consist of relatively large amounts of insoluble macromolecules. These organic particles serve the dual roles of colonizable substrata and potential energy substrates for some of these organisms. For instance, many of the cytophaga-like gliding bacteria are able to utilize complex polysaccharides such as cellulose, chitin, alginates, and agar (Reichenbach, 1992). Numerous other bacteria produce extracellular enzymes capable of hydrolyzing proteins, nucleic acids, and polysaccharides, thereby releasing soluble products that are metabolized by these organisms or by neighboring bacteria lacking this hydrolytic capability. The release of low-molecular-weight products by bacteria attached to insoluble substrates can elicit chemotactic responses toward the substratum by motile bacteria within the product diffusion zone (Malmcrona-Friberg et al., 1991). (See also Chapters 6 and 10, this volume.)

3.4.3.   Diversity in Bacterial Growth at Surfaces

Bott and Brock (1970) presented a direct demonstration of growth of bacteria on glass surfaces by irradiating controls with ultraviolet light to separate passive attachment from actual growth. Although continuous observations have been made of the behavioral and growth patterns of bacteria at surfaces by Lawrence et al. (1987), Szewzyk and Schink (1988), and Siebel and Characklis (1991), the substrates employed, glucose, gallic acid, and glucose, respectively, probably do not adsorb at surfaces and do not form CF. Davies and McFeters (1988) reported that cells of *Klebsiella oxytoca* attached to granular activated carbon particles had a growth rate 10 times higher than cells in liquid with glutamate, a substrate that adsorbed to the surface, but no difference was observed with glucose, a substrate that did not adsorb to granular activated carbon. Yet Pedrós-Alió and Brock (1983) found, using acetate as a substrate, that attached bacteria were larger (average vol. = 0.271 μm^3) than planktonic bacteria (average vol. = 0.118 μm^3), had a higher frequency of dividing cells, and were responsible for most of the acetate uptake. The larger attached cells suggest that many of the planktonic cells were starving (Morita, 1982) and, hence, remained small, whereas the attached cells were growing on substrate available at the surface.

3.4.3.1. Mother–Daughter or Shedding Cells
Starved cells of the marine *Vibrio* DW1 are more adhesive than unstarved cells, leading to the suggestion that adhesion may be a tactic in the survival strategy of these organisms under starvation conditions (Dawson et al., 1981). To test growth of starved cells at surfaces, the carbon, nitrogen (both as casamino acids equivalent to 2 mg organic $C L^{-1}$), and phosphorus components of a standard marine medium were reduced 1,000-fold in order to simulate an oligotrophic condition. Starved cells (7 days) of *Vibrio* DW1 were allowed to attach to the membrane of a dialysis microchamber (Duxbury, 1977), the medium flow was started and the behavior of the cells continuously monitored by time-lapse video recording (Kjelleberg et al., 1982). Viewing the videotape at normal speed revealed the following events: (1) after a variable lag period, the small starved cells (<0.5 μm diam.) grew to normal size (>1.0 μm length), presumably by utilizing substrate accumulating on the membrane surface (starved cells did not grow in the aqueous phase of the "oligotrophic" medium); (2) the growing cells attached to the substratum in a perpendicular orientation (Marshall and Cruickshank, 1973), and, as each growing cell began to divide, the daughter cells exhibited a rapid spinning motion prior to release into the aqueous phase (this indicated a form of physiological differentiation between the mother and daughter cells, as described by Dow and Whittenbury, 1980); (3) the mother cell, which remained attached to the substratum, repeated the processes of cellular growth and reproduction (Fig. 3.4A) at approximately 57 min intervals for some four to five cycles (Marshall, 1985); (4) many daughter cells returned to colonize the surface further; and (5) the extensive secondary colonization of the substratum resulted in nutrient uptake exceeding nutrient input to the surface, and, hence, the cells ceased dividing and reverted to typical small starvation forms. Lawrence and Caldwell (1987), using natural streamwater in a laminar flow chamber, reported similar surface colonization behavior, which they termed *shedding*.

A. "Mother-daughter" (Kjelleberg et al., 1982)

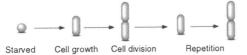

Starved   Cell growth   Cell division   Repetition

B. "Packing" (Lawrence and Caldwell, 1987)

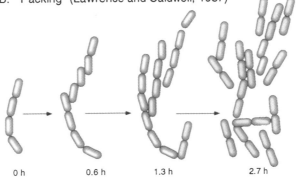

0 h   0.6 h   1.3 h   2.7 h

C. "Slow migration" (Power and Marshall, 1988)
   "Spreading"   (Lawrence and Caldwell, 1987)

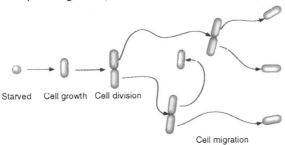

Starved   Cell growth   Cell division

Cell migration

D. "Non-adhesive" (Power and Marshall, 1988)

Drift from ⟶ Return of
surface   daughter cells

Starved   Cell growth   Early cell division

E. "Rolling" (Lawrence and Caldwell, 1988)

Fig. 3.4. Diversity in growth patterns in bacteria associated with surfaces: **(A)** Mother–daughter or shedding cells; **(B)** packing cells; **(C)** slow migratory or spreading cells; **(D)** reversibly adhering cells; **(E)** rolling cells.

3.4.3.2. Packing Cells

In their study of colonization from streamwater, Lawrence and Caldwell (1987) described bacteria that divided in a manner that resulted in the daughter cells aligning adjacent to each other. This eventually led to the formation of a monolayer of contiguous cells (Fig. 3.4B) and was termed a *packing maneuver.*

3.4.3.3. Slow Migratory or Spreading Cells

Studies of the behavior of *Pseudomonas* JD8 using time-lapse video and the Duxbury dialysis microchamber perfused with a medium lacking an energy source and CF of known composition (stearic acid adsorbed to the dialysis membrane as the sole energy source) have been reported by Power and Marshall (1988). Direct, progressive tracings from the TV screen clearly showed that adhering, small, starved cells grew to normal size (cellular growth), began to divide (cellular reproduction), and then the daughter cells migrated very slowly ($0.15 \pm 0.08 \ \mu m \ min^{-1}$) and at random for variable distances (5 to 50 times their cell length) across the surface before the next division occurred (Fig. 3.4C). This behavior, which is similar to that termed *spreading* by Lawrence and Caldwell (1987), may have resulted from the firm adhesion of cells to the hydrophobic fatty acid coated surface and the progressive utilization of substrate. This revealed the hydrophilic substratum below, allowing the cells to detach partly and move to adjacent fatty acid coated areas where the cells made contact until that substrate was utilized.

3.4.3.4. Reversibly Adhering or Rolling Cells

Although only reversibly attracted to the substratum, *Vibrio* MH3 was capable of metabolizing surface-bound stearic acid, the starved cells grew to normal size, and cellular division began near the surface, but, in all instances, the dividing cells returned to the aqueous phase (Fig. 3.4D) for the final production of daughter cells (Power and Marshall, 1988). That cell division proceeded to completion in the aqueous phase was confirmed by repeated observations of smaller cells returning to the surface and repeating the same cycle. The phenomenon described by Lawrence and Caldwell (1987) as *rolling,* that is, cells loosely attached and somersaulting across the surface as they grew and divided (Fig. 3.4E), similarly may arise as a result of the bacteria reversibly adhering to the surface.

3.4.3.5. Gliding Cells

Many cytophaga-like gliding bacteria are able to utilize complex polysaccharide substrata as an energy substrate (see section 3.4.2.2). In addition, they possess a unique form of adhesion (Stefan adhesion) that allows them to remain attached to the solid substrate in aqueous conditions and able to glide (translocate) across the surface (Humphrey et al., 1979). This characteristic certainly makes for efficient utilization of the solid substrates they metabolize.

Another factor that can modify the behavior of bacteria at surfaces is the adsorption of microbially produced surfactants to substrata. Starving cells of the marine *Vibrio* DW1 exhibit an increase in heat output and oxygen consumption and a decrease in cell volume at a surface (dialysis membrane) compared with the same batch of cells starving in the aqueous phase (Kjelleberg et al., 1982, 1983; Humphrey et al., 1983). Subsequently, Humphrey and Marshall (1984) found that the effects of the surface could be mimicked by a surfactant extracted from the membrane and by commercial surfactants. These authors urged caution in studies of bacterial activities at surfaces because of possible effects of adsorbed, microbially produced surfactants (Rosenberg, 1986) in modifying the behavior of attached bacteria. Aspects of bacterial behavior at surfaces have been reviewed recently by Korber *et al.* (1995).

Altered bacterial behavioral patterns also may result from surface-induced changes in gene expression (Dagostino et al., 1991; Davies et al., 1993), including differences in cell morphology and biofilm structure observed with the marine bacterium SW5 on hydrophobic and hydrophilic substrata (Dalton et al., 1994). One of many possible explanations for altered gene expression is a response to various nutrients accumulating at different surfaces (Marshall and Goodman, 1994; Goodman and Marshall, 1995).

Although Samuelsson and Kirchman (1991) determined growth rates of bacteria on surface-bound protein, these were determined by the uptake of radiolabeled substrate and not by direct microscopy. It is unfortunate that, to date, no attempt has been made to examine patterns of growth of bacteria on macromolecules adsorbed at surfaces. Researchers in this field are encouraged to use phase-contrast microscopy combined with either image analysis or time-lapse video recording in order to reveal greater diversity of growth behavior in bacteria scavenging surface-bound macromolecules.

3.5. DIVERSITY IN CELL DETACHMENT STRATEGIES

It would appear that many bacteria have developed a variety of strategies to alternate between the sessile and planktonic states in order to maximize their nutrient scavenging efficiency (Pedrós-Alió and Brock, 1983; Marshall, 1988, 1992b).

3.5.1. Direct Release of Daughter Cells

A very effective detachment adaptation results whenever bacteria attach to a surface in a perpendicular orientation by means of a hydrophobic pole of the cell (Marshall and Cruickshank, 1973). Cells attached in this manner do not divide by binary division, but the attached portion of the cell can be regarded as a mother cell that divides continually to give rise to a series of daughter cells (Fig. 3.4A). This mode of adhesion and, in general, reproduction has been found with *Escherichia coli* (Helmstetter and Cummings, 1963; Gilbert et al., 1989), *Leptothrix* (Brock and Conti, 1969), *Flexibacter* and *Hyphomicrobium* (Marshall and Cruickshank, 1973), *Caulobacter* and *Asticcaulis*

(Poindexter, 1981), *Rhodomicrobium* (Dow and Whittenbury, 1980), *Vibrio* DW1 (Kjelleberg et al., 1982), *Planktomyces* (Pedrós-Alió and Brock, 1983), and shedding cells from natural streamwater (Lawrence and Caldwell, 1987). A perpendicular orientation allows an efficient contact of the attached cell with the aqueous phase (Tyler and Marshall, 1967a) and provides an effective means of releasing daughter cells into the aqueous phase.

3.5.2. Enzymatic Degradation of Adhesive Polymers

Firm adhesion of bacteria to surfaces is often mediated by extracellular polymers bridging the bacterial cell and the surface (Fletcher and Floodgate, 1973; Marshall and Cruickshank, 1973), and it has been assumed that some bacteria detach from surfaces as a result of the degradation of these polymers by means of enzymes produced by the bacteria or by neighboring cells. Little direct evidence for the production of such extracellular enzymes at surfaces is available, but, recently, Boyd and Chakrabarty (1994) described the production of an alginate lyase by *Pseudomonas aeruginosa* that is responsible for cell detachment from surfaces.

Further studies are necessary to confirm the broader role of extracellular enzymes in the detachment of bacteria from surfaces.

3.5.3. Mechanical Rupture of Adhesive Polymers

Both the adhesive and cohesive strengths of bacterial adhesive polymers will vary considerably with bacterial type and with the substratum properties. For instance, Baier (1980) reviewed his work indicating lower adhesion rates and more ready detachment of bacteria on relatively low energy surfaces ($\gamma_c = 20$ to $24 \ mNm^{-1}$), a result confirmed in marine environments by Dexter et al. (1975; Dexter, 1979). Other bacteria adhere more effectively to intermediate or high energy surfaces (Pringle and Fletcher, 1983). Shear forces encountered in flowing water, wave motion, and other forms of turbulence, including that resulting from sonication, can result in the physical detachment of bacteria from surfaces (Duddridge et al. 1982; Powell and Slater, 1982). In most instances, the cohesive strength of the bacterial adhesive polymers will be less than their adhesive strength, resulting in breaks in the polymer chains to leave a polymer "footprint" on the surface (Neu and Marshall, 1991). These "footprints" make up part of the CF encountered by subsequent colonizing bacteria (Neu, 1992).

3.5.4. Chemical Alterations to Adhesive Polymers

Altered chemical environmental conditions in the vicinity of a surface may lead to a weakening in the adhesive or cohesive strength of adhesive polymers. The production of a chelating agent by some microorganisms may lead to the complexing of divalent bridging cations, particularly Ca^{2+} ions, resulting in the detachment of cells, such as found by Turakhia et al. (1983) with the commercial chelant EGTA. Other chemical changes near the surface may alter the configuration of adhesive molecules, rendering them susceptible to shear forces and leading to detachment.

3.5.5. Alterations to Substratum and/or CF Properties

Bacteria may detach from surfaces as a result of the excretion of surface-modifying products by the bacteria or by neighboring microorganisms. For instance, many bacteria produce surfactants that could displace hydrophobic bacteria from hydrophobic surfaces (Rosenberg, 1986) and can even alter the metabolic characteristics of attached bacteria (Humphrey and Marshall, 1984). Bacteria released in this manner may be transported to other colonizable sites or remain in the planktonic state.

The adsorption of CF to a surface changes the physicochemical properties of the surface and can modify the adhesive behavior of colonizing bacteria (see sections 3.2 and 3.3). In turn, the attached bacteria can alter the apparent surface properties by utilizing the adsorbed CF as an energy source. This is well illustrated by the study of the growth behavior of *Pseudomonas* JD8, where the substratum was a hydrophilic dialysis membrane but was rendered hydrophobic by the adsorption of stearic acid CF (Power and Marshall, 1988). Very slow movement of cells across the substratum surface was attributed to progressive utilization of the fatty acid CF in the vicinity of cells, partial detachment of cells from the original hydrophilic substratum, reattachment when cells again encountered more of the hydrophobic substrate, and a continued repetition of this cycle (Fig. 3.4C). The cycle ended when most of the hydrophobic substrate was consumed and the cells detached from the predominantly hydrophilic substratum (Power and Marshall, 1988).

3.5.6. Alterations to Bacterial Surface Properties

The outer chemical composition of a bacterial cell is never static and changes rapidly in response to changes in the external physicochemical environment (Wicken, 1985). An example of how changes in cell surface properties with time can effect detachment from a surface is the observation by Rosenberg et al. (1983) that *Acinetobacter calcoaceticus* attaches to oils and epithelial cells by means of hydrophobic thin fimbriae, but, later in the growth cycle, production of extracellular hydrophilic emulsan masks the fimbriae and results in detachment of cells from the surfaces. Similar phenomena have been found with the cyanobacterium *Phormidium* (Fattom and Shilo, 1984), whereby the sessile state of the organisms is characterized by a hydrophobic cell surface but, under certain conditions, a hydrophilic capsule is produced, leading to the cells detaching from the surface, and in *P. aeruginosa* biofilms, whereby cells released from the biofilm were constantly more hydrophilic than their adherent counterparts (Allison et al., 1990).

3.5.7. Reversibly Adhering Bacteria

Bacteria that do not adhere firmly to surfaces are capable of scavenging adsorbed CF at surfaces (see section 3.4.1). Power and Marshall (1988) demonstrated that small, starved cells of the nonadhesive *Vibrio* MH3 grew to normal size on utilization of surface-bound stearic acid and began the process of cellular reproduction while adjacent to the surface. The dividing cells then drifted from the surface, but it was assumed that the division cycle went to completion in the aqueous phase as numerous normal-sized cells were observed returning to near the surface.

3.5.8. Some Thoughts on the Relative Activities of Attached and Nonattached Bacteria

A great deal of attention has been given to the association of bacteria with particles and to the relative activities of attached and nonattached bacteria (see the excellent review on this subject by Fletcher, 1991). The statement by Pedrós-Alió and Brock (1983) that a simple division into sessile (attached) and planktonic (nonattached) is overly simplistic is conceptually sound in view of the diversity of detachment strategies detailed in sections 3.5.1 to 3.5.7. In addition, some aspects of the studies on distributions and activities of bacteria in relation to particle association suffer from faulty logic, such as (1) in many instances, the substrates employed to test metabolic activity are small, soluble molecules (glucose, in particular) that do not adsorb to surfaces and, in any case, would be rapidly utilized in the aqueous phase before encountering a particle or other solid surface; (2) attached and nonattached bacteria are normally separated by filtration procedures—the shear forces involved in this technique would ensure that reversibly attached bacteria would be classed as nonattached and, yet, they can scavenge surface-bound substrates (Hermansson and Marshall, 1985); and (3) the possession of a variety of mechanisms to ensure an interchange between the sessile and planktonic states means that, at any single sampling time, many nonattached cells will have recently left the surface from which they have gained an adequate supply of nutrients. This dynamic behavior of bacteria in natural habitats is well illustrated by the careful microscopic observations of Alldredge (1989) that demonstrate the movement of bacteria to and from the aqueous phase to suspended detrital aggregates (marine snow).

3.6. INORGANIC NUTRIENTS AND REACTIONS AT SURFACES

The discussion to this point has concentrated on access by bacteria to organic energy substrates at surfaces. Because most surfaces in nature possess or soon assume a net negative charge (Marshall, 1976), cations tend to be attracted to the vicinity of the surfaces, forming what is termed the *electrical double layer,* but anions are not entirely excluded from this zone, a process termed *negative adsorption* (see also Chapter 2, section 2.5.2.1, this volume). Some of the ions attracted to a surface are required as nutrients by all bacteria (Ca^{2+}, NH_4^+, Fe^{3+}, PO_4^{3-}, etc.), whereas others may serve as inorganic sources of energy for chemolithotrophic bacteria (NH_4^+, NO_2^-, Fe^{2+}, S^{2-}, etc.) or as alternate electron acceptors (NO_3^-, SO_4^{2-}, etc.).

3.6.1. Nitrification and Denitrification at Surfaces

Nitrification is a two-step oxidation of ammonia to nitrate, with the first step, the oxidation of ammonia to nitrite, carried out by *Nitrosomonas* and related genera and the second, the oxidation of nitrite to nitrate, by *Nitrobacter* and related genera. The nitrifying bacteria are typically found attached to surfaces in water conduits, wastewater treatment plants, deteriorating stone buildings, and, presumably, in soils (although see Stotzky, 1986). Positively charged ammonium ions are attracted to surfaces, and, in flowing systems, the *Nitrosomonas*-like bacteria should benefit by the availability of their main energy substrate at the surface. Nitrite ions, on the other hand, are negatively charged and should be repelled from a surface. A close association between the *Nitrosomonas*-like and *Nitrobacter*-like bacteria would provide ready access to the nitrite product of the former group, with minimal diffusional problems, for use as an energy substrate by the latter group. Studies on the growth of *Nitrosomonas* (Armstrong and Prosser, 1988; Allison and Prosser, 1993) and *Nitrobacter* (Keen and Prosser, 1987, 1988) at surfaces indicate that the kinetics of nitrification depend on the extent to which the surface is colonized. Initial growth is exponential, but becomes linear at high colonization densities as a result of a combination of reductions in pH and in oxygen and substrate diffusion.

Denitrification is the process whereby certain bacteria, in the absence of oxygen, utilize nitrate as an alternative electron acceptor and reduce it to nitrogen gas. This process is important in anaerobic zones in soils as it constitutes a loss of nitrogen normally available to plants. In wastewater treatment, where large amounts of organic nitrogen compounds are converted to ammonia and subsequently oxidized in nitrate, bacterial denitrification processes provide an economical means of removal of the potentially polluting nitrate. Biofilm formation (see section 3.7) by bacteria initially provides aerobic conditions suitable for nitrification to proceed, but oxygen diffusion becomes limiting in thick biofilms (Dalsgaard and Revsbech, 1992; de Beer et al., 1993a,b) and such conditions favor denitrification of any nitrate present.

3.6.2. Sulfur Oxidation at Surfaces

Bacteria whose sole source of energy may be reduced sulfur compounds, such as various species of *Thiobacillus,* often have to contend with solid forms of the substrate (elemental sulfur, metallic sulfides). In such cases, direct contact between the bacteria and the substrate is essential. *Thiobacillus thiooxidans* apparently attaches to the surface of elemental sulfur by an energy-dependent process involving thiol groups (Takakuwa et al., 1979). The means whereby the sulfur is transported across the cell membrane is uncertain (Kaplan and Rittenberg, 1962), although phosphatidylinositol may play a role as a wetting agent (Schaeffer and Umbreit, 1963). There is some evidence for enzymatically mediated attack by *T. ferrooxidans* on sulfide and/or metallic ions in mineral sulfide lattices (Tuovinen and Kelly, 1972). Interestingly, *T. ferrooxidans* simultaneously utilizes the oxidation of iron (II) and elemental sulfur when grown in liquid medium (Espejo et al., 1988), but it preferentially oxidizes the sulfide moiety rather than the iron when attached to pyrite (Wakao et al., 1984) or to chalcopyrite (Shrihari et al., 1991).

3.6.3. Iron and Manganese Oxidation at Surfaces

Bacteria inevitably are associated with iron (Lundgren and Dean, 1979) and manganese (Marshall, 1979) deposits found on surfaces in flowing aquatic systems. The oxidation of ferrous iron to the ferric form occurs chemically at neutral pH levels, and it is difficult to demonstrate that bacteria associated with iron deposition under these conditions gain energy from the oxidation process. At low pH (<3.0), *T. ferrooxidans* certainly utilizes the oxidation of iron (II) as its sole source of energy. Where the substrate is solid pyrite, however, the sulfide moiety tends to be utilized by the organism in preference to the iron (see section 3.6.2).

Manganous ions are more stable to autooxidation at pH 7.0 than ferrous ions, particularly when adsorbed to the insoluble manganese (IV) species (Marshall, 1979). It is uncertain that manganese bacteria gain energy from the oxidation of the manganese (II) species under most natural conditions, yet bacterial deposition of manganic oxides on many surfaces exposed to flowing water, such as in water reticulation and hydroelectric pipeline systems (Tyler and Marshall, 1967b), is a very regular occurrence. Bacterial oxidation of manganese does appear to be enhanced at surfaces (Marshall, 1980; Nealson and Ford, 1980), an observation that suggests altered gene expression at surfaces compared to the behavior of the same bacteria in the aqueous phase (Goodman and Marshall, 1995).

3.7. BIOFILM DEVELOPMENT—THE INEVITABLE CONSEQUENCE OF BACTERIAL ADHESION TO AND NUTRIENT ACCESS AT SURFACES

Biofilms develop as a result of continued colonization by and growth of microorganisms on surfaces. Some of the practical implications of biofilms formed on man-made structures are dealt with elsewhere in this volume (Chapter 4).

3.7.1. Biofilm Development and Structure

Although CF supports limited growth of the primary colonizing bacteria, sustained growth leading to biofilm formation is dependent on a continual replenishment of the nutrient supply by transport from the aqueous phase. This, in turn, is facilitated by flowing water, particularly under oligotrophic conditions (Characklis and Marshall, 1990). Beginning with a clean substratum, the various stages in the development of a biofilm (Marshall, 1992a) are the adsorption of molecules to the surface (CF), the adhesion of bacteria to the CF-modified surface (phase 1 of bacterial involvement), bacterial growth and reproduction (phase 2), and the development of a polymer matrix of bacterial origin surrounding the bacterial biomass (phase 3).

The old image of a biofilm consisting of a relatively uniform array of bacteria embedded in an equally uniform polymer matrix (Characklis and Marshall, 1990) has been put to rest by recent studies. The application of confocal laser microscopy, combined with the use of pH-sensitive fluorescent dyes (Lawrence et al., 1991; Caldwell et al., 1992) and with physical techniques such as oxygen microelectrodes (de Beer et al., 1993b; Stewart et al., 1993), has revealed that biofilms are discontinuous, with the bacterial biomass and surrounding polymer matrix existing as a series of columnar structures surrounded by water-filled spaces.

Despite the fact that this model of biofilm structure probably holds true for most biofilms, caution is urged in assuming that the model is universally applicable. Examples of filamentous biofilms in nature have been reported (Characklis, 1980), and even monocultures of normally rod-shaped bacteria, such as *Pseudomonas* sp. (McCoy and Costerton, 1982) and the marine bacterium SW5 (Dalton et al., 1994), can form filamentous biofilms.

3.7.2. Mass Transport of Nutrients, Metabolic By-Products, and Gases in Biofilms

There are important implications concerning nutrient access for biofilm bacteria in the modern concept of biofilm structure. Diffusion of oxygen and nutrients into biofilms and of carbon dioxide and metabolic by-products out of biofilms was thought to be a function of the hydrodynamics of the water phase and the barrier posed by a uniform biofilm. For instance, measurements of adenine nucleotide pools and adenylate energy charge in sections of quasi-steady-state *P. aeruginosa* biofilms indicate low microbial activity, especially in deeper levels of biofilms where oxygen diffusion should be limiting (Kinnement and Wimpenny, 1992). The discontinuities between the bacterial columnar structures, however, mean that void spaces are present, they are water filled, and there is definite water flow around the columnar biomass structures (Stoodley et al., 1994). Consequently, oxygen can freely diffuse to the base of a void (de Beer et al., 1993b; Lawrence et al., 1994), thereby dramatically increasing the interfacial area of biofilm exposed to oxygen compared with that anticipated by the old model of a biofilm. Potential nutrient diffusion into the biofilm, likewise, is increased, as is diffusion of carbon dioxide and metabolic by-products from the biomass to the aqueous phase.

It would appear that biofilm bacteria gain from continuous access to nutrients in flowing systems, and, as a result, they adapt to optimize the structure of the biofilms in order to facilitate the mass transport of nutrients, gases, and metabolic by-products into and out of the biofilms.

3.8. CONCLUSIONS

Although CF plays a definite role in the adhesion of bacteria to surfaces and in providing these bacteria with access to nutrients, very little is known of the precise composition of CF derived from different environments, the configuration of CF at different surfaces, the ways in which CF hinders or stimulates bacterial adhesion, and the overall availability as nutrient and energy sources in the mixture of molecules making up the CF.

The few studies made on the modes of growth of bacteria on CF have revealed some degree of diversity in cellular growth, reproduction, and detachment behavior. More extensive studies should demonstrate even greater diversity in these characteristics. It is certain that, following adhesion to a CF-coated surface, bacteria normally starving in the aqueous phase quickly exhibit cellular growth and progress to the stage of cellular reproduction in this specialized surface environment. Further input of nutrients transported from an aqueous phase flowing past the surface ensures continuous growth and polymer production, leading to the formation of mature biofilms.

Finally, the sequence beginning with nutrient access for bacteria at surfaces in the form of CF and ending with mature biofilms provides nature with a huge reservoir of bacterial biomass available at any time to ensure effective biodegradation of organic materials and the transformation of inorganic compounds. In a more modern context, these biofilms are also available for the biodegradation of xenobiotic compounds.

REFERENCES

Abramson HA, Moyer LS, Gorin MH (1942): Electrophoresis of Proteins and the Chemistry of Cell Surfaces. New York: Reinhold.

Alldredge L (1989): The significance of suspended detrital aggregates of marine snow as microhabitats in the pelagic zone of the ocean. In Hattori T, Ishida Y, Maruyama Y, Morita RY, Uchida A (eds): Recent Advances in Microbial Ecology. Tokyo: Japan Scientific Societies Press, pp 108–112.

Allison DG, Brown MRW, Evans E, Gilbert P (1990): Surface hydrophobicity and dispersal of *Pseudomonas aeruginosa* from biofilms. FEMS Microbiol Lett 71:101–104.

Allison SM, Prosser JI (1993): Ammonia oxidation at low pH by attached populations of nitrifying bacteria. Soil Biol Biochem 25:935–941.

Armstrong EF, Prosser JI (1988): Growth of *Nitrosomonas europaea* on ammonia-treated vermiculite. Soil Biol Biochem 20:409–411.

Baier RE (1980): Substrate influence on adhesion of microorganisms and their resultant new surface properties. In Bitton G, Marshall KC (eds): Adsorption of Microorganisms to Surfaces. New York: Wiley Interscience, pp 59–104.

Baier RE, Shafrin EG, Zisman WA (1968): Adhesion: Mechanisms that assist or impede it. Science 162:1360–1368.

Bartick G, Messerschmidt G (1984): Applications of cylindrical internal reflection for FTIR liquid sampling. Am Lab 14:56–61.

Blainey BL, Marshall KC (1991): The use of block copolymers to inhibit bacterial adhesion and biofilm formation on hydrophobic surfaces in marine habitats. Biofouling 4:309–318.

Bott TL, Brock TD (1970): Growth and metabolism of periphytic bacteria: Methodology. Limnol Oceanogr 15:333–342.

Boyd A, Chakrabarty AM (1994): Role of alginate lyase in cell detachment of *Pseudomonas aeruginosa*. Appl Environ Microbiol 60:2355–2359.

Brock TD, Conti F (1969): Electron microscope studies on *Leucothrix mucor*. Arch Mikrobiol 66:79–90.

Caldwell DE, Korber DR, Lawrence JR (1992): Confocal laser microscopy and computer image analysis in microbial ecology. Adv Microbial Ecol 12:1–67.

Characklis G (1980): Biofilm development and destruction. Electric Power Research Institute Report 902-1, USA (September, 1980).

Characklis WG, Marshall KC (eds) (1990): Biofilms. New York: Wiley Interscience.

Chet I, Asketh P, Mitchell R (1975): Repulsion of bacteria from marine surfaces. Appl Microbiol 30:1043–1045.

Cox PJ, Twigg GI (1974): Leptospiral motility. Nature (Lond) 250:260–261.

Dagostino L, Goodman AE, Marshall KC (1991): Physiological responses induced in bacteria adhering to surfaces. Biofouling 4:113–119.

Dalsgaard T, Revsbech NP (1992). Regulating factors of denitrification in trickling filter biofilms as measured with the oxygen/nitrous oxide microsensor. FEMS Microbiol Ecol 101:151–164.

Dalton HM, Poulsen K, Halasz P, Angles ML, Goodman AE, Marshall KC (1994): Substrata-induced morphological changes in a marine bacterium and their relevance to biofilm structure. J Bacteriol 176:6900–6906.

Davies DG, Chakrabarty AM, Geesey G (1993): Exopolysaccharide production in biofilms: Substratum activation of alginate expression by *Pseudomonas aeruginosa*. Appl Environ Microbiol 59:1181–1186.

Davies DG, McFeters GA (1988): Growth and comparative physiology of *Klebsiella oxytoca* attached to granular activated carbon and in liquid media. Microbial Ecol 15:165–175.

Dawson MP, Humphrey BA, Marshall KC (1981): Adhesion: A tactic in the survival strategy of a marine *Vibrio* during starvation. Curr Microbiol 6:195–199.

de Beer D, van den Heuvel JC, Ottengraph SPP (1993a): Microelectrode measurements of the activity distribution of nitrifying bacterial aggregates. Appl Environ Microbiol 59:573–579.

de Beer D, Stoodley P, Roe F, Lewandowski Z (1993b): Effects of biofilm structures on oxygen distribution and mass transport. Biotechnol Bioeng 43:1131–1138.

Dexter SC (1979): Influence of substratum critical surface tension on bacterial adhesion—in situ studies. J Colloid Interface Sci 70:346–354.

Dexter SC, Sullivan JD Jr, Williams J, Watson SW (1975): Influence of substratum wettability on the attachment of marine bacteria to various surfaces. Appl Microbiol 30:298–308.

Dow S, Whittenbury R (1980): Prokaryotic form and function. In Ellwood DC, Hedger JN, Latham MJ, Lynch JM Slater JH (eds): Contemporary Microbial Ecology. London: Academic Press, pp 391–417.

Duddridge JE, Kent CA, Laws JF (1982): Effect of surface shear on the attachment of *Pseudomonas fluorescens* to stainless steel under defined flow conditions. Biotechnol Bioeng 24:153–164.

Duxbury T (1977): A microperfusion chamber for studying the growth of bacterial cells. J Appl Bacteriol 42:247–251.

Einolf CW, Carstensen EL (1967): Bacterial conductivity in the determination of surface charge by microelectrophoresis. Biochim Biophys Acta 148:506–516.

Elwing H, Gölander C-G (1990): Protein and detergent interaction phenomena on solid surfaces with gradients in chemical composition. Adv Colloid Interface Sci 32:317–339.

Espejo T, Escobar B, Jedlicki E, Uribe P, Badilla-Ohlbaum R (1988): Oxidation of ferrous iron and elemental sulfur by *Thiobacillus ferrooxidans*. Appl Environ Microbiol 54:1694–1699.

Fattom A, Shilo M (1984): Hydrophobicity as an adhesion mechanism of benthic cyanobacteria. Appl Environ Microbiol 47:135–143.

Fletcher M (1976): The effects of proteins on bacterial attachment to polystyrene. J Gen Microbiol 94:400–404.

Fletcher M (1991): The physiological activity of bacteria attached to solid surfaces. Adv Microbial Physiol 32:53–85.

Fletcher M, Floodgate GD (1973): An electron-microscopic demonstration of an acidic polysaccharide involved in adhesion of a marine bacterium on solid surfaces. J Gen Microbiol 74:325–334.

Fletcher M, Loeb GI (1979): Influence of substratum characteristics on the attachment of a marine pseudomonad to solid surfaces. Appl Environ Microbiol 37:67–72.

Fletcher M, Marshall KC (1982): Bubble contact angle method for evaluating substratum interfacial characteristics and its relevance to bacterial attachment. Appl Environ Microbiol 44:184–192.

Gilbert P, Allison DG, Evans DJ, Handley PS, Brown MRW (1989): Growth rate control of adherent bacterial populations. Appl Environ Microbiol 55:1308–1311.

Gittens GJ, James AM (1963): Some physical investigations of the behavior of bacterial surfaces. VII. The relationship between zeta potential and surface charge as indicated by microelectrophoresis and surface conductance measurements. Biochim Biophys Acta 66:250–263.

Goodman AE, Marshall KC (1995): Genetic responses of bacteria at surfaces. In Lappin-Scott HM, Costerton JW (eds): Microbial Biofilms. Cambridge, England: Cambridge University Press, pp 80–98.

Gordon AS, Millero FJ (1985): Adsorption mediated decrease in the biodegradation rate of organic compounds. Microbial Ecol 11:289–298.

Griffith PC, Fletcher M (1991): Hydrolysis of protein and model dipeptide substrates by attached and nonattached marine *Pseudomonas* sp. strain NCIMB 2021. Appl Environ Microbiol 57:2186–2191.

Helmstetter CE, Cummings D (1963): An improved method for the selection of bacterial cells at division. Biochim Biophys Acta 82:608–610.

Hermansson M, Marshall KC (1985): Utilization of surface localized substrate by nonadhesive marine bacteria. Microbial Ecol 11:91–105.

Heukelekian H, Heller A (1940): Relation between food concentration and surface for bacterial growth. J Bacteriol 40:547–558.

Holmström C, Kjelleberg S (1994): The effect of external biological factors in settlement of marine invertebrate and new antifouling technology. Biofouling 8:147–160.

Humphrey BA, Dickson MR, Marshall KC (1979): Physicochemical and in situ observations on the adhesion of gliding bacteria to surfaces. Arch Microbiol 120:231–238.

Humphrey BA, Kjelleberg S, Marshall KC (1983): Response of marine bacteria under starvation conditions at a solid–water interface. Appl Environ Microbiol 45:43–47.

Humphrey BA, Marshall KC (1984): The triggering effect of surfaces and surfactants on heat output, oxygen consumption and size reduction of a starving marine *Vibrio*. Arch Microbiol 140:166–170.

Hunter KA, Liss PS (1982): Organic matter and the surface charge of suspended particles in estuarine waters. Limnol Oceanogr 27:322–335.

Jannasch HW (1973): Bacterial content of particulate matter in offshore surface waters. Limnol Oceanogr 18:340–342.

Kaplan IR, Rittenberg SC (1962): Fractionation of isotopes in relation to the problem of elemental sulfur transport by microorganisms. Nature (Lond) 194:1098–1099.

Keen GA, Prosser JI (1987): Interrelationship between pH and surface growth of *Nitrobacter*. Soil Biol Biochem 19:665–672.

Keen GA, Prosser JI (1988): The surface growth and activity of *Nitrobacter*. Microbial Ecol 15:21–39.

Kefford B, Kjelleberg S, Marshall KC (1982): Bacterial scavenging: Utilization of fatty acids localized at a solid–liquid interface. Arch Microbiol 133:257–260.

Kinnement SL, Wimpenny JW (1992): Measurements of the distribution of adenylate concentrations and adenylate energy charge across *Pseudomonas aeruginosa* biofilms. Appl Environ Microbiol 58:1629–1635.

Kirchman DL, Henry DL, Dexter SC (1989): Adsorption of proteins to surfaces in seawater. Marine Chem 27:201–217.

Kjelleberg S, Humphrey BA, Marshall KC (1982): The effect of interfaces on small starved marine bacteria. Appl Environ Microbiol 43:1166–1172.

Kjelleberg S, Humphrey BA, Marshall KC (1983): Initial phases of starvation and activity of bacteria at surfaces. Appl Environ Microbiol 46:978–984

Korber DR, Lawrence JR, Lappin-Scott, HM, Costerton (1995): Growth of microorganisms on surfaces. In Lappin-Scott HM, Costerton JW (eds): Microbial Biofilms. Cambridge, England: Cambridge University Press, pp 15–45.

Kristoffersen A, Rolla G, Sonju T, Jantzen E (1982): The organic film developed on metal surfaces exposed to seawater: Chemical studies. J Colloid Interface Sci 90:191–196.

Lawrence JR, Caldwell DE (1987): Behavior of bacterial stream populations within the hydrodynamic boundary layers of surface microenvironments. Microbiol Ecol 14:15–27.

Lawrence JR, Delaquis PJ, Korber DR, Caldwell DE (1987): Behavior of *Pseudomonas fluorescens* within the hydrodynamic boundary layers of surface microenvironments. Microbiol Ecol 14:1–14.

Lawrence JR, Korber DR, Hoyle BD, Costerton JW, Caldwell DE (1991): Optical sectioning of microbial biofilms. J Bacteriol 173:6558–6567.

Lawrence JR, Wolfaardt GM, Korber DR (1994): Determination of diffusion coefficients in biofilms by confocal laser microscopy. Appl Environ Microbiol 60:1166–1173.

Lorenz MG, Wackernagel W (1987): Adsorption of DNA to sand and variable degradation rates of adsorbed DNA. Appl Environ Microbiol 53:2948–2952.

Lundgren DG, Dean W (1979): Biogeochemistry of iron. In Trudinger PA, Swaine DJ (eds): Biogeochemical Cycling of Mineral-Forming Elements. Amsterdam: Elsevier, pp 211–251.

Maki JS, Rittschof D, Mitchell R (1992): Inhibition of larval barnacle attachment to bacterial films: An investigation of physical properties. Microbial Ecol 23:97–106.

Malmcrona-Friberg K, Blainey BL, Marshall KC (1991): Chemotactic response of a marine bacterium to breakdown products of an insoluble substrate. FEMS Microbiol Ecol 85:199–206.

Marshall KC (1971): Sorptive interactions between soil particles and microorganisms. In McLaren AD, Skujins J (eds): Soil Biochemistry. Vol 2. New York: Marcel Dekker Inc., pp 409–485.

Marshall KC (1976): Interfaces in Microbial Ecology. Cambridge, MA: Harvard University Press.

Marshall KC (1979): Biogeochemistry of manganese minerals. In Trudinger PA, Swaine DJ (eds): Biogeochemical Cycling of Mineral-Forming Elements. Amsterdam: Elsevier, pp 253–292.

Marshall KC (1980): The role of surface attachment in manganese oxidation by freshwater hyphomicrobia. In Trudinger PA, Walter MR, Ralph BJ (eds): Biogeochemistry of Ancient and Marine Environments. Canberra: Australian Academy of Science, pp 333–337.

Marshall KC (1985): Bacterial adhesion in oligotrophic habitats. Microbiol Sci 2:321–326.

Marshall KC (1988): Adhesion and growth of bacteria at surfaces in oligotrophic habitats. Can J Microbiol 34:503–506.

Marshall KC (1992a): Biofilms: An overview of bacterial adhesion, activity, and control at surfaces. ASM News 58:202–207.

Marshall KC (1992b): Planktonic versus sessile life of prokaryotes. In Balows A, Trüper HG, Dworkin W, Harder W, Schleifer K-H (eds): The Prokaryotes. A Handbook on the Biology of Bacteria: Ecophysiology, Isolation, Identification, Applications. Ed 2. New York: Springer-Verlag, pp 262–275.

Marshall KC, Blainey BL (1991): Role of bacterial adhesion in biofilm formation and biocorrosion. In Flemming H-C, Geesey GG (eds): Biofouling and Biocorrosion in Industrial Water Systems. Berlin: Springer-Verlag, pp 29–46.

Marshall KC, Cruickshank RH (1973): Cell surface hydrophobicity and the orientation of certain bacteria at interfaces. Arch Mikrobiol 91:29–40.

Marshall KC, Goodman AG (1994): Effects of adhesion on microbial cell physiology. Colloids Surfaces Biointerfaces 2:1–7.

Marshman NA, Marshall KC (1981a): Bacterial growth on proteins in the presence of clay minerals. Soil Biol Biochem 13:127–134.

Marshman NA, Marshall KC (1981b): Some effects of montmorillonite on the growth of mixed microbial cultures. Soil Biol Biochem 13:135–141.

McCoy WF, Costerton JW (1982): Fouling biofilm development in tubular flow systems. Dev Indust Microbiol 23:551–558.

Morita RY (1982): Starvation–survival of heterotrophs in the marine environment. Adv Microbiol Ecol 6:171–198.

Nealson KH, Ford J (1980): Surface enhancement of bacterial manganese oxidation: Implications for aquatic environments. Geomicrobiol J 2:21–37.

Neihof RA, Loeb GI (1972): The surface charge of particulate matter in seawater. Limnol Oceanogr 17:7–16.

Neihof R, Loeb G (1974): Dissolved organic matter in seawater and the electric charge of immersed surfaces. J Marine Res 32:5–12.

Neu TR (1992): Microbial "footprints" and the general ability of microorganisms to label interfaces. Can J Microbiol 38:1005–1008.

Neu TR, Marshall KC (1991): Microbial "footprints"—a new approach to adhesive polymers. Biofouling 3:101–112.

Neumann AW, Good RJ, Hope CJ, Sejpal M (1974): An equation of state approach to determine surface tensions of low energy solids from contact angles. J Colloid Interface Sci 49.291–304.

Nichols PD, Henson JM, Guckert JB, Nivens DE, White DC (1985): Fourier-transform infrared spectroscopic methods for microbial ecology: Analysis of bacteria, bacteria–polymer mixtures and biofilms. J Microbiol Methods 4:79–94.

Ørstavik D (1977): Sorption of streptococci to glass: effects of macromolecular solutes. Acta Pathol Microbiol Scand 85:47–53.

Owens NF, Gingell D, Rutter PR (1987): Inhibition of cell adhesion by a synthetic polymer adsorbed to glass shown under hydrodynamic stress. J Cell Sci 87:667–675.

Pedrós-Alió C, Brock TD (1983): The importance of attachment to particles for planktonic bacteria. Arch Hydrobiol 98:354–379.

Poindexter JS (1981): The caulobacters: Ubiquitous unusual bacteria. Microbiol Rev 45:123–179.

Powell MS, Slater NKH (1982): Removal rates of bacterial cells from glass surfaces by fluid shear. Biotechnol Bioeng 24:2527–2537.

Power K, Marshall KC (1988): Cellular growth and reproduction of marine bacteria on surface-bound substrate. Biofouling 1:163–174.

Pringle JH, Fletcher M (1983): Influence of substratum wettability on attachment of freshwater bacteria to solid surfaces. Appl Environ Microbiol 45:811–817.

Reichenbach H (1992): The Order Cytophagales. In Balows A, Trüper HG, Dworkin M, Harder W, Schleifer K-H (eds): The Prokaryotes. Ed 2. A Handbook on the Biology of Bacteria: Ecophysiology, Isolation, Identification, Applications. Ed 2. New York: Springer-Verlag, pp 3631–3675.

Robb ID (1984): Stereo-biochemistry and function of polymers. In Marshall KC (ed): Microbial Adhesion and Aggregation. Berlin: Springer-Verlag, pp 39–49.

Rosenberg E (1986): Microbial surfactants. CRC Crit Rev Biotechnol 3:109–132.

Rosenberg E, Gottlieb A, Rosenberg M (1983): Inhibition of bacterial adherence to epithelial cells and hydrocarbons by emulsan. Infect Immun 39:1024–1028.

Samuelsson M-O, Kirchman DL (1991): Degradation of adsorbed protein by attached bacteria in relationship to surface hydrophobicity. Appl Environ Microbiol 56:3643–3648.

Schaeffer WI, Umbreit WW (1963): Phosphatidyl-inositol as a wetting agent in sulfur oxidation by *Thiobacillus thiooxidans*. J Bacteriol 85:492–493.

Schneider RP, Marshall KC (1994): Retention of the gram-negative bacterium SW8 on surfaces—effects of microbial physiology, substratum nature and conditioning films. Colloids Surfaces Biointerfaces 2:387–396.

Shrihari KR, Gandhi KS, Natarajan KA (1991): Role of cell attachment in leaching of chalcopyrite mineral by *Thiobacillus ferrooxidans*. Appl Microbiol Biotechnol 36:278–282.

Siebel MA, Characklis WG (1991): Observations of binary population biofilms. Biotechnol Bioeng. 37:778–789.

Stark WH, Stadler J, McCoy E (1938): Some factors affecting the bacterial population of freshwater lakes. J Bacteriol 36:653–654.

Stewart PS, Peyton BM, Drury WJ, Murga R (1993): Quantitative observations of heterogeneities in *Pseudomonas aeruginosa* biofilms. Appl Environ Microbiol 59:327–329.

Stoodley P, de Beer D, Lewandowski Z (1994): Liquid flow in biofilm systems. Appl Environ Microbiol 60:2711–2716.

Stotzky G (1986): Influence of soil mineral colloids on metabolic processes, growth, adhesion, and ecology of microbes and viruses. In Interactions of Soil Minerals with Natural Organics and Microbes. Madison, WI: SSSA Spec. Publ. No 17, Soil Science Society of America, pp 305–427.

Szewzyk U, Schink B (1988): Surface colonization by and life cycle of *Pelobacter acidigallici* studied in a continuous flow microchamber. J Gen Microbiol 134:183–190.

Takakuwa S, Fujimori T, Iwasaki H (1979): Some properties of cell-sulfur adhesion in *Thiobacillus thiooxidans*. J Gen Appl Microbiol 25:21–29.

Touvinen OH, Kelly D (1972): Biology of *Thiobacillus ferrooxidans* in relation to the microbiological leaching of sulfide ores. Z Allg Mikrobiol 12:311–346.

Turakhia MH, Cooksey KE, Characklis WG (1983): Influence of a calcium-specific chelant on biofilm removal. Appl Environ Microbiol 46:1236–1238.

Tyler PA, Marshall KC (1967a): Form and function in manganese-oxidizing bacteria. Arch Mikrobiol 56:344–353.

Tyler PA, Marshall KC (1967b): Microbial oxidation of manganese in hydroelectric pipelines. Antonie van Leeuwenhoek J Microbiol Serol 33:171–183.

Wakao N, Mishina M, Sakurai Y, Shiota H (1984): Bacterial pyrite oxidation III. Adsorption of *Thiobacillus ferrooxidans* cells on solid surfaces and its effect on iron release from pyrite. J Gen Appl Microbiol 30:63–77.

Waksman SA, Carey CL (1935): Decomposition of organic matter in seawater by bacteria. 1. Bacterial multiplication in stored water. J Bacteriol 29:531–545.

Wicken AJ (1985): Bacterial cell walls and surfaces. In Savage DC, Fletcher M (eds): Bacterial Adhesion: Mechanisms and Physiological Significance. New York: Plenum Press, pp 45–70.

ZoBell CE (1943): The effect of solid surfaces upon bacterial activity. J Bacteriol 46:39–56.

ZoBell CE, Anderson DQ (1936): Observations on the multiplication of bacteria in different volumes of stored sea water and the influence of oxygen tension and solid surfaces. Biol Bull Woods Hole 71:324–342.

Zsolnay A, Little B (1983): Characterization of fouling films by pyrolysis and chemical ionization mass spectrometry. J Anal Appl Pyrol 4:335–341.

ADHESION TO BIOMATERIALS

MARC W. MITTELMAN

Departments of Surgery and Microbiology, University of Toronto, Toronto, Ontario M5G 2C4, Canada

4.1. INTRODUCTION

The application of implantable devices in medicine has increased dramatically over the past 20 years. Advances in materials engineering and surgical techniques—coupled with the demographics of an aging population—have led to annual increases of 7% to 15% in device implantations in North America alone. Implanted devices such as arterial grafts, hip joints, and dental implants have been approved for use on the basis of their host tissue biocompatibility and functional characteristics. For example, materials used in the total artificial heart are designed to ensure "blood compatibility" (nonthrombogenic) and structural integrity over a several-year time frame. Indeed, devices such as orthopedic pins, cardiac pacemakers, and vascular grafts are now designed for lifespans on the order of 10 to 20 years or more.

Despite these tremendous advances in the melding of the engineering and medical sciences, device-related bacterial infections are the single greatest challenge to the more widespread application of medical implants. Devices such as the total artificial heart and peritoneal dialysis catheters are so prone to colonization that their application is currently limited to relatively brief periods of time (Nose, 1992). Infection rates vary widely depending on the implant type and institution; however, rates of 1% to 2% for orthopedic implants and 100% for certain urinary catheters are often quoted in the literature (Dunn et al., 1993; Nickel, 1993; Stamm, 1991). The consequences of device-related infections are often catastrophic: For deep tissue and vascular implants that become colonized, amputation or death is seen as an outcome in approxi-

Bacterial Adhesion: Molecular and Ecological Diversity, pages 89–127
© *1996 Wiley-Liss, Inc.*

mately 50% of affected patients (Elliot and Faroqui, 1993). Considering that over 200,000 hip and knee implants, 350,000 vascular grafts, and 200,000,000 catheters are used annually in the United States alone (Ratner, 1993), the impact of device-related infections on patients and health care systems alike is staggering.

The vast majority of these infections are the result of bacterial colonization of device surfaces. Bacteria adhering to and colonizing medical devices exhibit a number of phenotypic and genotypic characteristics that are distinct from planktonic organisms and provide them with an adaptive advantage in an otherwise hostile environment replete with antibiotics and cellular and humoral antagonistic agents.

In marine, limnetic, riparian, and industrial aquatic ecosystems, the majority of microorganisms are associated with surfaces. The significance of sessile microorganisms in nutrient cycling, xenobiotic degradation, and biofouling activities was appreciated by environmental microbiologists long before medical microbiologists began to recognize the importance of adhesion as a virulence factor in a number of disease processes. While great strides have been made over the past 10 years in eliciting the molecular mechanisms of bacteria interactions with specific host tissue receptors (Chapters 7, 8, and 9, this volume), relatively little is known about the ecology of device-related infections. It is clear, however, that a number of similarities exist between bacterial adhesion to medical devices and other kinds of surfaces in both natural and man-made ecosystems.

This chapter will describe what is currently known about the sequence of events that leads to biomaterial-related infections. This material will, for the most part, be presented in the context of our current knowledge of "nonspecific" adhesion mechanisms as they have been elicited for a diverse array of substrata (Chapters 1 and 2, this volume).

4.2. DEVICE-RELATED INFECTIONS

4.2.1. Scope of Biomaterial Applications in Medicine

The applications for engineered materials in medicine have expanded dramatically over the past 20 years. The majority of medical devices have been and continue to be constructed of materials that are designed to meet the mechanical demands required by a given application. Table 4.1 provides an overview of the major types of medical devices along with the number implanted each year in North America. By some estimates (Ratner, 1993; Schoen, 1991), the number of surgical procedures to implant biomaterials is increasing by as much as 15% per year in North America alone. The tremendous growth in the biomaterials field has been motivated by three key factors: (1) improvements in both engineered materials and implantation (surgical) procedures; (2) demographics, which predict a longer lifespan in the developed world leading to

**TABLE 4.1. Prevalence of Medical Device
Implantation in North America**

Device	No. implanted/year[a]
Urinary catheters and stents	100,000,000
Breast prostheses	100,000
Hip and knee implants	150,000
Dental implants	20,000
Intraocular lenses	1,500,000
Contact lenses	18,000,000
Vascular grafts	350,000
Pacemakers	130,000
Peritoneal dialysis catheters	40,000
Heart valves	75,000
Vascular catheters and stents	150,000

[a] Figures for 1993. Sources: Bach and Motsch (1994); Egebo
et al. (1994); Gorman et al. (1994); Parras et al. (1994); Volkow
et al. (1994).

an increasingly aged population; and (3) improvements in the quality of health care delivery systems in the developing world. It is the combination of the diversity and number of implants used—coupled with the difficulty in treating biomaterial-related infections—that creates the need for a greater understanding of processes leading to colonization of biomaterials.

4.2.2. Selected Case Histories: Biomaterial-Centered Infections

The incidence of device-related bacterial infections is increasing with the more widespread application of implants in medicine. Infection problems linked to contamination of devices labeled as "sterile" are extremely rare. This is due to the effectiveness of current steam, dry heat, ethylene oxide, and γ-irradiation sterilization processes. One exception to this observation has been contamination of improperly sterilized xenograft tissues used in the manufacture of heart valves (Laskowski et al., 1977). In this case, the colonizing organisms were not killed by the "cold sterilant" protocol that was employed. This example contrasts, however, with the sources of bacteria that colonize the vast majority of implant-associated infections: (1) perioperative contamination, (2) exit site contamination for percutaneous devices, or (3) hematogenous spread from locations distal to the implant area. Each of these sources has common and distinct characteristics in terms of species diversity, severity of infection, and predisposing host conditions. As is discussed below, these characteristics are important determinants of bacterial colonization and biofilm formation leading to infection.

In one respect, device-related infections may be likened to accidents at nuclear power plants: low probability events with very severe consequences.

It is difficult to place a firm number on the probability that a given device will become infected. A number of factors can affect infection rate statistics; these include patient demographics (e.g., age, immune status), size of the institution performing implant procedures, and device design. The data in Table 4.2 may be taken as a broad indication of infection rates for various kinds of implanted devices. While a number of *in vitro* studies have shown that substratum surface properties influence bacterial adhesion (see below), there is no conclusive evidence that relates infection rates to materials of construction. The case histories cited in Table 4.2 suggest that the *application* is the single most important criterion influencing infection.

4.2.3. Genesis and Outcome of Biomaterial-Related Infections

Adhesion of one or more microorganisms to a foreign body is the first event in a series of both host and organismal reactions that leads to a biomaterial-related infection. Infection cannot initiate in the absence of the adhesion event. The effects of adhesion and subsequent fouling biofilm formation are similar for both biomaterials and other engineered surfaces. In a number of ways, the development of fouling biofilms on ship hulls, heat exchanger surfaces, and other engineered materials parallels the genesis and pathology of device-related infections. In both medical and aquatic ecosystems, otherwise clean surfaces are rapidly covered with an organic conditioning film. Bacteria adhere to this film and then produce extracellular polymeric substances (EPS), which can have a direct effect on both the substratum and the surrounding milieu. Finally, bacterial biofilms, whether on the surface of a heart valve or a ship's hull, are subject to attack by predators.

It is also interesting to note that there are deleterious consequences associated with the activities of these bacterial biofilm predators. In the body, complement activation, release of harmful cytokines, and production of mutagenic free radical species are all associated with biomaterial infections (Whitener et al., 1993). Similarly, the presence of a bacterial biofilm is required for the adhesion of some barnacle species and polychaete worm communities (Scheltema et al., 1981) to engineered surfaces in marine environments. The collateral damage caused by the response to development of a bacterial biofilm can thus greatly exceed that attributable to the activities of the bacteria alone.

The presence of bacteria on surfaces can lead to mechanical blockages, material deterioration, and surface contamination. Figure 4.1 provides some corresponding examples of fouling biofilms in both industrial and medical (host) environments. In each of these examples, a "fouling biofilm" is present that is composed of an organic conditioning film, bacterial cells, EPS, substratum reaction products, and entrained detrital material from the local environment. The role that these components play in bacterial adhesion to biomaterials and the genesis of infection is discussed below.

4.2.4. Host Responses to Implanted Biomaterials

The primary criterion for selection of medical implant materials of construction and design attributes has been biocompatibility with tissues in the vicinity of the implant site. Williams (1987) developed a consensus definition for biocompatibility as a material's ability to perform in a specific application with a suitable host response. There is an emerging trend to design materials that elicit specific biological activities (Langer et al., 1990; van Blitterswijk et al., 1991). These types of materials have been termed *biomimetic* (Ratner, 1993). The biocompatibility of a material for implantation into the body remains the most important factor in the selection of fabrication materials and device designs.

4.2.4.1. Biocompatibility

When a foreign body is implanted in an animal host, a number of reactions occur that are generally designed to wall-off and/or expel the invading material (Fig. 4.2). Migration of polymorphonucleocytes (PMN) and fibroblasts to the site occurs within minutes of implantation. The PMN merge to form multinucleated giant cells, and fibroblasts produce collagen and synthesize connective tissue *de novo*. Some materials, such as silicone, are highly reactive in this regard. For example, thick capsules are often observed surrounding explanted silicone breast implants. Virden et al. (1992) demonstrated that the presence of *Staphylococcus epidermidis* biofilms on silicone breast implants was associated with a significantly greater incidence of implants surrounded by contracted capsules versus those without contracted capsules (56% *vs.* 18%, $P < 0.05$). Although the host responses are similar, these types of inflammatory processes can occur in the absence of microorganisms. The presence of a fibrous capsule surrounding implanted materials can have two significant effects on bacterial colonization: (1) Decreased blood supply in the vicinity of the implant effectively impedes the transport of cellular and humoral factors to the interfacial region. In addition, transport of antimicrobials and other chemotherapeutic agents may be diffusion-limited as a function of capsule formation. (2) The capsular material itself may serve as a new substratum for colonization, in effect increasing the available surface area.

4.2.4.2. Host Responses to Colonized Biomaterials

While the widespread application of biomaterials in medicine has been a relatively recent innovation, it has been long recognized that "foreign bodies" in humans and animals are rapidly colonized by autochthonous and allochthonous bacteria. The implantation of a prosthetic device increases the risk of infection for many kinds of devices by at least a factor of fourfold (Christensen et al., 1989): The ability of microorganisms to adhere to biomaterials is certainly a key factor in the initiation of disease processes. However, local and systemic factors play an important role in the host's response to a colonized

TABLE 4.2. Case Summaries of Device-Related Infections

Device	Materials of construction	Incidence of infection[a]	References
Urinary catheters	Latex, polyurethane, silicone, fluoropolymers	7–10% increase in probability per day of indwelling catheter placement	Nickel (1993), Nickel and Costerton (1992), Nickel et al. (1985), Shand et al. (1985), Stamm (1991), Ganderton et al. (1992), Murakami et al. (1993), Stickler et al. (1993a,b)
Intravascular devices	Fluoropolymers, Dacron	0.5–1%	Beck-Sague and Jarvis (1989), Beck-Sague et al. (1990), Dickinson and Bisno (1989a), Elliot and Faroqui (1993), Passerini et al. (1992), Buchanan et al. (1994), Dasgupta et al. (1993), Bach and Bohrer (1993), Chew and Mendoza (1991), Egebo et al. (1994), Ena et al. (1992), Goldmann and Pier (1993), Gryn and Sacchetti (1992), Hammerberg et al. (1991), Johnson and Oppenheim (1992), Mermel et al. (1991), Poisson et al. (1992), Sterniste et al. (1994), Volkow et al. (1994)
Breast prostheses	Silicone	5%	Crespo et al. (1994), Virden et al. (1992)
Contact lenses	Polymethylmethacrylate (PMMA), various hydrogel polymers	0.05%–0.2%	Schein and Poggio (1990), Wilson et al. (1981), Stern (1990)
Orthopedic implants	Titanium, alumina, polyethylene	1%–4%	Nade (1990), Gristina et al. (1993), Chang and Merritt (1991)
Dental implants	Titanium, hydroxyapatite, PMMA	1%–5%	Haanaes (1990), Palenik et al. (1992)

Device	Material	Probability of infection[a]	References
Biliary stents	Silicone, titanium, stainless steel	Approaching 100% in pancreatic cancer patients	Hambraeus et al. (1990), Leung et al. (1992)
Central lines	Silicone	1%–12%	Dickinson and Bisno (1989b), Elliot and Faroqui (1993), Moonens et al. (1994), Pittet et al. (1994)
Prosthetic heart valves	Carbon, Dacron	1%–3%, nosocomial; 5%–24%, community	Grover et al. (1994), Karchmer (1991), Keys (1993), Wang (1993)
Pacemakers/defibrillators	Polyurethane, silicone	1%–10%	Gupta et al. (1994), Hammel et al. (1994), Pfeiffer et al. (1994)
Peritoneal dialysis catheters	Silicone	Average lifetime of catheter limited to approximately 19 patient-months	Burkhart (1988)
Voice prostheses	Silicone	Average lifetime of the prosthesis due to infection 4 months	Neu et al. (1993, 1994b)
Total artificial heart	Polyurethane and various polymers; titanium	Approaching 100% in first year	Nose (1992)

[a] Probability of infection within first year, unless otherwise noted.

(a)

Fig. 4.1. **(a)** Fouling of a mild steel pipeline leading to blockage. Photograph at scale. (Courtesy of Dr. D. White, University of Tennessee, Knoxville.) **(b)** Biliary stent fouling leading to occlusion. Bar = 1.5 μm. (Courtesy of Dr. J. Leung and K. Lam, University of California, Davis.) **(c)** Fouling of stainless steel exposed to freshwater. Bar = 2.0 μm. **(d)** *S. epidermidis* adhesion to and deterioration of medical-grade silicone. Bar = 1 cm. (Courtesy of Dr. S. Vas, University of Toronto, Ontario.) **(e)** Biofouling of PVC heat exchanger packing material. Photograph at scale. **(f)** Contamination of polyurethane catheter material with *S. epidermidis*. Bar = 0.75 μm.

foreign body. It is clear that bacterial colonization of biomaterials does not always lead to an intractable infection (Virden et al., 1992). While the mechanisms responsible for this type of "quiescent" colonization state are unclear, it is known that the immune response can be modulated by the presence of some kinds of biofilms.

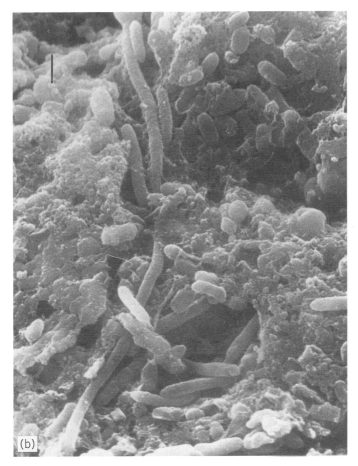

Fig. 4.1. *(Continued)*

Implanted tissue cage experiments have been used to study host reactions to various materials in the presence of infecting agents. The presence of a sterile foreign body decreases the minimum dose of bacteria required to establish an infection. In the case of *Staphylococcus aureus* and polymeric materials implanted in rat peritoneal cavities, a minimum infective subcutaneous dosage of 10^8 organisms was determined in the absence of a foreign body; in the presence of a sterile implant, the infective dosage was only 100 organisms (Christensen et al., 1989; Widmer et al., 1991). In addition to this altered host response to the presence of bacteria, device-associated bacteria possess an intrinsic resistance to topically or systemically applied antimicrobials. The combination of these two factors—host-mediated immune modulation and antimicrobial resistance—accounts for the often severe consequences of

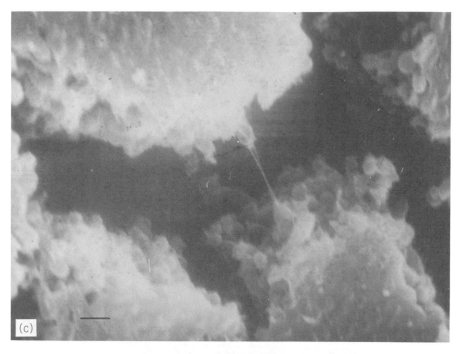

Fig. 4.1.   *(Continued)*

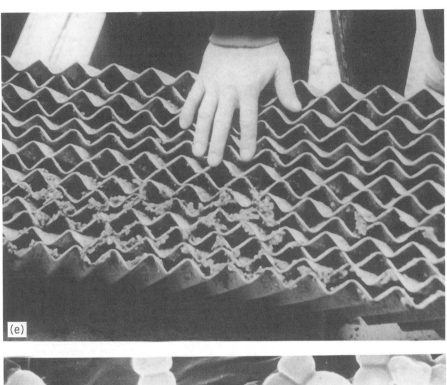

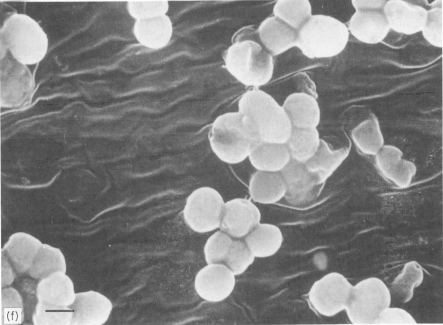

Fig. 4.1. *(Continued)*

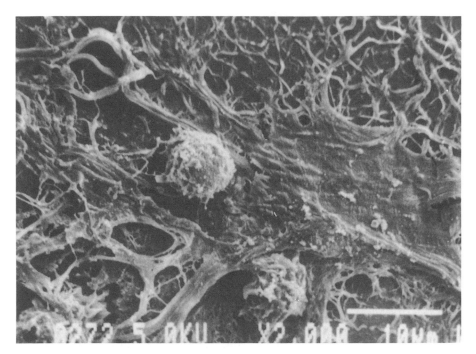

Fig. 4.2.   Macrophage and fibrin/fibrinogen adhering to the distal end of an explanted silicone peritoneal dialysis catheter. A macrophage is present at center left. The catheter substratum is obscured by fibrin/fibrinogen deposits. Bar = 10 μm. (Courtesy of Dr. F. Soboh, Toronto General Hospital, Toronto, Ontario.)

device-related bacterial infections (Gallimore et al., 1991; Gristina et al., 1993; Holmes and Evans, 1986; Jansen and Peters, 1993; Rigdon, 1975).

Finally, components of bacterial EPS can modulate the cellular response to biomaterial infections. Expression of EPS from *S. epidermidis* is not usually seen from clinical (bloodstream) isolates. However, strains recovered from explanted metallic and polymeric biomaterials possess significant amounts of an acidic polysaccharide (Deighton and Borland, 1993; Jones et al., 1992). *S. epidermidis* produces an EPS that elicits a negative chemotactic response in PMN (Stiver et al., 1988). *Pseudomonas aeruginosa* produces an EPS that fails to elicit an opsonic antibody response (Pier et al., 1994). This property is associated with chronic infections, including those associated with biomaterials and cystic fibrosis. Like *S. epidermidis, Candida* spp. appear to have increased slime production in association with vascular catheters (Branchini et al., 1994). It is unknown whether this EPS modulates host immune responses.

Interestingly, bacteria colonizing surfaces in marine ecosystems can act as either a stimulus (Kirchman et al., 1982) or an inhibitor (Maki et al., 1988) of marine invertebrate macrofoulants. Specific bacterial species have been characterized that appear either to promote or to retard barnacle and oyster formation on surfaces exposed to marine environments. Of 18 species of marine bacteria tested, 7 species were inhibitory, 10 showed no effect, and 1 was stimulatory for cyprid barnacle (*Balanus amphitrite*) attachment.

4.3. ADHESION MECHANISMS

In the context of this chapter, adhesion may be defined as the discrete association between a bacterium and a surface, or substratum. If a bacterium has adhered to a surface, energy is required to effect a separation. At the molecular level, bacteria adhere to surfaces by a combination of electrostatic and hydrophobic interactions (see Chapters 1 and 2, this volume). The nature of these interactions is dependent on a combination of physicochemical factors, including pH, temperature, ionic strength, ligand density, dipole moment, and charge density. The size and charge of bacteria are in the same range as colloidal solutes. Therefore bacteria can exist as lyophobic or lyophilic sols, association colloids, gels, or part of an emulsion. These physicochemical characteristics define (1) binding capacity, (2) binding strength, and (3) binding specificity for ligand–receptor interactions. The terms *specific* and *nonspecific,* as they are applied to bacterial adhesion mechanisms, have been used to differentiate stereospecific "lock and key" interactions from so-called physicochemical attractive forces, respectively. Regardless of whether specific adhesins are involved, the interactions that occur between bacteria and biomaterials can be described on the basis of the physicochemical factors listed above.

4.3.1. Adhesins and Biomaterials

Over the past 20 years, a tremendous effort has been made to characterize the adhesins that are associated with bacterial adhesion to and colonization of mucosal surfaces. These cell surface structures often have specific receptors on eukaryotic host cells. The paradigm for these types of "specific" adhesions is the type I fimbriae of *Escherichia coli, Klebsiella pneumoniae,* and *Salmonella typhimurium.* These fimbriae bind in a stereospecific manner to various glycosides and oligosaccharides of mannose (Ofek et al., 1977). Carbohydrate-binding proteins, or lectins, have been characterized from a number of human pathogens. Their specificity is often defined on the basis of the ability of one or more carbohydrates to inhibit adhesion. The role of lectins as adhesins in bacterial–tissue interactions is discussed elsewhere in this volume (Chapters 5, 8, and 9).

From an evolutionary standpoint, it is perhaps difficult to rationalize the involvement of pili or other "specific" adhesins as mediators of adhesion to biomaterials. However, several investigators have suggested that adhesins do play a role in bacterial attachment to engineered surfaces. A *S. epidermidis* biomaterial adhesin has been described that appears to mediate adhesion to polystyrene (Timmerman et al., 1991). Evidence for the involvement of a specific biomaterial adhesin was based on monoclonal antibody-blocking reactions using a whole-cell antibody to inhibit polystyrene adhesion. Muller and coworkers (1993a, b) provided evidence for a polysaccharide antigen involved in the attachment of coagulase-negative staphylococci (CNS) to silicone catheter material. Transposon mutants deficient in this polysaccharide antigen had a 10-fold decrease in their attachment to silicone surfaces. This finding was similar to results obtained with parent strains of *S. epidermidis* that did not produce the polysaccharide antigen. The same adhesin was, however, found in a small percentage of clinical CNS isolates that failed to adhere rapidly to catheter material under *in vitro* conditions. Another *S. epidermidis* strain, RP62A (ATCC 35984), was shown to produce a galactose-rich capsular polysaccharide adhesin that was implicated in biomaterial adhesion events (Christensen et al., 1990). Jansen et al. (1990) have provided some of the best data supporting involvement of an adhesin–receptor interaction with biomaterials. Surface lectins on *S. saprophyticus* S1 and *P. aeruginosa* (ATCC 27853) showing specificity for *N*-acetylgalactosamine/*N*-acetylglucosamine and *N*-acetylneuraminic acid, respectively, were identified using hemagglutination assays. They showed that attachment of these organisms to a polyether polyurethane preincubated in serum could be blocked by specific, competitive glycoconjugates.

Despite these findings, evidence for specific adhesins involved in biomaterial adhesion remains somewhat controversial. Indeed, one group showed that significantly greater numbers of nonpiliated, isogenic mutant *P. aeruginosa* cells than of a piliated parent strain adhered to unworn contact lenses (Fletcher et al., 1993). As is described below, experimental conditions can significantly influence the outcome of these kinds of experiments. Several investigators have shown that substratum characteristics as well as the components of conditioning films can significantly influence biofilm characteristics, including colonization rates. It is very likely that the outcomes of adhesin–biomaterial receptor studies would be influenced by host conditioning film constituents. For example, dental implants are rapidly coated with constituents of the salivary pellicle (Embery et al., 1986), including sialic acid. *S. sanguis* possesses a sialic acid-binding protein with lectin-like properties (Demuth et al., 1990). In addition, Ofek and Doyle (1994) have reviewed a number of putative receptor proteins and glycosides present in saliva and gingival crevicular fluid that could coat implanted devices.

It would also be of interest to characterize putative receptor sites on biomaterial surfaces. Demonstration of adhesin-mediated attachment to discrete biomaterial functional groups would make a fascinating story of adaptation among bacteria. It remains to be seen whether these functional groups would be recognized by adhesins on bacterial cell surfaces in the presence of host conditioning films.

The trend toward developing biomaterials with biospecific properties (blood compatibility, cell growth factors, immobilized enzyme delivery, and so forth) may result in the creation of conditions that are more *favorable* for the adhesion to and invasion of host cells. For example, *Yersinia* spp. bind to and penetrate host cells via an Arg-Gly-Asp (RGD)–dependent process (Ruoslahti and Pierschbacher, 1987). The invasin protein from this organism is located on the cell surface and binds to complementary RGD receptors on the host surface. It is possible that other bacteria utilize a similar process of cell surface recognition. The synthesis of RGD-containing peptides on biomaterial surfaces, while perhaps improving host cell integration (Massia and Hubbell, 1990), may also create new receptor sites for some bacteria. It is clear that biocompatibility alone is an insufficient criterion for the acceptance of a novel biomaterial.

4.3.2. Influence of Conditioning Films

Under *in vivo* conditions, conditioning films composed of proteinaceous and polysaccharide components in the blood, urine, bile, and saliva respiratory secretions rapidly coat implanted biomaterials (see also Chapter 3, section 3.2, this volume). A partial list of the components of these conditioning films is provided in Table 4.3. Christensen et al. (1989) have suggested that the conditioning film—rather than the underlying substratum—is the major determinant of bacterial adhesion to and colonization of biomaterials. However, this may be an oversimplification, since some investigators have shown that the orientation of proteinaceous conditioning film components can be influenced by substratum surface energetics (Uyen et al., 1990; van Loosdrecht and Zehnder 1990).

Conditioning films composed of the plasma proteins fibrin, fibrinogen, and fibrinectin are important mediators of staphylococcal adhesion to biomaterials. Cheung and Fischetti (1990) showed that both fibrin and fibrinogen significantly influenced *S. aureus* adhesion to polymeric catheters. Interestingly, *S. aureus* adhesion was blocked when fibrinogen-coated catheters were pretreated with goat antihuman fibrinogen but not fibrinectin antibody or nonimmune goat IgG. Prospective *in vivo* studies of *S. aureus* colonization have been conducted in patients with indwelling central venous catheters (Vaudaux et al., 1993). These studies demonstrated a statistically significant correlation between fibronectin concentrations and bacterial numbers on polyurethane and polyvinylchloride (PVC) catheter surfaces.

There is evidence for involvement of proteinaceous conditioning films in adhesion of CNS and *P. aeruginosa*. While plasma and serum albumin appear to inhibit CNS adhesion to polymers significantly, fibrinogen enhances adhesion to polyethylene, nylon, and PVC substrata (Carballo et al., 1991; Paulsson et al., 1993). The presence of host proteins on hydrogel contact lenses also appears to be an important factor in bacterial adhesion. *P. aeruginosa* adhesion to soft contact lenses increased by 45% in the presence of albumen, mucin, fibrinogen, or desialylated fibrinogen (Cook et al., 1993). Unlike the findings described above for *S. aureus* and *S. epidermidis,* however, there was little difference in the adhesion as a function of protein type. Albumen does not

TABLE 4.3. Constituents of Host Conditioning Films Influencing Bacterial Adhesion to Biomaterials

Biomaterial environment	Suspending fluids	Principle conditioning film components	References
Urinary tract	Urine, prostatic fluid, semen	Mucopolysaccharides, Tamm Horsfall proteins, glycoproteins	Reid et al. (1992)
Cardiovascular	Blood	Serum, albumins, fibrinogen, fibrinectin	Carballo et al. (1991), Cheung and Fischetti (1990), Fabrizius-Homan and Cooper (1991), Vaudaux et al. (1993)
Ophthalmic	Tears	Proteins (lysozyme), sialic acid, fibrin	Cook et al. (1993)
Oral cavity	Saliva	Mucopolysaccharides, glycoproteins, serum albumins	Embery et al. (1986), Glantz et al. (1991), Rosenberg (1991)
Biliary tract	Bile	Sterols, bile salts, mucopolysaccharides	Hambraeus et al. (1990), Jansen et al. (1993), Leung et al. (1992)
Subcutaneous/intervascular	Intervascular fluid	Fibrin	Bambauer et al. (1994), Paulsson et al. (1993)
Peritoneal cavity	Peritoneal fluid	Fibrin, serum albumins; dialysates	Dasgupta et al. (1993), Reid et al. (1994)
Respiratory tract	Respiratory secretions	Mucopolysaccharides	Neu et al. (1994a)

appear to inhibit *P. aeruginosa* adhesion to hydrogel polymers. In Figure 4.3, significantly greater numbers of *P. aeruginosa* are observed adhering to serum-coated than to uncoated silicone catheter surfaces, illustrating the importance of serum components in adhesion of bacteria of biomaterials.

4.3.3. Substratum and Bacterial Cell Surface Physicochemistry

In the search for surfaces that retard or otherwise inhibit bacterial colonization, a number of research efforts have focused on the physicochemical nature of the substrata. Although the initial stages of colonization often are influenced by surface topography and energy, longer term studies performed under *in vivo* or simulated *in vivo* conditions have failed to demonstrate conclusively a statistically significant effect. Interestingly, analogous findings have been observed with engineered materials that are exposed to various bulk-phase fluid environments. Pedersen (1990) studied biofilm development in a municipal drinking water system located in Goteborg, Sweden. They found that bacteria counts on hydrophilic stainless steel and hydrophobic PVC surfaces were not significantly different after 167 exposure days. In addition to the development of conditioning films, precolonization by commensal flora may be an important factor in the adhesion of putative pathogens. The role of primary colonizers in the adhesion of succeeding, pathogenic bacteria is discussed below.

4.3.3.1. Surface Topography

Surface topography appears to play an important role in the integration of some host tissues with biomaterials. For example, there is evidence that suggests that osseo-integration with both ceramic (Filiaggi et al., 1993) and titanium (Pilliar et al., 1993) implants is enhanced by a porous-type surface character. Short-term colonization studies performed in saline solutions did show increased attachment of *S. aureus* and *S. epidermidis* to sand-blasted polymer, composite (hydroxyapatite/poly[L-lactide]), and 316-type stainless steel relative to the unmodified surfaces (Verheyen et al., 1993). However, in a Japanese study performed under *in vivo* conditions, bacterial attachment was influenced by substrata characteristics 4 h following implantation of periodontal implants, but not after 48 h (Nakazato, 1990). Quirynen et al. (1993) found that both the bacterial species diversity and cell density were more significantly influenced by periodontal patients' dental status than by subgingival implant surface smoothness. Patel et al. (1990) also found that there was no effect of titanium surface roughness (1 μm *vs.* "sand-blasted" finish) on the numbers of adhering bacteria for two oral species.

4.3.3.2. Hydrophobicity and Nonspecific Interactions

There is a significant body of literature, including at least one book (Doyle and Rosenberg, 1990), that deals with hydrophobic interactions between bacterial cells and various kinds of surfaces (see also Chapter 1, section 1.3, and

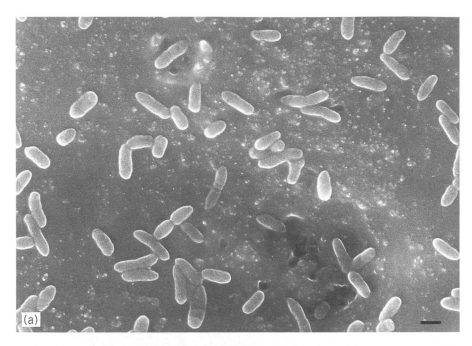

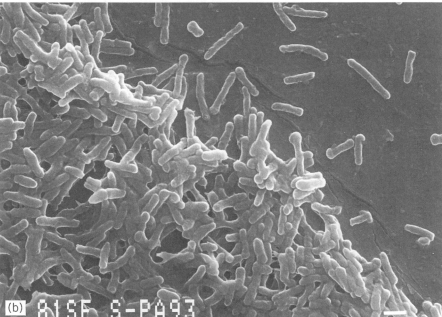

Fig. 4.3. Adhesion of *P. aeruginosa* cells to serum-coated **(a)** and uncoated **(b)** silicone catheter materials. Bacteria were allowed to colonize silicone sheet material in a flow chamber over a 48 h period. Bar = 1 μm.

Chapter 2, section 2.5.2.2, this volume). The "hydrophobic effect" refers to the proclivity of one nonpolar molecule for another nonpolar molecule over water. When two hydrophobic surfaces approach one another in an aqueous environment, intervening ordered layers of water are displaced. The increase in entropy, which is a consequence of this displacement reaction, creates energetically favorable conditions for adhesion (van Loosdrecht and Zehnder, 1990).

Hydrophobic interactions between bacteria and engineered or tissue surfaces have been termed *nonspecific* (Busscher and Weerkamp, 1987) to reflect the lack of dependence on "lock and key," stereospecific interactions. Bacteria do possess distinctly different cell wall constituents that contribute to their relative hydrophobicity. These so-called hydrophobins include aromatic amino acid groups, fatty acids, and mycolic acids. Similarly, engineered substrata possess discrete hydrophobic moieties, including aromatic and fluorine functionalities. Hydrophobic interactions, however, are balanced by the repulsive forces of the electrical double layer that surrounds both the bacteria and the substratum in an aqueous environment. Surface charge, therefore, plays a role in the so-called nonspecific adhesion process.

The DLVO theory (Derjaguin and Landau, 1941; Verwey and Overbeck, 1948) (see also Chapter 1, section 1.2.2, and Chapter 2, section 2.5.2.1, this volume) of colloid stabilization as applied by Marshall et al. (1971) holds that the separation between bacteria and adsorbents in an electrolyte solution is dependent on a balance between attractive and repulsive forces. The repulsive forces, mediated by like charges on both the bacterial and substratum surfaces, are balanced by attractive forces that include the hydrophobic effect and hydrogen bonding. This relationship can be described by the following equation:

$$V_t = V_r + V_a \tag{1}$$

where V_t is total energy of interactions, V_r is energy of repulsion, and V_a is energy of attraction.

Predictions of whether a given organism will adhere to a particular biomaterial surface are dependent not only on the net result of these complex interactions but also on the nature of the suspending host milieu and the physiological status of the bacterial community present. For example, the surface energetics of catheters are significantly influenced by sorption of urine (Reid et al., 1992) and dialysate (Reid et al., 1994) constituents. As with surface topography considerations, host status—rather than surface chemistry—appears to be the most important factor influencing bacterial adhesion to subgingival implants (Quirynen et al., 1994).

Adhesion to and colonization of biomaterials in biological systems is dependent on a suite of complex interactions (Fig. 4.4). Bacterial metabolism and cell surface structures are influenced by both the components of the host

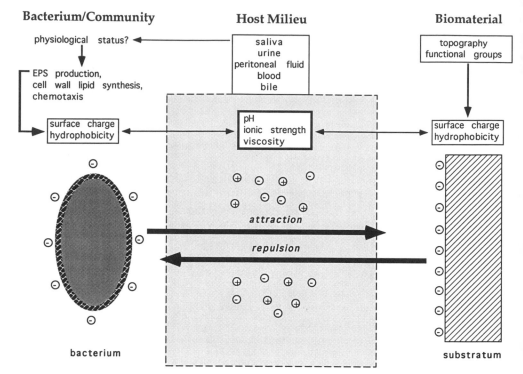

Fig. 4.4. Physicochemical interactions between bacterial cells and biomaterials in host environments.

milieu (e.g., saliva, blood, urine) and the physicochemical characteristics (e.g., surface charge, functional groups, steric factors) of the substratum. Conversely, bacterial extracellular polymer synthesis and short-chain organic acids can modify both the host microenvironment and the substratum characteristics, resulting in deterioration of biomaterials. For example, urease-producing *Proteus* spp. create conditions that encourage struvite-type encrustations of urinary catheters (Stickler et al., 1993a). A conceptual representation of the complex interactions that take place between the bacterium, substratum, and host milieu is presented in Figure 4.4.

4.4. MICROBIAL ECOLOGY OF BIOMATERIAL-ASSOCIATED BIOFILMS

The development of biofilm communities on biomaterials tends to be a function of the host species diversity and the nature of local host factors. In this sense, the ecology of biomaterial infections is subject to the same kinds of selective pressures as are natural marine, freshwater, and industrial fluid processing environments. As is the case in these ecosystems, implanted biomaterials can be colonized by autochthonous or allochthonous organisms.

4.4.1. Autochthonous and Allochthonous Colonizing Populations

Normal host flora are associated with the surfaces of the large intestine, oral cavity, and skin. Indeed, as is discussed elsewhere in this volume (see Chapters 8 and 9), these flora are necessary for normal health and play an important role in homeostasis. There is no evidence to support the existence of a "normal," autochthonous community structure in association with an implanted biomaterial. Gristina et al., (1993) and others have shown that colonization of devices, including orthopedic protheses and catheter surfaces, eventually leads to implant failure and/or pathological changes in host tissues. There have been reports, however, of "uninfected" implants with significant numbers of adherent bacteria. Virden et al. (1992) recovered silicone breast implants with significant numbers of adhered *S. epidermidis* cells. Although there were no overt signs of infection among the implant patients, contracted capsules were present in significantly higher numbers among the colonized patients (see section 4.2.4.1). Changes in host immune status, local nutritional factors, or environmental factors may effect a transition from this quiescent condition to a local or systemic response to infection.

4.4.1.1. Hematogenous Seeding of Biomaterials
With the exception of urinary catheter-related sepsis, percutaneous access device (PAD)-related infections are the most frequently encountered biomaterial-associated infection. However, the consequences associated with colonization of hematogenously seeded, totally implanted devices are far more severe. For example, the mortality rate associated with infected vascular catheters is on the order of 20%; the combined probability of either death or amputation is approximately 50% for infected vascular grafts (Elliot and Faroqui, 1993). Although the data are limited, it appears that the vast majority of vascular grafts, prosthetic heart valves, and orthopedic implants are colonized by bacteria via hematogenous spread. PAD are the major source of colonizing bacteria associated with so-called early-onset prosthetic valve endocarditis (Keys, 1993).

Oral microflora, including streptococci and *Actinobacillus* spp., are the most common etiological agents of all forms of bacterial endocarditis (Tan and Gill, 1992). Dental procedures and routine oral care can introduce these organisms into the bloodstream. Interestingly, CNS account for approximately 95% of the cases of prosthetic valve endocarditis (Whitener et al., 1993). The primary source for these organisms are also PAD as described above.

4.4.1.2. Exit-Site Infections
PAD, such as vascular access and peritoneal dialysis catheters, are usually fabricated from smooth silicone rubber. This material does not allow the skin

epithelial cells to adhere to it either mechanically or chemically. Following introduction of a standard PAD the epidermal cells begin to migrate, each seeking to surround itself completely with other epidermal cells. The granulation tissue, which forms near the skin surface, provides an ideal bed over which these cells can migrate under normal wound healing circumstances. The presence of a catheter across the skin at the exit site prevents contact of the epidermal cells with sister cells. This results in inward migration of the epidermal cells toward the subcutaneous tissues and the development of a wet sinus tract between the surface of the skin and the tubing. Necrotic epidermal cells and keratin line the sinus tract, creating an ideal environment for microbial colonization.

Despite the recent advances in aseptic handling techniques, antimicrobial therapy, and the understanding of bacterial biofilm principles, peritonitis is the most common and probably the most serious complication associated with peritoneal dialysis (PD) and is a major cause of morbidity (Burkart, 1988). Peritonitis rates fell from 2.4 episodes per year in 1979 to 0.8 episodes per year in 1989. Peritonitis may require hospital admission for treatment, and in approximately 20% of the patients surgery is required to remove the dialysis catheter if the infection is resistant to antibiotic therapy or is rapidly recurrent after cessation of therapy. The patient is transferred to hemodialysis followed by a second surgical procedure for insertion of a new catheter. Entry via the intraluminal route (i.e., through the lumen of the PD catheter) is thought to be caused by improper dialysis technique by the patient or personnel performing dialysis. Significant progress has been achieved by the modification of the techniques of handling the PD catheters and the improved connections to the catheters.

Clinical (Gorman et al., 1994) and laboratory (Dasgupta et al., 1993) experiences demonstrate that the subcutaneous portion of the catheter is usually colonized with endogenous organisms within 3 weeks of insertion of the catheter. The adhesion of these bacteria to the surface of PD catheters, followed by multiplication and spread of the organisms, leads to the formation of a dense microbial community. The presence of staphylococcal aggregates has been demonstrated in the dialysate effluent of patients with peritonitis (Reid et al., 1994). As with other biomaterial-centered infections, the presence of these aggregates is evidence of biofilm sloughing from the surface.

4.4.2. Succession and Climax Community Structure

As in natural ecosystems, succession plays an important role in the establishment of biomaterial-associated communities. Bacterial succession has been described for dental implants (Cowan and Busscher, 1993) and is an essential process in the biology of both healthy and diseased oral environments. Putative periodontal pathogens appear on implant surfaces at approximately the same time as on natural teeth (Quirynen et al., 1993). Both the successional order and climax community structures of dental implants are similar to those of natural teeth (Haanaes, 1990). Primary colonization by *S. epidermidis* nearly always precedes colonization by *Proteus* spp. Precolonization of catheter surfaces with *S. epidermidis* significantly increased subsequent colonization by *P. aeruginosa* and *Proteus mirabilis* to polymethyl methacrylate implant materials (Chang and Merritt, 1991).

4.4.3. Evolution of *S. epidermidis* in Device-Related Infections

Although approximately 20 different species of CNS have been isolated from normal human skin (Krieg and Holt, 1984), *S. epidermidis* is the etiological agent of the majority of PAD-related infections. CNS are the most common etiological agents of prosthetic valve endocarditis, yet are seen in only 5% of the cases of native valve infections (Whitener et al., 1993). This group of organisms was responsible for 10% of all intravenous line–related nosocomial bloodstream infections in a pairwise-matched case–control study (Pittet et al., 1994).

Contradictory reports of *S. epidermidis* adhesion to biomaterials are likely a function of differences in experimental methodology among different investigators. While *ex vivo* strains associated with biomaterials usually exhibit copious quantities of EPS, many *in vitro* isolates fail to produce significant amounts of EPS. Deighton and Borland (1993) have suggested that this finding is consistent with iron limitations, which are present in the *in vivo* implant environment. They reported on a "slime-negative" *S. epidermidis* isolate grown on tryptic soya broth that produced high concentrations of EPS after two passages through an iron-restricted medium. There appears to be a strong correlation between slime production and antibiotic resistance among *S. epidermidis* strains recovered from patients with septicemia/endocarditis (Kotilainen et al., 1991). However, the ability of clinical isolates to adhere to glass substrata in *in vitro* studies is not always correlated with either slime production (Ramirez de Arellano et al., 1994) or clinical outcome (Alexander and Richmond, 1987; Kotilainen, 1990). These findings explain, in part, the discrepancies between *in vivo* and *in vitro* observations. While standardization of laboratory adhesion assays may be difficult, these findings do highlight the importance of simulating the *in situ* environment in the conduct of biomaterial adhesion assays.

4.4.4. Iron and *S. epidermidis* Colonization

There is a significant body of evidence that suggests that iron limitations are an important environmental factor affecting EPS production and/or adhesion to surfaces (Anwar and Costerton, 1990; Deighton and Borland, 1993; Morck et al., 1991; Shand et al., 1985). Iron appears to be an important factor in controlling EPS production associated with PADs, such as central lines (Deighton and Borland, 1993). Lactoferrin is a high-affinity, iron-binding protein that is found in high concentrations in breast milk and in saliva. In the mouth, this protein has a dual protective function in that it sequesters iron and has a direct bactericidal effect on some streptococci (Malamud, 1985). Based on data from catheter-adhesion experiments conducted under iron-limiting conditions, EPS production in the oral environment may also be enhanced by sequestering agents such as lactoferrin.

4.5. PREVENTION AND TREATMENT OF DEVICE-RELATED INFECTIONS

4.5.1. Biofilm-Mediated Antimicrobial Resistance

Bacteria associated with surfaces exhibit a differential resistance to systemically or topically applied antimicrobial agents. A number of investigators have demonstrated that bacteria within biomaterial-associated biofilms can be several times more resistant to the bactericidal effects of antibiotics and other antimicrobials (Reid et al., 1993; Khoury et al., 1993). Nickel and Costerton (1992) described the resistance of a *P. aeruginosa* biofilm associated with urinary catheter material that was treated with 1,000 μg ml^{-1} tobramycin for 12 h. While significant numbers of viable cells remained within the biofilm despite treatment with this very high antibiotic concentration, planktonic (suspended, free-floating) organisms were completely killed by 50 μg ml^{-1} concentrations.

Bacteria within biofilms are also resistant to other antagonistic agents. For example, endemic strains of *S. aureus* isolated from poultry equipment were found to be eight times more resistant to chlorine than *S. aureus* strains isolated from normal skin (Bolton et al., 1988). The major phenotypic difference between these strains was the extensive EPS associated with the poultry equipment isolates and their ability to form macro-clumps. Other workers (Anderson et al., 1990) have shown that *P. aeruginosa* survived within biofilms associated with PVC piping after 7 days exposure to iodophor and phenolic antimicrobial solutions. They suggested that these organisms survived within EPS "masses" on the interior walls of the piping. Low levels of sodium hypochlorite in the range of 0.5 to 5 ppm were inhibitory only to biofilms associated with stainless steel surfaces; concentrations exceeding 50 ppm were required for bacterial inactivation under process control conditions (Caldwell, 1990).

4.5.2. Mechanisms

Giwercman et al. (1991) showed that β-lactamase production was induced in *P. aeruginosa* biofilms. These workers suggested that β-lactamase production could be an important factor in the persistence of this organism in chronic infections despite antibiotic intervention. As was noted above, *S. epidermidis,* which is one of the most common etiological agents of device-related infections, produces copious quantities of EPS *in vivo*. Extracted extracellular polymer from *S. epidermidis* was found to decrease the efficacy of glycoside antibiotics (Farber et al., 1988). Hoyle et al., (1990, 1992) and others have suggested that EPS is an important diffusional barrier to antibiotic penetration. This reduction in the rate of antibiotic penetration into bacterial cells may provide the cells with sufficient time to switch on the expression of antibiotic degrading enzymes. Other groups have shown that "older" biofilms can significantly reduce antibiotic efficacy, suggesting changes in membrane permeability and/or porin structure (Anwar et al., 1992). EPS expression does appear to be a common theme among biofilm populations resistant to antimicrobials. This may be a critical factor in other disease processes involving adherent bacterial populations. For example, cystic fibrosis isolates of *P. aeruginosa* up-expressed gene products involved with alginate biosynthesis (Davies et al., 1993). Since relatively little is known about either the community structure or metabolic status and activity of biomaterial-associated bacteria, the mechanism(s) of biofilm antimicrobial resistance have not been clearly elucidated.

Attenuated total reflectance Fourier-transform infrared spectroscopy (ATR-FTIR) has been utilized to evaluate both antibiotic penetration and biofilm interactions at the molecular level (Suci et al., 1994). Biofilms were developed for varying periods of time on germanium crystals within circle-type ATR cells from *P. aeruginosa* continuous cultures. Ciprofloxacin penetration was followed on a real-time basis. The presence of a *P. aeruginosa* biofilm created a diffusion barrier for this antimicrobial, resulting in a 17 min lag in the time required to establish bulk-phase equilibrium concentrations at the surface (Fig. 4.5).

Although the acute local inflammatory changes and bacteremia may temporarily be controlled with large doses of antibiotics, they frequently recur when therapy is discontinued. Repeated cycles of inflammation followed by antibiotic therapy result in a chronic inflammatory process associated with significant tissue destruction and extensive scarring in the vicinity of the device. Direct examination of devices associated with chronic infections often demonstrate the presence of a thick biofilm composed of both host and bacterial components (Neu et al., 1994a).

4.5.3. Prevention of Bacterial Colonization

4.5.3.1. Silver Coatings

A number of topical treatments have been developed to address the problem of exit-site infections. Success in controlling PD catheter infections in long-term rabbit models has been reported (Dasgupta et al., 1994). Silver-coated catheters have been used to prevent exit-site infections associated with chronic venous access (Groeger et al., 1993), urinary (Takeuchi et al., 1993), and peritoneal dialysis (Mittelman et al., 1994) catheters in human clinical trials. However, long-term studies sometimes fail to demonstrate a significant reduction in the number or severity of exit-site infections (Babycos et al., 1993).

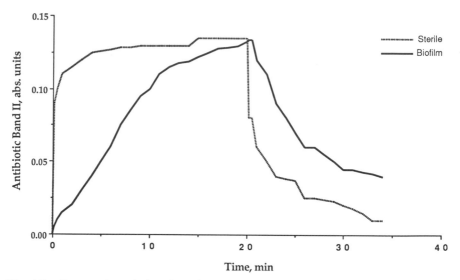

Fig. 4.5. Penetration of ciprofloxacin through a 48 h *P. aeruginosa* biofilm adapted from Suci et al. (1994). Absorbance of the antibiotic II band was followed using ATR-FTIR. The dotted line represents a sterile control germanium crystal surface. Antibiotic addition was stopped at 20 min.

In addition, bacterial resistance to silver can develop over time and carries with it the risk of emerging multiple antibiotic resistance (Silver and Misra, 1988; Summers et al., 1993) among colonizing organisms (Silver and Misra, 1988). Despite potential questions concerning long-term efficacy and the development of resistant populations, *in vitro* and clinical *in vivo* data suggest that silver-impregnated catheters are effective in reducing the incidence of exit-site infections.

4.5.3.2. Antibiotic Coatings

Various antibiotics have been used to coat the surfaces of catheters in an ionic bonding process. Trooskin and coworkers (1985, 1987) described a method by which catheter surfaces were soaked in antibiotic solutions prior to their implantation. Duran et al. (1992) covalently bonded a photoactivated hydrogel onto the surface of silicone materials. Vancomycin was then immobilized within the hydrogel matrix. In both of the above studies, most of the antibiotic was released within a very short period of time (days). Short-term studies with silver-sulfadiazine– and chlorhexidine-coated polyurethane vascular access catheters did reduce the incidence of exit-site infections in a rat model (Bach et al., 1994). Fewer intravenously injected *S. epidermidis* cells adhered to rifampicin-treated than to untreated Dacron vascular grafts in sheep (D'Addato et al., 1994). Although the results of these *in vivo* animal studies were promising, they were all conducted over relatively short periods of time following the introduction of the biomaterial. Biomaterial-related infections can only be controlled when bacterial colonization is prevented for an extended period of time (weeks), enabling complete host tissue integration as described above.

4.5.4. Emerging Treatment Strategies

4.5.4.1. Drug Delivery Systems for Biomaterials

In addition to the diffusion-controlled drug delivery systems described by Duran and coworkers (1992), there exist several more sophisticated *in situ* drug delivery systems. These include polymeric liposomes (Stefely et al., 1988; Hope and Wong, 1995), bioadhesives (Hui and Robinson, 1985; Longer et al., 1985), bioerodable polymers (Heller, 1988; Mathiowitz et al., 1989), chemical and physical stimuli responsive polymers (Hsieh et al. 1983; Makino et al., 1990), and polymer drugs (Ouchi et al., 1990; Takemoto and Inaki, 1988). Applications of these materials have included the delivery of antitumor drugs in cancer therapy (Ouchi et al., 1990), insulin for diabetics (Makino et al., 1990), and antimicrobial drugs for vascular grafts (Karch et al., 1993). In their studies on vascular grafts, Karch et al. (1993) combined gentamicin with a fibrin sealant, which was then coated onto Dacron surfaces. This bioerodable system takes advantage of the degradative features of the biopolymer fibrin to release the gentamicin. The degradation process depends on the hydrolysis of amide bonds and dissolution of the fibrin network. Following implantation in a porcine model, gentamicin release was elevated for the first few days, but decreased significantly shortly thereafter. In addition, over 50% of the specimens containing the fibrin–gentamicin matrix were found to be infected upon retrieval. To date, polymer-based antimicrobial delivery systems have failed to demonstrate *in vitro* or *in vivo* efficacy over extended periods of time.

4.5.4.2. Chlorhexidine

Perhaps the most successful agent thus far applied to biomaterials for the prevention and treatment of infective biofilms has been the bisbiguanide chlorhexidine. Chlorhexidine has a broad spectrum of activity against gram-positive and gram-negative bacteria as well as a variety of fungi, including *Candida* spp. In addition to its bacteriostatic and bactericidal properties, chlorhexidine tends to bind very avidly to mucosal tissues, tooth surface, and dental implant materials (Sodhi et al., 1992). Chlorhexidine coatings on dental implant surfaces have demonstrated excellent *in vivo* activity against *S. mutans, Porphyromonas gingivalis,* and other dental pathogens (Bach et al., 1994). In addition, local tissue integration with the implants is not adversely affected by the presence of this compound (Buchard et al., 1991).

4.5.4.3. Surface-Active Agents

One approach to overcoming the intrinsic antimicrobial resistance of biofilm bacteria would be to enhance the penetration of agents through the biofilm matrix. Protamine is known to play a significant role in natural host defenses against infection and to exert antimicrobial activities (Richards, 1976). Parsons and coworkers (1981) have shown that protamine sulfate penetrates and disrupts the protective glycosaminoglycan layer lining the urinary bladder. Synergistic activity with the fluoroquinolone ciprofloxacin has also been demonstrated (Soboh et al., 1995). The efficacy of vancomycin against *S. aureus* was enhanced by protamine sulfate in an infected rat model (Teichman et al., 1994). Proteolytic agents, used in combination with antibiotics, may also enhance the efficacy of systemically applied antimicrobials against biofilm organisms (Selan et al., 1993).

4.5.4.4. *"Natural" Antifoulants*

Many sessile and sedentary plants and animals have hard, nondesquamating surfaces and are therefore subject to the same biofouling pressures as are engineered surfaces. The "natural" antifouling properties associated with some of these nonfouling surfaces have been the subject of several research programs sponsored by the U.S. Navy and other organizations affected by biofouling activities. Extracts from Gorgonian coral (Keifer et al., 1986; Vrolijk et al., 1990), eel grass (Harrison and Chan, 1980), and marine sponges (Sears et al., 1990) have all been employed as so-called natural antifoulants in marine coating formulations.

Although one or more of these compounds may hold promise as antifouling agents, perhaps with applications as biomaterial additives, natural compounds suffer from three major disadvantages in this regard: (1) They are often available in limited quantities, (2) the compounds are frequently difficult to synthesize *de novo* and possess multiple chiral centers, and (3) their range of application is often limited in terms of both species selectivity and environmental conditions. The same considerations should be applied to emerging antimicrobial coatings applied to biomaterial surfaces. Although the concept of a "natural products" antimicrobial compound may hold an aesthetic appeal, there is no evidence that these compounds are either safer or more effective than those synthesized *de novo*.

4.6. CONCLUSIONS

That bacteria are able to adhere to and colonize biomaterial substrata in various hostile host environments is an indication of their great adaptability. Although the scientific application of biomaterials is a very recent development, they became a niche for a number of clinically significant bacteria. *S. epidermidis,* in particular, has evolved into perhaps the most significant biomaterial-associated infecting agent. By virtue of its extracellular polymeric substances, this organism not only modulates host immune responses, but appears to impede antimicrobial transport to labile cellular components. Biomaterial-related infections are a significant problem in human medicine, directly responsible for hundreds of thousands of infections and tens of thousands of deaths each year. However, relatively little is known about the mechanisms of adhesion to engineered substrata, the cellular and humoral immune system response to colonized biomaterials, or the mechanisms responsible for bacterial biofilm-mediated antibiotic resistance. Modifications of substratum physicochemistry have generally failed to prevent bacterial adhesion; this could be predicted from similar experiences with modified surfaces in marine, freshwater, and industrial fluid-handling systems. The keys to effective prevention appear to be in (1) enhancing the wound-healing process in the vicinity of implants to form a "biological barrier" to invading organisms and (2) mobilizing the host's immune response to combat more effectively colonizing bacteria on biomaterial surfaces. In the former case, there is a significant amount of research currently in progress to develop "biomimetic" surfaces to facilitate the wound-healing process, enabling the host to win what Gristina (1987) has called the "race to the surface."

ACKNOWLEDGMENTS

The support of Bayer Healthcare Canada, the Natural Sciences and Engineering Research Council of Canada, and the Department of Surgery, University of Toronto, is gratefully acknowledged.

REFERENCES

Alexander W, Richmond D (1987): Lack of correlation of slime production with pathogenicity in continuous ambulatory peritoneal dialysis peritonitis caused by coagulase negative staphylococci. Diagn Microbiol Infect Dis 8:215–222.

Anderson RL, Holland BW, Carr JK, Bond WW, Favero MS (1990): Effect of disinfectants on pseudomonads colonized on the interior surface of PVC pipes. Am J Public Health 80:17–21.

Anwar H, Costerton JW (1990): Enhanced activity of combination of tobramycin and piperacillin for eradication of sessile biofilm cells of *Pseudomonas aeruginosa*. Antimicrob Agents Chemother 34:1666–1671.

Anwar H, Strap JL, Costerton JW (1992): Establishment of aging biofilms: possible mechanism of bacterial resistance to antimicrobial therapy. Antimicrob Agents Chemother 36:1347–1351.

Babycos CR, Barrocas A, Webb WR (1993): A prospective randomized trial comparing the silver-impregnated collagen cuff with the bedside tunneled subclavian catheter. J Parenter Enter Nutr 17:61–63.

Bach A, Bohrer H (1993): Infections caused by intravascular catheters [in German]. Anasthesiol Intensivmed Notfallmed Schmerzther 28:404–414.

Bach A, Bohrer H, Motsch J, Martin E, Geiss HK, Sonntag HG (1994): Prevention of bacterial colonization of intravenous catheters by antiseptic impregnation of polyurethane polymers. J Antimicrob Chemother 33:969–978.

Bach A, Motsch J (1994): Prevention of infections caused by intravascular catheters—A review of the literature [in German]. Infusions Ther Transfusionsmed 21:104–114.

Bambauer R, Mestres P, Pirrung KJ, Sioshansi P (1994): Scanning electron microscopic investigation of catheters for blood access. Artif Org 18:272–275.

Beck-Sague CM, Jarvis WR (1989): Epidemic bloodstream infections associated with pressure transducers: A persistent problem. Infect Cont Hosp Epidemiol 10:54–59.

Beck-Sague CM, Jarvis WR, Brook H, Culver DH, Potts A, Gay E, et al. (1990): Epidemic bacteremia due to *Acinetobacter baumannii* in five intensive care units. Am J Epidemiol 132:723–733.

Bolton KJ, Dadd CER, Mead GC, Waites WM (1988): Chlorine resistance of strains of *Staphylococcus aureus* isolated from poultry processing plants. Appl Microbiol 6:31–34.

Branchini ML, Pfaller MA, Rhine-Chalberg J, Frempong T, Isenberg HD (1994): Genotypic variation and slime production among blood and catheter isolates of *Candida parapsilosis*. J Clin Microbiol 32:452–456.

Buchanan WE, Quinn MJ, Hasbargen JA (1994): Peritoneal catheter colonization with *Alternaria:* Successful treatment with catheter preservation. Peritoneal Dialysis Int 14:91–92.

Burchard WB, Cobb CM, Drisko CL, Killow WJ (1991): Effects of chlorhexidine and stannous fluoride on fibroblast attachment to different implant surfaces. Int J Oral Maxillofac Implants 6:418–426.

Burkart JM (1988): Comparison of peritonitis rates using standard spike vs. Y sets in CAPD. ASAIO-Trans 34:433–436.

Busscher HJ, Weerkamp AH (1987): Specific and non-specific interactions in bacterial adhesion to solid substrata. FEMS Microbiol Rev 46:165–173.

Caldwell DR (1990): Analysis of biofilm formation: Confocal laser microscopy and computer image analysis. 77th Annu Meet Int Assoc Milk Food Environ Sanitarians, pp 11–16.

Carballo J, Ferreiros CM, Criado MT (1991): Importance of experimental design in the evaluation of the influence of proteins in bacterial adherence to polymers. Med Microbiol Immunol 180:149–155.

Chang CC, Merritt K (1991): Effect of *Staphylococcus epidermidis* on adherence of *Pseudomonas aeruginosa* and *Proteus mirabilis* to polymethyl methacrylate (PMMA) and gentamicin-containing PMMA. J Orthop Res 9:284–288.

Cheung AL, Fischetti VA (1990): The role of fibrinogen in staphylococcal adherence to catheters in vitro. J Infect Dis 161:1177–1186.

Chew S, Mendoza M (1991): Bacterial contamination of three-way taps—One Australian hospital's experience. Aust J Adv Nurs 8:15–18.

Christensen GD, Baddour LM, Hasty DL, Lowrance GH, Simpson WA (1989): Microbial and foreign body factors in the pathogenesis of medical device infections. In Bisno AL, Waldvogel FA (eds): Infections Associated With Indwelling Medical Devices. Washington, DC: American Society for Microbiology, pp 59–70.

Christensen GD, Barker LP, Mawhinney TP, Baddour LM, Simpson WA (1990): Identification of an antigenic marker of slime production for *Staphylococcus epidermidis*. Infect Immun 58:2906–2911.

Cook AD, Sagers RD, Pitt WG (1993): Bacterial adhesion to protein-coated hydrogels. J Biomat Appl 8:72–89.

Cowan M, Busscher HJ (1993): Flow chamber study of the adhesion of *Prevotella intermedia* to glass after preconditioning with mutans streptococcal species: Kinetics and spatial arrangement. Microbios 73:135–144.

Crespo L, Eberlein TJ, O'Connor N, Hergrueter CA, Pribaz JJ, Eriksson E (1994): Postmastectomy complications in breast reconstruction. Ann Plastic Surg 32: 452–456.

D'Addato M, Curti T, Freyrie A, Agus GB, Bertini D, Biasi G (1994): Prevention of early graft infection with rifampicin-bonded Gelseal grafts: A multicentre experimental study. Cardiovasc Surg 2:254–258.

Dasgupta MK, Kowalewaska-Grochowska K, Costerton JW (1993): Biofilm and peritonitis in peritoneal dialysis. Peritoneal Dialysis Int 13:322–325.

Dasgupta MK, McKay S, Olsen M, Costerton JW (1994): Silver coated peritoneal catheter reduces colonization by *Staphylococcus aureus* in a rabbit model of peritoneal dialysis. Peritoneal Dialysis Int 14:586.

Davies DG, Chakrabarty AM, Geesey GG (1993): Exopolysaccharide production in biofilms: Substratum activation of alginate gene expression by *Pseudomonas aeruginosa*. Appl Environ Microbiol 59:1181–1186.

Deighton M, Borland R (1993): Regulation of slime production in *Staphylococcus epidermidis* by iron limitation. Infect Immun 61:4473–4479.

Demuth DR, Golub EE, Malamud D (1990): Streptococcal–host interactions: Structural and functional analysis of a *Streptococcus sanguis* receptor for a human salivary glycoprotein. J Biol Chem 265:7120–7126.

Derjaguin BV, Landau L (1941): Theory of the stability of strongly charged lyophobic soils of the adhesion of strongly charged particles in solutions of electrolytes. Acta Physicochim 14:633–662.

Dickinson GM, Bisno AL (1989a): Infections associated with indwelling devices: Concepts of pathogenesis; infections associated with intravascular devices. Antimicrob Agents Chemother 33:597–601.

Dickinson GM, Bisno AL (1989b): Infections associated with indwelling devices: Infections related to extravascular devices. Antimicrob Agents Chemother 33:602–607.

Doyle RJ, Rosenberg M (1990): Microbial Cell Surface Hydrophobicity. Washington, DC: American Society for Microbiology.

Dunn DS, Raghavan S, Volz RG (1993): Gentamicin sulfate attachment and release from anodized Ti-6Al-4V orthopedic materials. J Biomed Mat Res 27:895–900.

Duran LW, Marcy JA, Josephson MW (1992): Antimicrobial coatings on medical devices. Proceedings of meeting on Surfaces in Biomaterials, Minneapolis, May 1992, pp 37–41.

Egebo K, Toft P, Christensen EF, Steensen P, Jakobsen CJ (1994): Contamination of central venous catheters: Use of infusion lines does not increase catheter contamination. J Hosp Infect 26:105–109.

Elliot TSJ, Faroqui MH (1993): Infections and intravascular devices. Br J Hosp Med 17:498–503.

Embery G, Heaney TG, Stanbury JB (1986): Studies on the organic polyanionic constituents of human acquired dental pellicle. Arch Oral Biol 31:623–625.

Ena J, Cercenado E, Martinez D, Bouza E (1992): Cross-sectional epidemiology of phlebitis and catheter-related infections. Infect Cont Epidemiol 13:15–20.

Fabrizius-Homan DJ, Cooper SL (1991): A comparison of the adsorption of three adhesive proteins to biomaterial surfaces. J Biomat Sci 3:27–47.

Farber BF, Kaplan MH, Clogston AG (1988): *Staphylococcus epidermidis* extracted slime inhibits the antimicrobial action of aminoglycocide antibiotics. J Infect Dis 161:37–40.

Filiaggi MJ, Pilliar RM, Coombs NA (1993): Post-plasma-spraying heat treatment of the HA coating/Ti-6A1-4V implant system. J Biomed Mat Res 27:191–198.

Fletcher EL, Weissman BA, Efron N, Fleiszig SM, Curcio AJ, Brennan NA (1993): The role of pili in the attachment of *Pseudomonas aeruginosa* to unworn hydrogel contact lenses. Curr Eye Res 12:1067–1071.

Gallimore B, Gagnon RF, Subang R, Richards GK (1991): Natural history of chronic *Staphylococcus epidermidis* foreign body infection in a mouse model. J Infect Dis 164:1220–1223.

Ganderton L, Chawla J, Winters C, Wimpenny J, Stickler D (1992): Scanning electron microscopy of bacterial biofilms on indwelling bladder catheters. Eur J Clin Microbiol Infect Dis 11:789–796.

Giwercman B, Jensen ET, Hoiby N, Kharazmi A, Costerton JW (1991): Induction of β-lactamase production in *Pseudomonas aeruginosa* biofilm. Antimicrob Agents Chemother 35:1008–1010.

Glantz POJ, Baier RE, Attstrom R, Meyer AE, Gucinski H (1991): Comparative clinical wettability of teeth and intraoral mucosa. J Adhesion Sci Technol 5:401–408.

Goldmann DA, Pier GB (1993): Pathogenesis of infections related to intravascular catheterization. Clin Microbiol Rev 6:176–192.

Gorman SP, Adair CG, Mawhinney WM (1994): Incidence and nature of peritoneal catheter biofilm determined by electron and confocal laser scanning microscopy. Epidemiol Infect 112:551–559.

Gristina AG, Giridhar G, Gabriel BL, Naylor PT, Myrvik QN (1993): Cell biology and molecular mechanisms in artificial device infections. Int J Artif Org 16:755–763.

Groeger JS, Lucas AB, Coit D, LaQuaglia M, Brown AE, Turnbull A, Exelby P (1993): A prospective, randomized evaluation of the effect of silver impregnated subcutaneous cuffs for preventing tunneled chronic venous access catheter infections in cancer patients. Ann Surg 218:206–210.

Grover F, Cohen DJ, Oprian C, Henderson WG, Sethi G, Hammermeister KE (1994): Determinants of the occurrence of and survival from prosthetic valve endocarditis. Experience of the Veterans Affairs Cooperative Study on Valvular Heart Disease. J Thorac Cardiovasc Surg 108:207–214.

Gryn J, Sacchetti A (1992): Emergencies of indwelling venous catheters. Am J Emerg Med 10:254–257.

Gupta S, Preve CD, Shaheen K, Wilkens E, Smith DJ, Kirsh MM, Bolling SF, Rees RS (1994): Wound complications and treatment of the infected implantable cardioverter defibrillator generator. J Cardiol Surg 9:372–373.

Haanaes HR (1990): Implants and infections with special reference to oral bacteria. J Clin Periodontol 17:516–524.

Hambraeus A, Laurell G, Nybacka O, Whyte W (1990): Biliary tract surgery: A bacteriological and epidemiological study. Acta Chir Scand 156:155–162.

Hammel D, Scheld HH, Block M, Breithardt G (1994): Nonthoracotomy defibrillator implantation: A single-center experience with 200 patients. Ann Thorac Surg 58:321–326.

Hammerberg O, Bialkowska-Hobrzanska H, Gopaul D (1991): Isolation of *Agrobacterium radiobacter* from a central venous catheter. Eur J Clin Microbiol Infect Dis 10:450–452.

Harrison PG, Chan AT (1980): Inhibition of the growth of micro-algae and bacteria by extracts of eelgrass (*Zostera marina*) leaves. Mar Biol 61:21–26.

Heller J (1988): Chemically self-regulated drug delivery systems. J Controlled Release 8:111.

Holmes CJ, Evans R (1986): Biofilm and foreign body infection in CAPD-associated peritonitis. Peritoneal Dialysis Bull 6:168–177.

Hope MJ, Wong KW (1995): Liposomal formulation of ciprofloxacin. In Shek PN (ed): Liposomes in Biomedical Applications. Singapore: Harwood Academic, pp 121–134.

Hoyle BD, Alcantara J, Costerton JW (1992): *Pseudomonas aeruginosa* biofilm as a diffusion barrier to piperacillin. Antimicrob Agents Chemother 36:2054–2056.

Hoyle BD, Jass J, Costerton JW (1990): The biofilm glycocalyx as a resistance factor. J Antimicrob Chemother 26:1–5.

Hsieh DS, Rhine WD, Langer R (1983): Zero-order controlled-released polymer matrices for micro- and macromolecules. J Pharm Sci 17:17–22.

Hui HW, Robinson JR (1985): Ocular delivery of progesterone using a bioadhesive polymer. Int J Engl Ed August:196.

Jansen B, Beuth J, Ko HL (1990): Evidence for lectin-mediated adherence of *S. saprophyticus* and *P. aeruginosa* to polymers. Int J Med Microbiol 272:437–442.

Jansen B, Goodman LP, Ruiten D (1993): Bacterial adherence of hydrophilic polymer-coated polyurethane stents. Gastrointest Endosc 39:670–673.

Jansen B, Peters G (1993): Foreign body associated infection. J Antimicrob Chemother 32:69–75.

Johnson A, Oppenheim BA (1992): Vascular catheter-related sepsis: Diagnosis and prevention. J Hosp Infect 20:67–78.

Jones JW, Scott RJ, Morgan J, Pether JV (1992): A study of coagulase-negative staphylococci with reference to slime production, adherence, antibiotic resistance patterns and clinical significance. J Hosp Infect 22:217–227.

Karch M, Forgione L, Haverich A (1993): The efficacy of controlled antibiotic release for prevention of polyethylene terephthalate-(Dacron)-related infection in cardiovascular surgery. Clin Mat 13:149–154.

Karchmer A (1991): Prosthetic valve endocarditis: A continuing challenge for infection control. J Hosp Infect 18:355–356.

Keifer PA, Rhinehart KL, Hooper IR (1986): Renilla foulins, antifouling diterpenes from the sea pansy *Renilla reniformis* (Octocorallia). J Org Chem 5:4450–4454.

Keys T (1993): Early-onset prosthetic valve endocarditis. Cleve Clin J Med 60:455–459.

Khoury AE, Bruce AW, Reid G, Soboh F, Davidson D, Mittelman MW (1993): Susceptibility of bacterial biofilms to ciprofloxacin, tobramycin, ceftazidime, and trimethoprim/sulfamethoxazole under dynamic-flow conditions. Proceedings of 18th International Congress of Chemotherapy, Stockholm.

Kirchman D, Graham S, Reish D, Mitchell R (1982): Bacteria induce settlement and metamorphosis of *Janua* (*Dexiospora*) *brasiliensis* (Grube). J Exp Mar Biol Ecol 56:153–163.

Kotilainen P (1990): Association of coagulase-negative staphylococcal slime production and adherence with the development and outcome of adult septicemias. J Clin Microbiol 28:2779–2785.

Kotilainen P, Nikoskelainen J, Huovinen P (1991): Antibiotic susceptibility of coagulase-negative staphylococcal blood isolates with special reference to adherent, slime-producing *Staphylococcus epidermidis* strains. Scand J Infect Dis 23:325–332.

Krieg NR, Holt JG (1984): Bergey's Manual of Systematic Bacteriology. Baltimore: Williams & Wilkins.

Langer R, Cima LG, Tamada JA, Wintermantel E (1990): Future directions in biomaterials. Biomaterials 11:738–745.

Laskowski LF, Marr JJ, Spernoga JF, Frank NJ, Barner HB, Kaiser G, Tyras DH (1977): Fastidious mycobacteria grown from porcine prosthetic-heart-valve cultures. N Engl J Med 297:101–102.

Leung JW, Lau GT, Sung JJ, Costerton JW (1992): Decreased bacterial adherence to silver-coated stent material: An *in vitro* study. Gastrointest Endosc 38:338–340.

Longer MA, Ch'ng HS, Robinson JR (1985): Bioadhesive polymers as platforms for oral controlled drug delivery. III. Oral delivery of chlorothiazide using a bioadhesive polymer. J Pharm Sci 74:406.

Maki JS, Rittschof D, Costlow JD, Mitchell R (1988): Inhibition of attachment of larval barnacles, *Balanus amphritrite,* by bacterial surface films. Mar Biol 97:199–206.

Makino K, Mark E, Okano T, Kim SW (1990): A microcapsule self-regulating delivery system for insulin. J Controlled Release 12:235.

Malamud D (1985): Influence of salivary proteins on the fate of oral bacteria. In Mergenhagen SE, Rosan B (eds): Molecular Basis of Oral Microbial Adhesion. Washington, DC: ASM, pp 117–124.

Marshall KC, Stout R, Mitchell R (1971): Mechanisms of the initial events in the sorption of marine bacteria to surfaces. J Gen Microbiol 68:337–348.

Massia SP, Hubbell JA (1990): Covalent surface immobilization of Arg-Gly-Asp– and Tyr-Ile-Gly-Ser-Arg–containing peptides to obtain well-defined cell-adhesive substrates. Anal Biochem 187:292–301.

Mathiowitz E, Ron E, Mathiowitz G, Langer R (1989): Surface morphology of bioerodible poly(anhydrides). Polym Prepr 30:460.

Mermel LA, McCormick RD, Springman SR, Maki DG (1991): The pathogenesis and epidemiology of catheter-related infection with pulmonary artery Swan-Ganz catheters: A prospective study utilizing molecular subtyping. Am J Med 91:197S–205S.

Mittelman MW, Davidson D, Vas S, Khoury AE (1994): *In vitro* studies of silver-coated catheter antimicrobial efficacy. Ann Conf Peritoneal Dialysis, Orlando, FL.

Moonens F, el-Alami S, Gossum AV, Struelens MJ, Serruys E (1994): Usefulness of Gram staining of blood collected from total parenteral nutrition catheter for rapid diagnosis of catheter-related sepsis. J Clin Microbiol 32:1578–1579.

Morck DW, Ellis BD, Domingue PA, Olson ME, Costerton JW (1991): *In vivo* expression of iron regulated outer-membrane proteins in *Pasteurella haemolytica*-A1. Microb Pathol 11:373–378.

Muller E, Hubner J, Gutierrez N, Takeda S, Goldmann DA, Pier GB (1993a): Isolation and characterization of transposon mutants of *Staphylococcus epidermidis* deficient in capsular polysaccharide/adhesin and slime. Infect Immun 61:551–558.

Muller E, Takeda S, Shiro H, Goldmann D, Pier GB (1993b): Occurrence of capsular polysaccharide/adhesin among clinical isolates of coagulase-negative staphylococci. J Infect Dis 168:1211–1218.

Murakami S, Igarashi T, Tanaka M, Tobe T, Mikami K (1993): Adherence of bacteria to various urethral catheters and occurrence of catheter-induced urethritis [in Japanese]. Hinyokika Kiyo Acta Urol Japonica 39:107–111.

Nade S (1990): Infection after joint replacement—What would Lister think? Med J Aust 152:394–397.

Nakazato G (1990): Studies on plaque formed on implants [in Japanese]. Gifu Shika Gaikkai Zasshi 17:131–151.

Neu TR, DeBoer CE, Verkerke GJ, Schuttes HK, Rakhorst G, Mei HCVD, Busscher HJ (1994a): Biofilm development in time on a silicone voice prosthesis—A case study. Microb Ecol Health Dis 7:27–33.

Neu TR, Mei HCVD, Busscher HJ, Dijk F, Verkerke GJ (1993): Biodeterioration of medical-grade silicone rubber used for voice prostheses: A SEM study. Biomaterials 14:459–464.

Neu TR, Verkerke GJ, Herrmann IF, Schutte HK, Mei HCVD, Busscher HJ (1994b): Microflora on explanted silicone rubber voice prostheses: Taxonomy, hydrophobicity and electrophoretic mobility. J Appl Bacteriol 76:521–528.

Nickel JC (1993): A practical approach to urinary tract infections. Can J Diagn 10:64–80.

Nickel JC, Costerton JW (1992): Bacterial biofilms and catheters: A key to understanding bacterial strategies in catheter-associated urinary tract infections. Can J Infect Dis 3:261–267.

Nickel JC, Ruseska I, Wright JB, Costerton JW (1985): Tobramycin resistance of *Pseudomonas aeruginosa* cells growing as a biofilm on urinary catheter material. Antimicrob Agents Chemother 27:619–624.

Nose Y (1992): Is a totally implantable artificial heart realistic. Artif Org 16:19–42.

Ofek I, Doyle RJ (eds) (1994): Bacterial Adhesion to Cells and Tissues. New York: Chapman & Hall, pp 195–238.

Ofek I, Mirelman D, Sharon N (1977): Adherence of *Escherichia coli* to human mucosal cells mediated by mannose receptors. Nature 265:623–625.

Ouchi T, Kobayshi H, Banda T (1990): Design of poly(a-malic acid)-5-FU conjugate exhibiting antitumor activity. Br Polym J 23:221.

Palenik CJ, Behnen MJ, Setcos JC, Miller CH (1992): Inhibition of microbial adherence and growth by various glass ionomers in vitro. Dent Mat 8:16–20.

Parras F, Ena J, Bouza E, Guerrero MC, Moreno S, Galvez T, Cercenado E (1994): Impact of an educational program for the prevention of colonization of intravascular catheters. Infect Cont Hosp Epidemiol 15:239–242.

Parsons CL, Stauffer C, Schmidt JD (1981): Impairment of antibacterial effect of bladder surface mucin by protamine sulfate. J Infect Dis 144:180–184.

Passerini L, Lam K, Costerton JW, King EG (1992): Biofilms on indwelling vascular catheters. Crit Care Med 20:665–673.

Patel M, Drake D, Keller J (1990): Bacterial adhesion to titanium implant surfaces—Development of an *in vitro* model. J Dent Res (abstract). 69:369.

Paulsson M, Kober M, Freij-Larsson C, Stollenwerk M, Wesslen B, Ljungh A (1993): Adhesion of staphylococci to chemically modified and native polymers, and the influence of preadsorbed fibronectin, vitronectin and fibrinogen. Biomaterials 14:845–853.

Pedersen K (1990): Biofilm development on stainless steel and PVC surfaces in drinking water. Water Res 24:239–243.

Pfeiffer D, Jung W, Fehske W, Korte T, Manz M, Moosdorf R, Luderitz B (1994): Complications of pacemaker-defibrillator devices: Diagnosis and management. Am Heart J 127:1073–1080.

Pier GB, DesJardin D, Grout M, Garner C, Bennett SE, Pekoe G, et al. (1994): Human immune response to *Pseudomonas aeruginosa* mucoid exopolysaccharide (alginate) vaccine. Infect Immun 62:3972–3979.

Pilliar RM, Lee JM, Davies JE (1993): Interface zone: Factors influencing its structure for cementless implants. In Morrey BF (ed): Biological, Material, and Mechanical Considerations of Joint Replacement. New York: Raven Press, pp 225–235.

Pittet D, Tarara D, Wenzel RP (1994): Nosocomial bloodstream infection in critically ill patients. J Am Med Assoc 271:1598–1601.

Poisson DM, Touquet S, Bercault N, Arbeille B (1992): Electron-microscopic description of accretions occurring on tips of infected and non-infected central venous catheters. Intensive Care Med 18:464–468.

Quirynen M, Mei HCVD, Bollen CM, Bossch LHVD, Doornbusch GI, Steenberghe DV, Busscher HJ (1994): The influence of surface-free energy on supra- and subgingival plaque microbiology. An *in vivo* study on implants. J Periodont 65:162–167.

Quirynen M, Mei HCVD, Bollen CM, Schotte A, Marechal M, Doornbusch GI, Naert I, Busscher HJ, Steenberghe DV (1993): An *in vivo* study of the influence of the surface roughness of implants on the microbiology of supra- and subgingival plaque. J Dent Res 72:1304–1309.

Ramirez de Arellano E, Pascual A, Martinez-Martinez L, Perea EJ (1994): Activity of eight antibacterial agents on *Staphylococcus epidermidis* attached to Teflon catheters. J Med Microbiol 40:43–47.

Ratner BD (1993): New ideas in biomaterials science—A path to engineered biomaterials. J Biomed Mat Res 27:837–850.

Reid G, Khoury AE, Preston CA, Costerton JW (1994): Influence of dextrose dialysis solutions on adhesion of *Staphylococcus aureus* and *Pseudomonas aeruginosa* to three catheter surfaces. Am J Nephrol 14:37–40.

Reid G, Tieszer C, Foerch R, Busscher HJ, Khoury AE, Bruce AW (1993): Adsorption of ciprofloxacin to urinary catheters and effect on subsequent bacterial adhesion and survival. Coll Surf B Biointerfaces 1:9–16.

Reid G, Tieszer C, Foerch R, Busscher HJ, Khoury AE, Mei HCVD (1992): The binding of urinary components and uropathogens to a silicone latex urethral catheter. Cells Mat 2:253–260.

Richards GK (1976): Resistance to infection. In Freedman SO, Gold P (eds): Clinical Immunology. New York: Harper and Row, pp 65–77.

Rigdon RH (1975): Tissue reaction to foreign materials. Crit Rev Food Sci Technol 7:435.

Rosenberg M (1991): Basic and applied aspects of microbial adhesion at the hydrocarbon: water interface. Crit Rev Microbiol 18:159–173.

Ruoslahti E, Pierschbacher MD (1987): New perspectives in cell adhesion: RGD and integrins. Science 238:491–497.

Schein OD, Poggio EC (1990): Ulcerative keratitis in contact lens wearers, incidence and risk factors. Cornea 9:55–58.

Scheltema RS, Williams IP, Shaw MA, London C (1981): Gregarious settlement by the larvae of *Hydroides dianthyus* (Polychaeta: Serpulidae). Mar Ecol Prog Ser 5:69–74.

Schoen FJ (1991): Biomaterials science, medical devices, and artificial organs. Synergistic interactions for the 1990s. ASAIO-Trans 37:44–48.

Sears MA, Gearhart DJ, Tittschof D (1990): Antifouling agents from marine sponge *Lissodendoryx isodictyalis* Carter. J Chem Ecol 16:791–799.

Selan L, Berlutti F, Passariello C, Comodi-Ballanti MR, Thaller MC (1993): Proteolytic enzymes: A new treatment strategy for prosthetic infections? Antimicrob Agents Chemother 37:2618–2621.

Shand GH, Anwar H, Kadurugamuwa J, Brown MRW, Silverman SH, Melling J (1985): *In vivo* evidence that bacteria in urinary tract infections grow under iron-restricted conditions. Infect Immun 48:35–39.

Silver S, Misra TK (1988): Plasmid-mediated heavy metal resistances. Annu Rev Microbiol 42:717–743.

Soboh F, Zamboni AC, Davidson D, Khoury AE, Mittelman MW (1995): Interaction of *Pseudomonas aeruginosa* biofilm with ciprofloxacin and protamine sulfate. Antimicrob Agents Chemother 39:1281–1286.

Sodhi RNS, Grad HA, Smith DC (1992): Examination by x-ray photoelectron spectroscopy of the adsorption of chlorhexidine on hydroxyapatite. J Dent Res 71:1493–1497.

Stamm WE (1991): Catheter-associated urinary tract infections: Epidemiology, pathogenesis, and prevention. Am J Med 91:65–71.

Stefely JS, Markowitz MA, Regen SL (1988): Permeability characteristics of lipid bilayers from lipoic acid derived phosphatidylcholines: Comparison of monomeric, cross-linked, and non-cross linked polymerized membranes. J Am Chem Soc 110:7463.

Stern GA (1990): *Pseudomonas* keratitis and contact lens wear: The lens/eye is at fault. Cornea 9:36–38.

Sterniste W, Vavrik K, Lischka A, Sacher M (1994): Effectiveness and complications of percutaneous central venous catheters in neonatal intensive care [in German]. Klin Padiatr 206:18–21.

Stickler D, Ganderton L, King J, Nettleton J, Winters C (1993a): *Proteus mirabilis* biofilms and the encrustation of urethral catheters. Urol Res 21:407–411.

Stickler DJ, King JB, Winters C, Morris SL (1993b): Blockage of urethral catheters by bacterial biofilms. J Infect 27:133–135.

Stiver HG, Zachidniak Z, Speert DP (1988): Inhibition of polymorphonuclear leukocyte chemotaxis by the mucoid exopolysaccharide of *Pseudomonas aeruginosa.* Clin Invest Med 11:247–252.

Suci PA, Mittelman MW, Yu FP, Geesey GG (1994): Investigation of ciprofloxacin penetration into *Pseudomonas aeruginosa* biofilms. Antimicrob Agents Chemother 38:2125–2133.

Summers AO, Wireman J, Vimy MJ, Lorscheider FL, Marshall B, Levy SB, Bennett S, Billard L (1993): Mercury released from dental "silver" fillings provokes an increase in mercury- and antibiotic-resistant bacteria in oral and intestinal floras of primates. Antimicrob Agents Chemother 37:825–834.

Takemoto K, Inaki Y (1988): New aspects of the chemical modification on the nucleic acid analogs. Acta Polym 39:33.

Takeuchi H, Hida S, Yoshida O, Ueda T (1993): Clinical study on efficacy of a Foley catheter coated with silver-protein in prevention of urinary tract infections [in Japanese]. Hinyokika Kiyo Acta Urol Japonica. 39:293–298.

Tan S, Gill G (1992): Selection of dental procedures for antibiotic prophylaxis against infective endocarditis. J Dent 20:375–376.

Teichman JMH, Abraham VE, Stein PC, Parsons CL (1994): Protamine sulfate and vancomycin are synergistic against *Staphylococcus epidermidis* prosthesis infection *in vivo.* J Urol 152:213–216.

Timmerman CP, Fleer A, Besnier JM, Graaf LD, Cremers F, Verhoef J (1991): Characterization of a proteinaceous adhesin of *Staphylococcus epidermidis* which mediates attachment to polystyrene. Infect Immun 59:4187–4192.

Trooskin SZ, Donetz AP, Baxter J, Harvey RA, Greco RS (1987): Infection-resistant continuous peritoneal dialysis catheters. Nephron 46:263–267.

Trooskin SZ, Donetz AP, Harvey RA, Greco RS (1985): Prevention of catheter sepsis by antibiotic bonding. Surgery 97:547–551.

Uyen HMW, Schakenraad JM, Sjellema J, Noordmans J, Jongebold WL, Stokroos I, Busscher HJ (1990): Amount and different surface structure of albumin adsorbed to solid substrata with different wettabilities in a parallel plate flow cell. J Biomed Mat Res 24:1599–1614.

van Blitterswijk CA, Bakker D, Hesseling SC, Koerten HK (1991): Reactions of cells at implant surfaces. Biomaterials 12:187–193.

van Loosdrecht MCM, Zehnder AJB (1990): Energetics of bacterial adhesion. Experientia 46:817–822.

Vaudaux P, Pittet D, Haeberli A, Lerch PG, Morgenthaler JJ, Proctor RA, Waldvogel FA, Lew DP (1993): Fibronectin is more active than fibrin or fibrinogen in promoting *Staphylococcus aureus* adherence to inserted intravascular catheters. J Infect Dis 167:633–641.

Verheyen CC, Dhert WJ, Blieck-Hogervorst JM, Reijden TJvd, Petit PL, Groot KD (1993): Adherence to a metal, polymer and composite by *Staphylococcus aureus* and *Staphylococcus epidermidis.* Biomaterials 14:383–391.

Verwey EJW, Overbeek JTG (1948): Theory of the Solubility of Lyophobic Colloids. Amsterdam: Elsevier.

Virden C, Dobke MK, Stein P, Parsons CL, Frank DH (1992): Subclinical infection of the silicone breast implant surface as a possible cause of capsular contracture. Aesthetic Plastic Surg 16:173–179.

Volkow P, Sanchez-Mejorada G, Vega SLdl, Vazquez C, Tellez O, Baez RM, et al. (1994): Experience of an intravenous therapy team at the Instituto Nacional de Cancerologia (Mexico) with a long-lasting, low-cost silastic venous catheter. Clin Infect Dis 18:719–725.

Vrolijk NH, Targett NM, Baier RE, Meyer AE (1990): Surface characterization of two gorgonian coral species: Implications for a natural antifouling defense. Biofouling 2:39–54.

Wang D (1993): Multiple organ failure after valve replacement [in Chinese]. Chung-Hua Wai Ko Tsa Chih 31:653–656.

Whitener C, Caputo GM, Weitekamp MW, Karchmer AW (1993): Endocarditis due to coagulase-negative staphylocci. Microbiologic, epidemiologic, and clinical considerations. Infect Dis Clin North Am 7:81–96.

Widmer AF, Wiestner A, Frei R, Zimmerli W (1991): Killing of nongrowing and adherent *Escherichia coli* determines drug efficacy in device-related infections. Antimicrob Agents Chemother 35:741–746.

Williams DF (1987): Definitions in Biomedicals. New York: Elsevier.

Wilson LA, Schlitzer AB, Ahearn DG (1981): *Pseudomonas* corneal ulcers associated with soft contact lens wear. Am J Ophthalmol 92:546–554.

5

ADHESION IN
THE RHIZOSPHERE

ANN G. MATTHYSSE

Department of Biology, University of North Carolina, Chapel Hill, North Carolina 27599-3280

5.1. INTRODUCTION: THE ROOT AND THE RHIZOSPHERE

The rhizosphere is generally considered to be that part of the soil that is in intimate association with plant roots and whose composition is influenced by the uptake of chemicals by the plant roots and by the release of various compounds by the roots into the soil. Some soil bacteria may find themselves in the rhizosphere by mere chance, but many soil bacteria are found preferentially in the vicinity of plant roots. Some of these bacteria adhere to the root or even invade the root in symbiotic or pathogenic associations.

The general structure of a root is shown in Figure 5.1. Roots grow just behind the tip in a region filled with dividing cells that is called the *meristem.* Covering the meristem is the root cap, whose cells have their origin at the bottom of the root meristem. The cells of the root cap may be sloughed off or removed by abrasion during the growth of the root through the soil. These cells may survive for some time in the soil, influencing the composition of the rhizosphere (Hawes and Brigham, 1992). The root cap and the root epidermis for some distance above the cap may be covered with a layer of mucilage. Just above the root meristem is an area where the cells of the root are elongating and differentiating. The cells in the interior of the root differentiate into vascular tissue that is surrounded by the pericycle (a layer of meristematic cells) and the cortex. The epidermal cells in this region form projections called *root hairs.* These projections are filled with cytoplasm and separated from the soil by the

Bacterial Adhesion: Molecular and Ecological Diversity, pages 129–153
© 1996 Wiley-Liss, Inc.

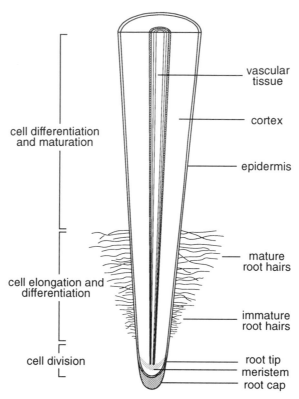

cell differentiation
and maturation

cell elongation and
differentiation

cell division

vascular
tissue

cortex

epidermis

mature
root hairs

immature
root hairs

root tip
meristem
root cap

Fig. 5.1. Diagram of the tip of a root showing structures involved in bacterial adhesion to the root. The root grows down through the soil from the tip. As the distance of cells from the growing tip increases, the cells begin to elongate and differentiate. Most uptake of water and solutes occurs in the region of the root hairs. (Drawing by Susan Whitfield and Ann G. Matthysse.)

plasma membrane and the plant cell wall. As the root grows and the epidermal cells bearing root hairs become further removed from the root tip, the root hairs become mature and cease to elongate. Eventually the root hairs are lost from the epidermal cells. Secondary roots are initiated in the pericycle and grow through the cortex and epidermis to emerge through a crack or wound in the epidermis.

The root may secrete various compounds, such as B vitamins and flavonoids, into the soil. The exact compounds released vary with the species of plant and the growth conditions. With the exception of studies on the interaction between rhizobia and legumes, there is little information on the regulation of this secretion or whether it is influenced by microorganisms.

5.2. DESCRIPTIONS OF ADHESION OF VARIOUS TYPES OF BACTERIA TO ROOTS

Many different species of bacteria are found in the rhizosphere in association with roots. The interactions of two groups of soil bacteria with roots have been examined in considerable detail: the rhizobia and their symbiotic interaction with legumes and the agrobacteria and their pathogenic interaction with dicots. In addition, the association of various species of *Pseudomonas, Klebsiella, Enterobacter,* and *Azospirillum* with plant roots has been examined.

5.2.1. The Adhesion of Rhizobia to Legume Roots

Members of the Rhizobiaceae are capable of living free in the soil. However, under conditions of nitrogen limitation the bacteria can form a symbiotic association with the roots of leguminous plants resulting in the formation of nitrogen-fixing root nodules. These bacteria are divided into three genera: *Rhizobium, Bradyrhizobium,* and *Azorhizobium.* For the purposes of this chapter the members of all three genera will be referred to as *rhizobia.* Each species or biovar of *Rhizobium* generally forms nodules on only a limited number of legume species or cultivars. At one time it was thought that the basis of this specificity was the ability (or lack thereof) of the bacteria to adhere to root hairs of the host plant. However, the specificity of the adherence of rhizobia to legume roots is in some dispute. Dazzo et al. (1976) found that *R. leguminosarum* bv. *trifolii* adhered preferentially to clover roots, which they nodulate. Smit et al. (1986, 1989) observed that the adherence of *R. leguminosarum* bv. *viciae* to pea roots, which they nodulate, was nonspecific. They found that *R. leguminosarum* bv. *trifolii,* which does not nodulate pea, and indeed even *Agrobacterium tumefaciens,* which forms tumors in wound sites, could bind to pea root hairs. In examining the adherence of *R. meliloti* to alfalfa roots, which they nodulate, Wall and Favelukes (1991) found that about 10% of the inoculated bacteria adsorbed to the roots and that binding of about half of these bacteria could not be competed by a thousand-fold excess of *R. leguminosarum* bv. *trifolii,* which does not nodulate alfalfa. They concluded that *R. meliloti* bound specifically to alfalfa. Ho et al. (1986) found that *Bradyrhizobium japonicum,* which nodulates soybean, bound to soybean tissue culture cells and that other rhizobia that did not nodulate soybean did not adhere to soybean cells.

5.2.1.1. Nod Factors Legume roots release a number of compounds, including flavonoids, into the soil. The particular compounds released depend on the legume species. Rhizobia are chemotactic to the compounds released by legume roots. Each species of *Rhizobium* possesses a system for sensing the flavonoids produced by the plant species with which it interacts. These flavonoids stimulate the expression of a set of *Rhizobium* genes, the *nod* genes. Recent reviews of signaling between the plant and the bacteria and the role of *nod* genes can be found in Fisher and Long (1992) and Vijn et al. (1993). A diagram of the initial interactions between rhizobia and a legume root is shown in Figure 5.2.

The *nod* genes can be grouped into the common *nod* genes, which are expressed in most species of rhizobia, and the host-specific *nod* genes, which are involved in the interaction of a particular *Rhizobium* species or biovar with a particular plant species. Common *nod* genes encode enzymes required to synthesize a chitin oligosaccharide (a short oligomer of *n*-acetylglucosamine). Host-specific *nod* genes encode enzymes that add substitutions such as acetate, sulfate, or fatty acyl side chains to this oligosaccharide. Also included in the *nod* genes are a membrane transport system, a calcium-binding protein with homology to hemolysin, and regulatory and sensory systems. The transcription of the *nod* genes is activated by the product of the *nodD* gene(s). Bacterial nod factors may be released into the soil or bound to the surface of the bacterium presumably by the fatty acyl side chain. The nod factors cause a number of responses in the host plant, including root hair deformations, cortical cell divisions, and the expression of some of the early plant proteins involved in nodule formation.

At some time during this complicated chemical conversation between the bacterium and the plant host the bacteria bind to the surface of immature root hairs. An electron micrograph of *R. meliloti* bound to an alfalfa root hair is shown in Figure 5.3. The relationship between this adhesion and the expression of *nod* genes is unclear. Nod factors cause root hairs to become deformed and in some cases to form a characteristic morphology termed "shepherd's crooks." The bacteria synthesize cellulose fibrils that link them together to form aggregates on the tips of the root hairs and in the curl of the shepherd's crooks (Napoli et al., 1975; Smit et al., 1987). The rhizobia then enter the root hair via an invagination of the plant cell plasma membrane to form an infection thread. The infection thread grows into the root cortex where the bacteria, still surrounded by a membrane derived from the plant plasma membrane, are released into vacuoles in the cytoplasm of the cortical cells. During this process the plant cells continue to divide to form a nodule. The end result is a nodule on the root that contains cortical cells with vacuoles filled with bacteroids that are capable of fixing atmospheric nitrogen. The bacteroids rely on the plant for both their energy and their carbon supply.

The role and mechanism of bacterial adhesion in the development of the legume root nodule remain uncertain. As mentioned above, early studies indicated that rhizobia bound only to the legume species on which they were able to form nodules, and the suggestion was made that host specificity of the rhizobia was determined by bacterial adhesion to the root hairs of only those plants that the bacteria could nodulate. However, the discovery of nod factors and the role of the host-specific modifications in the structure of the nod factors, as well as the role of the host-specific flavonoids in the induction of the genes for the synthesis of nod factors, has cast doubt on the role of specific

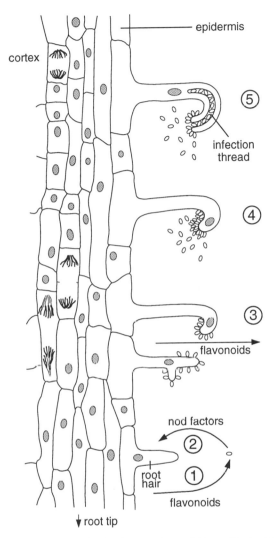

Fig. 5.2. Diagram of the stages in the interaction of *R. meliloti* with an alfalfa root. The bacteria are found living in the soil and are chemotactic to substances released from the root. Included in these substances are flavonoids specific for the alfalfa root such as luteolin (step 1). The bacteria respond to the flavonoids by the activation of *nod* genes and the release of nod factors (step 2). The plant responds to the nod factors by cortical cell divisions and by release of additional compounds, including more flavonoids. The young root hairs become branched and/or curl. The bacteria adhere to these root hairs (step 3). Increasing numbers of bacteria adhere to the tip of the root hair, and the bacteria begin to synthesize cellulose fibrils and to form aggregates (step 4). The infection thread forms by an invagination of the plant cell plasma membrane, and the bacteria grow down the infection thread (step 5). (Drawing by Susan Whitfield and Ann G. Matthysse.)

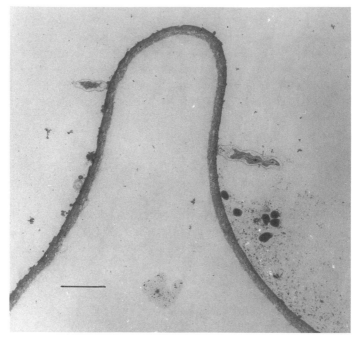

Fig. 5.3. Transmission electron micrograph of *R. meliloti* attached to the surface of an alfalfa root hair. Two bacteria are shown, one on each side of the root hair surface. The bacterium is a mutant that can make nod factors and adhere to root hair tips, but cannot form infection threads. Bar = 1 μm. (Courtesy of Ann Hirsch.)

bacterial adhesion in determining the host range of rhizobia. As yet there are no studies of the role, if any, of *nod* genes and nod factors in bacterial adhesion.

5.2.1.2. *Plant and Bacterial Lectins*

The earlier models of the adhesion of rhizobia to root hairs involved the adhesion of the bacteria via a surface polysaccharide, either a capsule or a lipopolysaccharide, to a lectin produced by the plant root. In the case of clover and *R. trifolii*, the root lectin is reported to be released into the soil and to serve as a bridge between polysaccharides on the root hair surface and on the surface of the bacterium (Dazzo et al., 1982a). The bacterial polysaccharide is localized at the polar ends of the bacterium (Dazzo et al., 1982b). The hapten for the pea lectin, 3-*O*-methyl-D-glucose, inhibited the attachment of *R. leguminosarum* to pea root hair tips only if the bacteria were grown with limiting manganese. When carbon was the growth-limiting factor, bacterial adherence was no longer inhibited by the hapten. Manganese-limited bacteria were more effective at forming infection threads than carbon-limited cells (Kijne et al., 1988). In a recent version of the lectin model of adhesion, it is proposed that *B. japonicum* produces a lectin, Bj38, localized at one pole of the bacterial surface (Loh et al., 1993).

This lectin binds the bacteria to the surface of immature root hairs. In both models the bacterial polysaccharide or lectin protein is produced only at limited times during bacterial growth, thus accounting for the observation that the ability of the bacteria to adhere to root hairs varies with the growth phase of the bacteria (Dazzo et al., 1984; Ho et al., 1994).

No nonadhesive mutants of *R. trifolii* are available, but mutants that are reduced in adhesion to root hairs of *B. japonicum* do exist (Ho et al., 1994). At low bacterial concentrations these mutants adhere to root hairs almost as well as the wild-type parent. When higher bacterial concentrations were examined it was found that the number of the wild-type bacteria that bound increased as the bacterial inoculum increased. The mutant bacteria showed no such increase in bacterial binding. Both the mutant and wild-type bacteria cause the formation of nitrogen-fixing nodules on soybean roots, but the mutants induce only about 10% to 20% as many nodules as the wild type. These bacterial mutants lack detectable Bj38 on their surface. Since the mutants were isolated by chemical mutagenesis, the gene(s) altered in the mutants are unknown. It is unclear whether these mutants are reduced in their ability to bind directly to the root hair surface or whether they are reduced in their ability to form aggregates. In the case of *A. tumefaciens* (discussed below), mutants that are unable to synthesize cellulose fibrils and thus to form bacterial aggregates are reduced in virulence (Minnemeyer et al., 1991). These *A. tumefaciens* mutants show normal adhesion at low bacterial inoculum concentrations and short incubation times, but are reduced in adherence to plant cells compared with the wild type at high inoculum densities or longer incubation times (Matthysse, 1983).

Preincubation of *B. japonicum* with root extracts or soybean lectin increases the effectiveness of the bacteria in inducing nodules, decreases the time between the addition of the bacteria to the roots and bacterial binding to root hairs, and increases the competitiveness of the bacteria in binding to root hairs (Lodeiro and Favelukes, 1995). To be effective, the preincubation must last at least 12 hours, suggesting that the lectin is eliciting some response from the bacteria. Preincubation of *R. meliloti* with alfalfa, but not clover, root extract also increased the number of bacteria that bound to root hairs. The active factor (which had a high molecular weight) appeared to be removed from the solution by the bacteria (Wall and Favelukes, 1991).

The question of the involvement of plant lectins in the specificity of root nodule formation by rhizobia has also been examined by experiments in which the plant host is modified. Peas cannot be nodulated by *R. leguminosarum* bv. *trifolii*, which forms nitrogen-fixing nodules on clover. However, when the clover lectin gene was introduced into pea roots using *Agrobacterium rhizogenes*, the resulting transgenic pea roots could form nodules when inoculated with *R. leguminosarum* bv. *trifolii* (Diaz et al., 1989). A similar experiment using lotus and soybean has recently been reported. The bacteria that nodulate lotus and soybean, *R. loti* and *B. japonicum*, respectively, are very different. When the soybean lectin gene is introduced into lotus, the transgenic plants can form nodules when inoculated with *B. japonicum.* These nodules do not contain bacteroids and do not fix nitrogen (van Rhijn et al., 1995).

Thus it appears that plant lectins do play a role in the host specificity of the interaction between rhizobia and legumes. Whether the lectins interact directly with the nod factors or whether they play some direct or indirect role in bacterial adhesion to root hair tips is unclear at present. It is also uncertain how these experiments relate to earlier experiments reporting that added soybean lectin had no effect on the adhesion of *B. japonicum* to soybean roots (Pueppke, 1984) or how they relate to the effects of preincubation of bacteria with lectins. Thus, at the present time, the role of lectins in bacterial adhesion and nodule formation is quite unclear.

5.2.1.3. *Bacterial Fimbriae* Fimbriae have been found to be involved in many instances of adherence of bacteria to animal cells. For this reason the possible role of fimbriae in the adherence of rhizobia to root hairs has been examined. No fimbrial adhesions have been found in the case of *R. leguminosarum,* although cellulose fibrils that are involved in the second stage of bacterial adhesion have occasionally been mistaken for fimbriae (Smit et al., 1987). In the case of *B. japonicum* some evidence for the involvement of fimbriae in adhesion exists (Vesper and Bauer, 1986). These bacteria produce fimbriae at certain growth phases that coincide with those times during which the bacteria show the ability to bind to soybean root hairs. Of the bacteria that bound, 25% were tightly bound; the remaining 75% were loosely bound and could be removed by gentle washing (Vesper and Bauer, 1985). The extremely loose binding of most of the bacteria may account for differences in the estimates of binding by various researchers. The loose binding could be inhibited by galactose but not by other sugars tested, as if a lectin was involved. Antibody to fimbriae was able to block bacterial adhesion to roots. There is no information on the relation of this possible fimbrial lectin and the bacterial surface Bj38 described by Ho et al. (1986).

5.2.1.4. *Bacterial and Plant Surface Proteins* Another possible candidate for involvement in the adhesion of rhizobia to root hairs is a small bacterial surface protein that binds calcium (Smit et al., 1989). This protein has been named *rhicadhesin.* It can be purified from extracts of the bacterial surface. When added to the culture medium rhicadhesin inhibits the adhesion of *R. leguminosarum* bv. *vicae* to pea root hairs. The protein is acidic, with a molecular weight of about 14,000. Purified rhicadhesin binds calcium, and calcium is reported to be required for the adhesion of *R. leguminosarum* bv. *viciae* to pea root hairs. The researchers believe that calcium is involved in the binding of rhicadhesin to the bacterial surface. A protein similar to rhicadhesin was isolated from several other rhizobia and from *A. tumefaciens* (Smit et al., 1992).

R. leguminosarum bv. *viciae* may bind to a protein on the root hair surface. A 32,000 molecular weight glycoprotein that blocked the ability of rhicadhesin to inhibit bacterial attachment to root hairs was isolated from pea roots (Swart et al., 1994). The end-terminal amino acid sequence was determined and showed no homology with known proteins. The protein may contain an RGD (Arg-Gly-Asp) sequence, since an RGD-containing peptide can also block the inhibition of bacterial attachment by free rhicadhesin. Obviously this research is still preliminary. The assay of the blocking of an inhibition of bacterial attachment is complex and subject to many possible artifacts. The role of calcium and of rhicadhesin in the attachment of rhizobia to root hairs is uncertain. Ho et al. (1988) found that EGTA did not inhibit the adhesion of B. japonicum to soybean roots, although 50 mM NaCl did inhibit adhesion. Caetano-Anolles et al. (1989) found that divalent cations were required for the adsorption of R. meliloti to alfalfa roots, and Lodeiro et al. (1995) found divalent cations to be required for the specific adsorption of R. leguminosarum bv. phaseoli to bean roots. In the absence of bacterial mutants that fail to produce rhicadhesin, the role of this protein in adherence of rhizobia to root hairs remains intriguing but uncertain.

5.2.2. The Adhesion of Agrobacteria to Host Cells

Agrobacteria are gram-negative bacteria, which, like rhizobia, are capable of living free in the soil or entering into an association with higher plants. However, unlike rhizobia, association with agrobacteria is detrimental to the host plant and results in the formation of crown gall tumors (A. tumefaciens) or hairy roots (A. rhizogenes). Agrobacteria are unable to infect plants through an intact epidermis and require a wound for entry. Such wounds occur frequently in root tissue due to abrasion from soil particles and are also formed naturally at the site of emergence of lateral roots. Wounds are also frequent at the region where the plant emerges from the soil. This part of the plant is referred to as the *crown* of the plant, hence the name of the disease, *crown gall*. The mechanism of tumorigenesis or hairy root formation by these bacteria involves the transfer of DNA sequences (the T DNA) from the bacterium to host plant cells where the T DNA becomes integrated into the plant chromosomes. Genes in the T DNA contain eukaryotic promoters and encode enzymes involved in the synthesis of the plant growth hormones auxin and cytokinin. In addition, they encode enzymes for the synthesis and secretion of unique small-molecular-weight products, the opines. Different strains of agrobacteria encode the enzymes for the synthesis of different opines. In general, the bacterial strain that carries the genes for the synthesis of a particular opine also carries elsewhere in its DNA the genes for the metabolism of the same opine. Thus the bacteria can convert the plant cells into factories that are making and secreting opines that the bacteria can then use a source of carbon and energy (in some cases in which the opine is an amino acid derivative, the bacteria can also use the opine as a nitrogen source). For more information on tumor formation by A. tumefaciens, see reviews by Ream (1989) and by Binns and Thomashow (1988).

Agrobacteria and rhizobia are closely related. *R. meliloti* and biotype 1 *A. tumefaciens* share more than 65% DNA homology, and their chromosomal maps are superimposable. Rhizobia carry sym plasmids on which are located the *nod* and *nif* genes required for the formation of nitrogen-fixing nodules. Agrobacteria carry tumor-inducing plasmids (Ti plasmids) on which are located the T DNA, genes required for the transfer of the T DNA to plant cells (*vir* genes), and genes involved in the catabolism of opines. In most strains of *A. tumefaciens* the genes required for adhesion to plant cells are carried on the bacterial chromosome (Matthysse, 1986; Robertson et al., 1988).

5.2.2.1. Nonattaching Bacterial Mutants Several nonattaching mutants of *A. tumefaciens* are known. In every case these mutants are avirulent, suggesting that bacterial adhesion is required for the transfer of T DNA from the bacterium to the plant host cell (Matthysse, 1986). Theoretically, nonattaching bacterial mutants can be of three types: The site that the bacterium uses to bind to the host cell may be altered or missing, the entire bacterial surface may be so altered that the molecules involved in bacterial adhesion are no longer exposed in the proper configuration on the cell surface, or the bacterium may overproduce some other surface component such as exopolysaccharide and so mask the molecules involved in adhesion. Three of the nonattaching mutants of *A. tumefaciens* (ChvA, ChvB, and PscA) appear to involve substantial nonspecific alterations in the bacterial surface. The *chvA* and *chvB* genes are involved in the synthesis of β-1,2-glucans, which are localized in the periplasmic space and seem to aid the bacterium in its resistance to low external osmotic pressures (Douglas et al., 1982, 1985). ChvA and ChvB mutants are avirulent on some, but not all, host plants. In addition to the lack of β-1,2-glucans, the bacteria show reduced motility and overproduction of the acidic exopolysaccharide. The *pscA* or *exoC* gene encodes the β-glucosyltransferase required for the synthesis of UDP-glucose (Thomashow et al., 1987; Cangelosi et al., 1987). Not surprisingly, these mutants show a very slow growth rate and are altered in the synthesis of lipopolysaccharide and exopolysaccharide. They fail to adhere to plant cells and are avirulent. Another group of mutants that fails to adhere to plant cells is the Att mutants (Matthysse, 1987). The *att* mutations are clustered in a region of about 22 kb on the bacterial chromosome. Most of the mutants show normal motility and exopolysaccharide synthesis, although some of them fail to synthesize cellulose. All of them are avirulent. The cross-talk between the bacterium and the legume host seen with rhizobia may also occur between agrobacteria and their dicot plant hosts. This is suggested by the fact that the ability of some Att mutant bacteria to adhere to plant tissue culture cells can be restored by the addition to the assay-mix of filter-sterilized medium conditioned by incubation of the wild-type bacteria with tissue culture cells (Matthysse, 1994).

5.2.2.2. Bacterial Surface Proteins Although the presence of a rhicadhesin protein similar to that found in the rhizobia has been reported in agrobacteria, as well as the ability of externally added rhicadhesin to complement ChvB mutants of *A. tumefaciens* (Swart et al., 1993), the role of this protein in adhesion of agrobacteria is uncertain. The *chvB* gene encodes a protein required for the synthesis of β-1,2-glucan; however, the effects of the lack of this enzyme are pleiotropic and thus difficult to analyze.

Agrobacteria infect wound sites in dicotyledonous plants. The bacteria generally adhere, and transfer DNA, to dividing cells in the wound. In many respects the growth of cells in plant tissue culture resembles a prolonged wound response. Agrobacteria adhere to cells in wound sites and to tissue culture cells. The bacteria do not generally transform epidermal cells and their adhesion to these cells may differ from adhesion to cells in wound sites. Calcium is not required for the binding of agrobacteria to tissue culture cells, and thus the role of a calcium-binding protein in the adhesion of these bacteria to wound site cells remains problematic (Gurlitz et al., 1987). The gene for rhicadhesin has not been identified, and no mutants of rhicadhesin in *A. tumefaciens* are available at present.

5.2.2.3. Two-Step Attachment Attachment of *A. tumefaciens* to plant cells appears to be a two-step process (see Figs. 5.4 and 5.5). In the first step the bacteria bind loosely to a receptor on the plant cell surface. It is this step that is blocked in *chvA*, *chvB*, *pscA*, and *att* gene mutants. This step is required for virulence as witnessed by the fact that all of these mutants are avirulent. At this first stage of adhesion the bacteria can be removed from the host cell

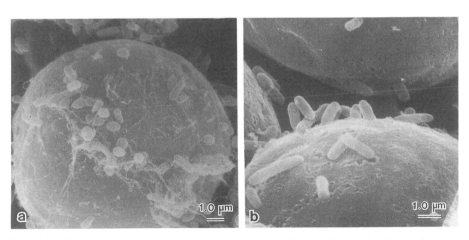

Fig. 5.4. Scanning electron micrographs of *A. tumefaciens* bound to the surface of carrot cells. **(a)** Wild-type bacteria. Note the fibrils and the formation of bacterial aggregates including the large aggregate that has pulled away from the cell surface in the upper part of the picture. **(b)** Cellulose-minus mutant. Note the lack of fibrils and bacterial aggregates. (Reproduced from Matthysse, 1983, with permission of the publisher.)

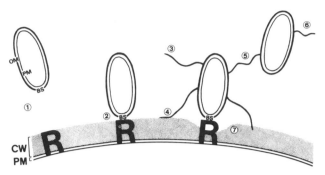

Fig. 5.5. Diagram of the steps in the adhesion of *A. tumefaciens* to suspension culture cells. The free bacterium has a plasma membrane (PM) and an outer membrane (OM) with an exposed binding site (BS). The surface of the plant cell has a plasma membrane (PM) covered by a cell wall (CW). A receptor (R) is located on the plant cell surface (step 1). The bacterium binds to the plant cell by an interaction of the bacterial binding site and the plant cell receptor (step 2). Substances coming from the plant cell cause the bacteria to begin to elaborate cellulose fibrils (step 3). Some of these fibrils anchor the bacteria firmly to the plant cell (step 4). Other fibrils entrap additional free bacteria (step 5), which are in turn induced by proximity to the plant to elaborate cellulose fibrils (step 6). Repetition of this process results in the formation of large aggregates of bacteria. The bacteria may also alter the plant cell surface in the vicinity of the receptor (step 7). (Reproduced from Matthysse et al., 1981, with permission of the publisher.)

by vortexing of tissue culture cells or by gentle water washing of wound sites on leaves (Matthysse, 1986). In the next step some signal coming from the host cell causes the bacteria to elaborate cellulose fibrils. These fibrils anchor the bacteria tightly to the host cell surface, and they can no longer be removed by shear forces. These cellulose fibrils also extend into the surrounding medium and entrap additional bacteria, forming large aggregates of bacteria bound to the host cell surface (Matthysse et al., 1981). This second step in bacterial attachment is not absolutely required for virulence (Matthysse, 1983). Mutants that are unable to elaborate cellulose fibrils are still capable of inducing tumors on host plants, but the virulence of such mutants is severely reduced compared with the wild-type bacteria. Depending on the particular cellulose mutant being examined, between 10 and 1,000 times more mutant than wild-type bacteria are required to induce tumors at 50% of the inoculated sites on a *Bryophyllum* leaf (Minnemeyer et al., 1991). Cellulose-minus mutants are also very easy to remove from wound sites by gentle water washing and probably would rarely form tumors in the natural world (Sykes and Matthysse, 1986).

5.2.2.4. Plant Receptors for Bacterial Adhesion The plant receptor to which the bacteria adhere appears to be a protein. Plant cell wall preparations and the pectin-rich fraction of the cell wall can block the adherence of agrobacteria to tissue culture cells (Neff and Binns, 1985). These preparations are no longer effective if they are treated with proteases before they are added to the incubation medium. Brief treatments of carrot cells with trypsin prevent the adhesion of agrobacteria (Gurlitz et al., 1987). Such treatments do not kill the carrot cells, which recover after 6 h and are again able to bind the bacteria. Treatment of the carrot cells with dilute solutions of Triton also prevents the adhesion of the bacteria. The proteins removed by this Triton treatment can be examined using polyacrylamide gel electrophoresis. Less than 15 proteins appear to be found in such extracts (Gurlitz et al., 1987).

Several proteins have been described that are localized at the surface of plant cells. Among them is a vitronectin-like protein that is involved in the adhesion of the marine algae *Fucus* to coastal rocks (Wagner et al., 1992). Although a vitronectin-like protein has been detected on the surface of cells from several species of plants, including tobacco and soybean (Sanders et al., 1991), the gene has not been cloned and the purified protein is not available from any plant source. Thus, at the present time, experiments involving the use of vitronectin or antivitronectin antibodies must rely on the use of the human protein. Human vitronectin or antibodies to human vitronectin added to incubation mixtures of carrot cells and *A. tumefaciens* inhibit the adhesion of the bacteria to the plant cells (Wagner and Matthysse, 1992). In addition, wild-type *A. tumefaciens* bind radioactive vitronectin, while *att* gene mutants fail to show this binding. Two of the proteins extracted from carrot suspension culture cells with Triton react with antihuman vitronectin antibodies on Western blots. These results suggest that the plant receptor to which agrobacteria bind may be a vitronectin-like protein. What relationship, if any, this protein has to the RGD-containing protein reported to be involved in the adhesion of rhizobia to pea roots is uncertain.

5.2.3. Adhesion of Pseudomonads to Roots

The soil pseudomonads whose interactions with plants have been studied can be divided into two groups: those that are pathogenic and those whose presence seems to promote the growth of the plants with which they are associated. The majority of the pathogenic pseudomonads studied cause disease on aerial parts of plants rather than on roots. These bacteria may be found in the soil, but their association with the surface of leaves appears to be the most significant in terms of disease. The evidence suggests the *Pseudomonas syringae* pathovars such as *phaseolicola* and *syringae* bind to the surface of leaves of plants on which they cause disease by means of fimbriae. *P. syringae* pv *phaseolicola* binds to the surface of bean leaves in the vicinity of the stomata. The bacterium enters the leaf through the stomata, so this localization appears to be helpful to the bacteria. The binding is believed to be mediated by pili (Romantschuk and Bamford, 1986). Mutant bacteria that are unable to make pili have been isolated. They are no longer able to adhere to bean leaves and are avirulent if they are sprayed on the surface of susceptible bean leaves. If the bean leaves are inoculated by injecting the bacteria into the intercellular spaces, then the mutant and wild-type bacteria are equally capable of causing disease. The binding of *P. syringae* pv *syringae* to bean leaves appears also to involve pili, but the binding shows a different specificity (Nurmiaho-Lassila et al., 1991). Instead of being clustered around the stomata, the bacteria are found distributed over the surface of the leaf. The distribution of these two pathovars of *P. syringae* correlates well with their presumed modes of entry: *phaseolicola* via the stomata and *syringae* via wounds made in the epidermis by nucleation of ice crystal formation by the bacteria.

The role of bacterial adhesion after potentially pathogenic bacteria have entered the plant has been much discussed. Potentially pathogenic bacteria inoculated into nonhost or resistant plants give rise to a resistant reaction called the *hypersensitive response* (HR). In many cases bacteria that have elicited an HR are observed bound to the surface of plant cells at the site of inoculation. The role of this binding in the HR has been controversial. However, there are now several cases in which the HR can be elicited by purified proteins or other molecules from the bacteria, suggesting that bacterial adhesion is not a required component of the HR in all cases. Whether bacterial adhesion plays a role in the HR to some plant pathogens remains to be determined.

Pseudomonas species that associate with plant roots are widespread in the soil. Some of these bacteria seem to promote the growth of the plants with which they associate. Because of their potential agricultural importance it is the adhesion of these strains of pseudomonads that has been studied the most carefully. Some of these bacteria possess pili whose presence correlates with the ability of the bacteria to adhere to roots (Vesper, 1987). While not absolutely required for bacterial adhesion in most cases, calcium and magnesium seem to promote adhesion (James et al., 1985). Most of these bacteria are loosely bound, but a small percentage may be tightly bound (Buell and Anderson, 1992). Mutational changes in the surface charge or hydrophobicity of the bacteria do not alter their adhesion (de Weger et al., 1989). Mutants of *P. putida* that are altered in their ability to adhere to and colonize bean roots have been obtained (Anderson et al., 1988). The wild-type bacteria are agglutinated by a glycoprotein from the root surface. The nonadhesive bacteria fail to react with this protein. The gene that is mutated in the nonagglutinable bacteria has been cloned and sequenced (Buell and Anderson, 1992). Unfortunately, the sequence does not suggest how it acts to bind bacteria to the root surface. Factors that agglutinate pseudomonads appear to be widespread in the roots of crop plants. However, there is no correlation between the ability of the factors from a particular plant to agglutinate a bacterium and the ability of that bacterium to colonize the plant's roots (Glandorf et al., 1994).

5.2.4. Adhesion of Other Rhizosphere Bacteria to Roots

The adhesion of several other types of bacteria to roots has been examined. Most prominent among them are the free-living bacteria that are capable of nitrogen fixation. These bacteria have been studied because of their importance in reducing the need for fertilizer application and in increasing crop yields. *Klebsiella* and *Enterobacter* species appear to adhere to root hairs of various grasses, including *Poa pratensis, Trifollium pratense,* and *Festuca rubra,* using fimbriae (Haahtela and Korhonen, 1985). Klebsiella species use both type 1 and type 3 fimbriae. *Enterobacter* species use type 1 fimbriae. Presumably these fimbriae bind to the appropriate sugars on the surface of the root hairs, but the actual molecules to which they bind have not been determined. There does not appear to be any particular host specificity in these associations. However, there is some cell specificity in that bacteria were observed to adhere to root hairs, but not to the remainder of the epidermis (Haahtela et al., 1988). In the case of *P. pratensis* the plant responded by forming larger roots with more root hairs (some of which were curled or branched) and a shorter zone of elongation (Haahtela et al., 1986). To what extent these alterations were due directly to an interaction of the plant with the bacterium, rather than to the increased nitrogen supply provided by the bacterium, is uncertain.

Another free-living nitrogen-fixing bacterium, *Azospirillum brasilense,* was found to adhere to root hairs of tomato, pepper, and cotton (Bashan et al., 1991). In this case the bacteria bound both to the root hairs and to the epidermal cells of the elongation zone. The bacteria appeared to form clusters held together by fibrils.

Two species of plant-associated *Erwinia, E. carotovora* and *E. rhupontici,* and *Xanthomonas campestris* pv. *hyacinthi* have been reported to possess fimbriae that may be involved in adherence to plant surfaces (Korhonen et al., 1987; Romantschuk, 1992).

5.3. THE EFFECTS OF SURFACE CHARACTERISTICS ON BACTERIAL ADHERENCE

The effects of changes in surface hydrophobicity or charge on bacterial adhesion have only been examined using mutants of *Pseudomonas* species. The results suggest that neither of these factors determines the ability of these bacteria to adhere to roots (de Wegner et al., 1989). The adherence of *P. fluorescens* to radish roots and of *B. japonicum* to soybean root hairs appeared to have no hydrophobic component (James et al., 1985; Pueppke, 1984).

Mutants with altered lipopolysaccharide (LPS) have been examined in both *A. tumefaciens* and *R. meliloti* (Metts et al., 1991; Clover et al., 1989). In both cases the adherence of the bacteria was unaffected by alterations in LPS. Inhibitors of LPS biosynthesis also have no effect on the ability of *A. tumefaciens* to bind to carrot cells (Goldman et al., 1992). The introduction of the plasmid pSA into *A. tumefaciens* prevents the bacteria from adhering to plant cells and causes them to become avirulent (Loper and Kado, 1979). The altered phenotype of bacteria carrying pSA includes alterations in LPS (New et al., 1983), but the results just described suggest that it is unlikely that it is the alteration in LPS that renders the bacteria nonadherent.

The role of charge interactions in bacterial adherence in the rhizosphere is more complicated. Neither ionic strength nor the presence or absence of divalent cations affects the adherence of *A. tumefaciens* biotype 1 to tissue culture cells (Sykes and Matthysse, 1988). However, the adherence of *A. tumefaciens* biotypes 2 and 3 is sensitive to ionic strength and is best at ionic strengths below 0.1 M. The binding of *P. tolaasii* to barley roots requires divalent cations, and the binding of *P. fluorescens* to radish roots was promoted by divalent cations (James et al., 1985). The adhesion of *B. japonicum* to soybean cells is inhibited by 50 mM NaCl and by EDTA (Ho et al., 1988). However, EGTA has no effect on bacterial adhesion, suggesting that calcium is not required. The adherence of *R. leguminosarum* bv. *vicae* to pea root hairs requires the presence of calcium, and a calcium-binding protein, rhicadhesin, has been implicated in this process (Smit et al., 1989).

Thus, general hydrophobic interactions appear to play little role in the adhesion of bacteria to roots. The role of charge interactions and the role of divalent cations, particularly calcium, appear to depend on the particular organism being examined and may differ between biotypes of the same bacterial species.

5.4. THE SIGNIFICANCE OF ADHESION

5.4.1. The Need to Adhere in Various Situations

Bacteria in the rhizosphere do not have the same need to adhere tightly to surfaces in order to avoid being removed as do bacteria in the ocean or inside structures such as pipelines or in the intestine or lungs. In those cases that have been examined, the bacteria found in association with roots in the rhizosphere are generally chemotactic to roots or root exudates. Once the bacteria have reached the root surface they probably do not need to adhere tightly to the roots in order to maintain this association. Water washing during rain storms may cause the removal of some bactera, particularly those associated with the root near the soil surface. Those bacteria associated with the root tip or the region of elongation will need to maintain this association as the root grows through the soil. However, the association of many of the soil bacteria, such as pseudomonads and azosprilla, with plants is often a loose adherence rather than a tight one. Bacteria that invade the plant to cause disease or to form symbiotic associations are likely to need to adhere more strongly to the root surface than bacteria that remain on the exterior of the root. The symbiotic rhizobia form tight associations with the root hair prior to initiating an infection thread. Agrobacteria adhere to the plant surface prior to forming whatever bridge may be used to transfer DNA from the bacterium to the plant host cell. The need for some intimate interaction between the bacterial and plant cell surface seems obvious in these cases. Bacteria that live on the surface of leaves or any of the other aerial parts of the plant may need to adhere strongly to the plant surface to avoid being removed by wind and rain.

5.4.2. Adherence to Specific Cell Types Versus General Adherence

Most of the bacteria that adhere to the root surface adhere to specific cell types or structures. Rhizobia and bradyrhizobia adhere preferentially to root hairs. *B. japonicum* adheres in the greatest numbers to emerging root hairs of a particular age range (Vesper and Bauer, 1985). *B. japonicum* adherence to mature root hairs or to epidermal cells is much less. When *R. leguminosarum* bv. *trifolii* and bv. *viciae* are inoculated onto their host plants, more bacteria are observed to adhere to the tips than to the sides of root hairs (Dazzo et al., 1984; Smit et al., 1986). *Klebsiella* species bind to the root hairs of grasses and not to the remainder of the root epidermis (Haahtela et al., 1986). *A. brasilense* also binds to root hairs of pepper (Bashan et al., 1991). However, when incubated with tomato or soybean roots the bacteria bind to the epidermal surface. If inoculated onto cotton, these bacteria bind to both root hairs and the epidermal surface. Whether this unusual group of binding abilities of *A. brasilense* is due to a single mechanism or to several distinct types of adhesion mechanism is unknown. *P. syringae* pv. *phaseolicola* and *X. campestris* pv. *hyacinthi* bind to leaves in the vicinity of stomata (Romantschuk and Bamford, 1986; Romantschuk, 1992; van Doorn et al., 1994). In both of these cases, purified pili from the bacteria have been shown to have the same cell-binding preference as the bacteria. Not all pseudomonads show such cell specificity, however. *P. syringae* pv. *syringae* binds to the entire epidermis of the leaf without any preference for the stomata (Nurmiaho-Lassila et al., 1991). The different adhesion sites of these bacteria may be related to their different modes of entry into the plant. *P. syringae* pv. *phaseolicola* and *X. campestris* pv. *hyacinthi* enter the leaves through the stomata. *P. syringae* pv. *syringae* enters the leaves through wounds that it forms in the leaf by catalyzing the formation of ice crystals on the leaf surface at temperatures at which such crystals would ordinarily not form.

Agrobacteria that require a wound to infect the plant show little specificity in the cell types to which they bind. When wounded surfaces inoculated with agrobacteria are examined in the microscope, many different cell types within the wound are found to have adherent bacteria (Graves et al., 1988). These bacteria also bind to tissue culture cells and to root cap cells (Matthysse and Gurlitz, 1982; Hawes and Pueppke, 1987). In general the binding of agrobacteria to intact epidermal tissues and to the surface of plant embryos is reduced compared with bacterial binding to suspension culture cells (Matthysse and Gurlitz, 1982). Comparisons of bacterial adhesion to intact surfaces and tissue culture cells with adherence to cells in wound sites are hard to make, since it is difficult to obtain accurate quantitative measures of adherence in wound sites that are not confused by trapping of bacteria in torn tissues.

5.5. THE ADHESION PROCESS

5.5.1. Two-Step Adherence: Loose Reversible and Then Tight Irreversible

Adherence of *A. tumefaciens* to plant cells in culture or in a wound is a two-step process (Matthysse, 1983). Mutants that are unable to carry out each step are available and have been used to clarify the roles of the initial adherence and the subsequent cellulose fibril-mediated adherence (Figs. 5.4 and 5.5). In the first step the bacteria bind loosely to the host cell surface and can be removed by shear forces such as water washing (Matthysse, 1983). In the second step the bacteria synthesize cellulose fibrils that bind them tightly to the host cell surface. The bacteria can then be removed only by digesting the bacterial or host cell wall. *R. leguminosarum* bv. *trifolii* and *leguminosarum* and *B. japonicum* have been observed to undergo a similar two-step binding process (Dazzo et al., 1984; Smit et al., 1987; Ho et al., 1988). In the first step the binding is reversible. During the first hour after the addition of *B. japonicum* to soybean cells the bacteria could be removed by washing. After 4 h it was difficult to remove the bacteria. Cellulose fibrils that bind the bacteria to the root hair surface and aid in the formation of bacterial aggregates have been observed with *R. leguminosarum* bv. *trifolii* and *viciae*.

Fibrils whose nature is unknown but that appear to play a role in adherence to plant cells have been observed in *P. putida* and *P. tolaasii* binding to the surface of *Agaricus bisporus* mycelium (Rainey, 1991), *P. syringae* pv. *syringae* binding to apple tissues (Mansvelt and Hattingh, 1989), and *A. brasilense* binding to tomato, cotton, and pepper roots (Bashan et al., 1991). Thus, it appears that this type of two-step binding may be widespread in bacteria that adhere to plant surfaces. One possible advantage of this type of binding is that the bacteria are not irreversibly committed to adherence at the first encounter with the plant surface.

5.5.2. Signal Exchange and Adherence

Before bacteria can adhere to the root surface they must encounter it. Many soil bacteria are chemotactic to root exudates that may contain sugars, amino acids, and vitamins, as well as compounds that are produced by only some species of plants, an example being the production of flavonoids by legume species. In the case of most of the bacteria that adhere to roots, we do not know whether they sense the presence of the root or whether they simply bump into it and stick to it. In the case of the interaction of rhizobia with roots, our knowledge is much more detailed (Fisher and Long, 1992; Vijn et al., 1993). Each biovar of *Rhizobium* nodulates only a specific host range of legumes. Legume roots release flavonoids and other compounds into the soil. The particular flavonoids released depend on the species of legume. Rhizobia sense the flavonoids from the particular species of plant that they nodulate, and this induces the expression of the *nod* genes. The product of the *nod* genes is an oligosaccharide, a nod factor, whose structure differs depending on the biovar of *Rhizobium*. The appropriate legume responds to the nod factors by cortical cell divisions, root hair deformations, and release of additional flavonoids and possibly other compounds. At roughly this point in the conversation, the bacteria adhere to the root hair tips. At any stage, if the two partners do not communicate successfully the interaction does not progress and the bacteria do not commit themselves by adhering to the plant. This protects both the bacterium and the plant from proceeding with an interaction in which the other partner is not the desired organism or in which the physiological state of the other partner is not suitable.

No other bacterial plant interaction is known to have such elaborate signaling mechanisms. In the case of *A. tumefaciens* there are three steps in the interactions of the bacteria with the plant host cells that appear to involve signaling between the bacteria and the plant host. Certain nonadherent mutants of *A. tumefaciens* can be complemented by the addition of filter-sterilized medium from the incubation of wild-type bacteria with carrot cells (Matthysse, 1994). This suggests that there are molecules released into the medium that are required for the initial binding of these bacteria to host cells. Once the bacteria have bound to the plant cells the bacteria begin to elaborate cellulose fibrils. It is thought that some signal from the plant triggers the fibril synthesis (Matthysse, 1983). The chemical nature of the compounds involved in these first two signals is unknown. Finally, acetosyringone is released from the wound in the plant and induces the expression of the *vir* genes, which are required for the transfer of the T DNA from the bacteria to the plant cell (Ream, 1989).

5.5.3. Is Adherence a One-Way Street? Can Bound Bacteria Be Released?

The question of the fate of adherent bacteria is obviously of some importance in assessing the role of bacterial adherence in the rhizosphere. In the case of the rhizobia, the fate of the bacteria is to enter the plant and become bacteroids inside the cells of the nodule. This process does not seem likely to be reversible under natural conditions at any high frequency. Agrobacteria that are tightly bound to the root surface at wound sites could be released if they were to digest the cellulose fibrils that hold them together in large aggregates. Bacteria that are loosely bound to the root surface by structures such as fimbriae could easily lose their fimbriae and swim away from the root. Whether this happens and if so under what conditions are unknown.

5.6. DOES IT PAY FOR EVERYONE TO ADHERE?

One of the common observations in all systems involving bacterial adherence to roots is that only a fraction of the bacteria adhere. About 10% of the inoculated *R. meliloti* bind to alfalfa roots; the percent inoculated bacteria bound is similar for *R. leguminosarum* bv. *trifolii* binding to clover roots (Caetano-Anolles and Favelukes, 1986). In the case of *B. japonicum* about 2% of the added bacteria bind to soybean cells (Ho et al., 1988). Between 0.1% and 1% of *P. fluorescens* bind to radish roots (James et al., 1985). About 10% of the inoculated *P. putida* bind to bean roots (Anderson et al., 1988). Less than 10% of the added *Klebsiella* or *Enterobacter* species bind to roots of grasses (Haahtela et al., 1988). *A. tumefaciens* shows a higher percentage of inoculated cells binding than do the other bacteria, but, with the bacterial strain with the greatest attachment, only 60% of the added bacteria bind to the plant cell surface (Matthysse et al., 1981).

This observation that only a small fraction of the total bacterial population adheres to the plant cell surface raises the question of why the remainder of the bacteria do not bind. The answer does not appear to lie in a lack of plant sites for bacterial adhesion. With both *A. tumefaciens* and *B. japonicum* the percentage of the bacterial inoculum bound does not change as the investigator increases the number of added bacteria until very high bacterial concentrations are reached (on the order of 10^8 bacteria/ml). In many cases it has been observed that only a fraction of the bacterial population possesses molecules or structures believed to be involved in adherence. For example, only 70% of *B. japonicum* have Bj38 (Loh et al., 1993), and only a small fraction of the bacteria have fimbriae (Vesper and Bauer, 1986); about 60% of *E. carotovora* have fimbriae (Korhonen et al., 1987). Although nothing is known about the molecular mechanism of the regulation of the ability to adhere to the plant, in the case of *A. tumefaciens* if the bacteria that fail to adhere to carrot cells are collected and again mixed with carrot cells they continue to fail to adhere. However, if this population of nonadherent bacteria is allowed to grow for a few hours, the percentage of the bacteria that adhere to carrot cells returns to that initially observed (Matthysse, unpublished observation). This distribution in the population of bacteria that will and will not adhere to the plant surface may represent a survival strategy in which only a fraction of the bacterial population can be committed to any one site.

The need for agrobacteria and rhizobia to adhere to the plant in order to transfer T DNA or to initiate infection thread formation is obvious. Other types of bacteria that are more loosely associated with the root surface may obtain nutrients from the association. It appears reasonable for all of these organisms that the bacterial population as a whole could derive an advantage from the adherence of only a fraction of its members at various sites on the root.

5.7. CONCLUSIONS

As yet our information about the adhesion of bacteria from the rhizosphere to roots is very fragmentary. Bacteria residing in this environment have to make complex distinctions between adherence to intact or wounded plant surfaces, which may represent growing plant roots, and dead or dying parts of plants such as leaves or twigs that have dropped off the aerial part of the plant. The rhizosphere contains many small niches with widely varying microenvironments. Thus it is not surprising that in general rhizosphere bacteria appear to commit only a fraction of their population to adhering to a surface at any one time and that some rhizosphere bacteria engage in prolonged conversations with the plant, designed to determine if the root they have contacted belongs to a living healthy plant of the right type. Only then do these bacteria enter into an intimate relationship with the root. The two-step nature of the adherence process (first, a loose, reversible association and, second, a tight binding) also reflects the cautious approach of the bacteria to committing themselves to adherence to the plant.

This view of adherence in the rhizosphere reflects studies that have focused mainly on rhizobia and agrobacteria. As our knowledge of the actions of other bacteria in proximity to the root increases, the picture will surely change. Rhizosphere interactions are of major agricultural importance, but our information is not sufficient to allow us to manage them for our benefit. It is hoped that we will soon gain this knowledge.

REFERENCES

Anderson AJ, Habibzadegah-Tari P, Tepper CS (1988): Molecular studies on the role of a root surface agglutinin in aderence and colonization by *Pseudomonas putida*. Appl Environ Microbiol 54:375–380.

Bashan Y, Levanony H, Whitmoyer RE (1991): Root surface colonization of non-cereal crop plants by pleomorphic *Azospirillum brasilense* Cd. J Gen Microbiol 137:187–196.

Binns AN, Thomashow MF (1988): Cell biology of *Agrobacterium* infection and transformation of plants. Annu Rev Microbiol 42:575–606.

Buell CR, Anderson AJ (1992): Genetic analysis of the *aggA* locus involved in agglutination and adherence of *Pseudomonas putida,* a beneficial fluorescent pseudomonad. Mol Plant Microbe Interact 5:154–162.

Caetano-Anolles G, Favelukes G (1986): Quantitation of adsorption of rhizobia in low numbers to small legume roots. Appl Environ Microbiol 52:371–376.

Caetano-Anolles A, Lagares A, Favelukes G (1989): Adsorption of *Rhizobium meliloti* to alfalfa roots: Dependence on divalent cations and pH. Plant Soil 177:67–74.

Cangelosi GA, Hung L, Puvanesarajah V, Stacey G, Ozga DA, Leigh JA, Nester EW (1987): Common loci for *Agrobacterium tumefaciens* and *Rhizobium meliloti* exopolysaccharide synthesis and their roles in plant interactions. J Bacteriol 159:2086–2091.

Clover RH, Kieber J, Signer ER (1989): Lipopolysaccharide mutants of *Rhizobium meliloti* are not defective in symbiosis. J Bacteriol 171:3961–3967.

Dazzo FB, Napoli CA, Hubbell DH (1976): Adsorption of bacteria to roots as related to host specificity in the *Rhizobium*–clover symbiosis. Appl Environ Microbiol 32:166–171.

Dazzo FB, Truchet GL, Kijne JW (1982a): Lectin involvement in root hair tip adhesion as related to the *Rhizobium trifolii*–clover symbiosis. Plant Physiol 56:143–147.

Dazzo FB, Truchet GL, Sherwood JE, Hrabak M, Gardiol AE (1982b): Alteration of the trifoliin A-binding capsule of *Rhizobium trifolii* 0403 by enzymes released from clover roots. Appl Environ Microbiol 44:478–490.

Dazzo FB, Truchet GL, Sherwood JE, Hrabak EM, Abe M, Pankratz SH (1984): Specific phases of root hair attachment in the *Rhizobium trifolii*–clover symbiosis. Appl Environ Microbiol 48:1140–1150.

de Wegner LA, van Loosdrecht MCM, Klassen HE, Lugtenberg B (1989): Mutational changes in physicochemical cell surface properties of plant-growth-stimulating *Pseudomonas* spp. do not influence the attachment properties of the cells. J Bacteriol 171:2756–2761.

Diaz CL, Melchers LS, Hooykaas PJJ, Lugtenberg BJJ, Kijne JW (1989): Root lectin as a determinant of host-plant specificity in the *Rhizobium*–legume symbiosis. Nature 338:579–581.

Douglas CJ, Halperin W, Nester EW (1982): *Agrobacterium tumefaciens* mutants affected in attachment to plant cells. J Bacteriol 152:1265–1275.

Douglas CJ, Staneloni RJ, Rubin RA, Nester EW (1985): Identification and genetic analysis of an *Agrobacterium tumefaciens* chromosomal virulence region. J Bacteriol 161:850–860.

Fisher RF, Long SR (1992) *Rhizobium*–plant signal exchange. Nature 357:655–660.

Glandorf DCM, van der Sluis I, Anderson AJ, Bakker PAHM, Schippers B (1994): Agglutination, adherence, and root colonization by fluorescent pseudomonads. Appl Environ Microbiol 60:1726–1733.

Goldman RC, Capobianco JO, Doran CC, Matthysse AG (1992): Inhibition of lipopolysaccharide synthesis in *Agrobacterium tumefaciens* and *Aeromonas salmonicida*. J Gen Microbiol 138:1527–1533.

Graves AE, Goldman SL, Banks SW, Graves ACF (1988): Scanning electron microscope studies of *Agrobacterium tumefaciens* attachment to *Zea mays, Gladiolus* sp., and *Triticum aestivum*. J Bacteriol 170:2395–2400.

Gurlitz RHG, Lamb PW, Matthysse AG (1987): Involvement of carrot cell surface proteins in attachment of *Agrobacterium tumefaciens*. Plant Physiol 83:564–568.

Haahtela K, Korhonen TK (1985): In vitro adhesion of N_2-fixing enteric bacteria to roots of grasses and cereals. Appl Environ Microbiol 49:1186–1190.

Haahtela K, Laakso T, Korhonen TK (1986): Associative nitrogen fixation by *Klebsiella* spp.: Adhesion sites and inoculation effects on grass roots. Appl Environ Microbiol 52:1074–1079.

Haahtela K, Laakso T, Nurmiaho-Lassila E-L, Ronkko R, Korhonen TK (1988): Interactions between N_2-fixing enteric bacteria and grasses. Symbiosis 6:139–150.

Hawes MC, Brigham LA (1992): Impact of root border cells on microbial populations in the rhizosphere. Adv Plant Pathol 8:119–148.

Hawes MC, Pueppke SG (1987): Correlation between binding of *Agrobacterium tumefaciens* by root cap cells and susceptibility of plants to crown gall. Plant Cell Rep 6:287–290.

Ho S-C, Shahnaz M-H, Wang JL, Schindler M (1986): Endogenous lectin from cultured soybean cells: Isolation of a protein immunologically cross-reactive with seed soybean agglutinin and analysis of its role in binding of *Rhizobium japonicum*. J Cell Biol 103:1043–1054.

Ho S-C, Wang JL, Schindler M, Loh JT (1994): Carbohydrate binding activities of *Bradyrhizobium japonicum* III. Lectin expression, bacterial binding, and nodulation efficiency. Plant J 5:873–884.

Ho S-C, Ye W, Schindler M, Wang JL (1988): Quantitative assay for binding of *Bradyrhizobium japonicum* to cultured soybean cells. J Bacteriol 170:3882–3890.

James DW Jr, Suslow TV, Steinback KR (1985): Relationship between rapid, firm adhesion and long-term colonization of roots by bacteria. Appl Environ Microbiol 50:392–397.

Kijne JW, Smit G, Diaz CL, Lugtenberg BJJ (1988): Lectin-enhanced accumulation of manganese-limited *Rhizobium leguminosarum* cells on pea root hair tips. J Bacteriol 170:2994–3000.

Korhonen TK, Kalkkinen N, Haahtela K, Old DC (1987): Characterization of type 1 and mannose-resistant fimbriae of *Erwinia* spp. J Bacteriol 169:2281–2283.

Lodeiro AR, Favelukes G (1995): Enhanced adsorption to soybean roots, infectivity, and competition for nodulation of *Bradyrhizobium japonicum* pretreated with soybean lectin. Tenth International Congress of Nitrogen Fixation, abstract.

Lodeiro AR, Lagares A, Martinez EN, Favelukes G (1995): Early interactions of *Rhizobium leguminosarum* bv. *phaseoli* and bean roots: Specificity in adsorption and its requirement of Ca^{2+} and Mg^{2+} ions. Appl Environ Microbiol 61:1571–1579.

Loh JT, Ho S-C, de Feuter AW, Wang JL, Schindler M (1993): Carbohydrate binding activities of *Bradyrhizobium japonicum*: Unipolar localization of the lectin Bj38 on the bacterial cell surface. Proc Natl Acad Sci USA 90:3033–3037.

Loper JA, Kado CI (1979): Host range conferred by the virulence-specifying plasmid of *Agrobacterium tumefaciens*. J Bacteriol 139:591–596.

Mansvelt EL, Hattingh MJ (1989): Scanning electron microscopy of invasion of apple leaves and blossoms by *Pseudomonas syringae* pv. *syringae*. Appl Environ Mircrobiol 55:535–538.

Matthysse AG (1983): Role of bacterial cellulose fibrils in *Agrobacterium tumefaciens* infection. J Bacteriol 154:906–915.

Matthysse AG (1986): Initial interactions of *Agrobacterium tumefaciens* with plant host cells. CRC Crit Rev Microbiol 13:281–307.

Matthysse AG (1987): Characterization of nonattaching mutants of *Agrobacterium tumefaciens*. J Bacteriol 169:313–323.

Matthysse AG (1994): Conditioned medium promotes the attachment of *Agrobacterium tumefaciens* strain NT1 to carrot cells. Protoplasma 183:131–136.

Matthysse AG, Gurlitz RHG (1982): Plant cell range for attachment of *Agrobacterium tumefaciens* to tissue culture cells. Physiol Plant Pathol 21:381–387.

Matthysse AG, Holmes KV, Gurlitz RHG (1981): Elaboration of cellulose fibrils by *Agrobacterium tumefaciens* during attachment on carrot cells. J Bacteriol 145:583–595.

Metts J, West J, Doares S, Matthysse AG (1991): A new class of chromosomal avirulent mutants of *Agrobacterium tumefaciens* which fail to induce *vir* genes. J Bacteriol 173:1080–1087.

Minnemeyer SL, Lightfoot R, Matthysse AG (1991): A semi-quantitative bioassay for relative virulence of *Agrobacterium tumefaciens* strains on *Bryophyllum daigremontiana*. J Bacteriol 173:7723–7724.

Napoli C, Dazzo F, Hubbell D (1975): Production of cellulose fibrils by *Rhizobium*. Appl Microbiol 30:123–131.

Neff NT, Binns AN (1985): *Agrobacterium tumefaciens* interaction with suspension-cultured tomato cells. Plant Physiol 77:35–42.

New PB, Scott JJ, Ireland CR, Farrand SK, Lippincott BB, Lippincott JA (1983): Plasmid pSa causes loss of LPS-mediated adhesion in *Agrobacterium*. J Gen Microbiol 129:3657–3660.

Nurmiaho-Lassila E-L, Rantala E, Romantschuk M (1991): Pilus-mediated adsorption of *Pseudomonas syringae* to the surface of bean leaves. Macron Microscopia Acta 22:71–72.

Pueppke SG (1984): Adsorption of slow- and fast-growing rhizobia to soybean and cowpea roots. Plant Physiol 75:924–928.

Rainey PB (1991): Phenotypic variation of *Pseudomonas putida* and *P. tolaasii* affects attachment to *Agaricus bisporus* mycelium. J Gen Microbiol 137:2769–2779.

Ream W (1989): *Agrobacterium tumefaciens* and interkingdom genetic exchange. Annu Rev Phytopathol 27:583–618.

Robertson JL, Holliday T, Matthysse AG (1988): Mapping of *Agrobacterium tumefaciens* chromosomal genes affecting cellulose synthesis and bacterial attachment to host cells. J Bacteriol 170:1408–1411.

Romantschuk M (1992): Attachment of plant pathogenic bacteria to plant surfaces. Annu Rev Phytophatol 30:225–243.

Romantschuk M, Bamford DH (1986): The causal agent of halo blight in bean, *Pseudomonas syringae* pv. *phaseolicola,* attaches to stomata via pili. Microbial Pathogenesis 1:136–148.

Sanders L, Wang C-O, Walling L, Lord E (1991): A homolog of the substrate adhesion molecule vitronectin occurs in four species of flowering plants. Plant Cell 3:629–635.

Smit G, Kijne JW, Lugtenberg BJJ (1986): Correlation between extracellular fibrils and attachment of *Rhizobium leguminosarum* to pea root hair tips. J Bacteriol 168:821–827.

Smit G, Kijne JW, Lugtenberg BJJ (1987): Involvement of both cellulose fibrils and a Ca^{2+}-dependent adhesin in the attachment of *Rhizobium leguminosarum* to pea root hair tips. J Bacteriol 169:4294–4301.

Smit G, Logman TJJ, Boerrigter MET, Kijne JW, Lugtenberg BJJ (1989): Purification and partial characterization of the *Rhizobium leguminosarum* biovar *vicae* Ca^+-dependent adhesin, which mediates the first step in attachment of the cells of the family *Rhizobiacea* to plant root hair tips. J Bacteriol 171:4054–4062.

Smit G, Swart S, Lugtenberg BJJ, Kijne JW (1992): Molecular mechanisms of attachment of *Rhizobium* bacteria to plant roots. Mol Microbiol 6:2897–2903.

Swart S, Smit G, Lugtenberg BJJ, Kijne JW (1993): Restoration of attachment, virulence and nodulation of *Agrobacterium tumefaciens chvB* mutants by rhicadhesin. Mol Microbiol 10:597–605.

Swart S, Logman TJJ, Smit G, Lugtenberg BJJ, Kijne JW (1994): Purification and partial characterization of a glycoprotein from pea (*Pisum sativum*) with receptor activity for rhicadhesin, an attachment protein of Rhizobiaceae. Plant Mol Biol 24:171–183.

Sykes LC, Matthysse AG (1986): Time required for tumor induction by *Agrobacterium tumefaciens*. Appl Environ Microbiol 52:597–598.

Sykes L, Matthysse AG (1988): Differing attachment of biotypes I, II, and III of *Agrobacterium tumefaciens* to carrot suspension culture cells. Phytopathology 78:1322–1326.

Thomashow MF, Karlinsky JE, Marks JR, Hurlburt RE (1987): Identification of a new virulence locus in *Agrobacterium tumefaciens* that affects polysaccharide composition and plant cell attachment. J Bacteriol 169:3209–3216.

van Doorn J, Boonekamp PM, Oudega B (1994): Partial characterization of fimbriae of *Xanothomonas campestris* pv. *hyacinthi*. Mol Plant Microbe Interact 7:334–344.

van Rhijn P, Scambry J, Lim P, Hirsch AM (1995): Transfer of soybean seed lectin gene to lotus enables *Bradyrhizobium japonicum* to trigger cell divisions in the transgenic lotus plants. Tenth International Congress of Nitrogen Fixation, abstract.

Vesper SJ (1987): Production of pili (fimbriae) by *Pseudomonas fluorescens* and correlation with attachment to corn roots. Appl Environ Microbiol 53:1397–1405.

Vesper SJ, Bauer WD (1985): Characterization of *Rhizobium* attachment to soybean roots. Symbiosis 1:139–162.

Vesper SJ, Bauer WD (1986): Role of pili (fimbriae) in attachment of *Bradyrhizobium japonicum* to soybean roots. Appl Environ Microbiol 52:134–141.

Vijn I, das Neves L, van Kammen A, Franssen H, Bisseling T (1993): Nod factors and nodulation in plants. Science 260:1764–1765.

Wagner VT, Brian L, Quatrano RS (1992): Role of a vitronectin-like molecule in embryo adhesion of the brown alga *Fucus*. Proc Natl Acad Sci USA 89:3644–3648.

Wagner VT, Matthysse AG (1992): Involvement of a vitronectin-like protein in attachment of *Agrobacterium tumefaciens* to carrot suspension culture cells. J Bacteriol 174:5999–6003.

Wall LG, Favelukes G (1991): Early recognition in the *Rhizobium meliloti*–alfalfa symbiosis: Root exudate factor stimulates root adsorption of homologous rhizobia. J Bacteriol 173:3492–3499.

6

THE CELLULOSOME: A CELL SURFACE ORGANELLE FOR THE ADHESION TO AND DEGRADATION OF CELLULOSE

EDWARD A. BAYER
ELY MORAG

Department of Membrane Research and Biophysics, The Weizmann Institute of Science, Rehovot, 76100 Israel

YUVAL SHOHAM

Department of Food Engineering and Biotechnology, Technion - Israel Institute of Technology, Haifa, 32000 Israel

JOSÉ TORMO

Department of Molecular Biophysics and Biochemistry, Bass Center for Molecular and Structural Biology, Yale University, New Haven, Connecticut 06520

RAPHAEL LAMED

Department of Molecular Microbiology and Biotechnology, Tel Aviv University, Ramat Aviv, 69978 Israel

6.1. INTRODUCTION

The adhesion of cellulose-degrading bacteria to cellulose is the primary event that precedes the degradation process. In *Clostridium thermocellum*—the intensely studied, anaerobic, thermophilic, cellulolytic bacterium—adhesion to cellulose is accomplished by means of a discrete multifunctional, multicomponent cell surface protein complex, known as the *cellulosome* (Bayer et al., 1983; Lamed and Bayer, 1988a,b; Lamed et al., 1983b). The cellulosome is not only responsible for the primary property of adhesion, but also furnishes the bacterium with sufficient catalytic potential for the efficient solubilization of recalcitrant cellulosic substrates. The major extracellular end product of the degradation process is the disaccharide cellobiose, which is then taken up and assimilated by the cell (Lamed and Bayer, 1991; Strobel et al., 1995).

The cellulosome consists of numerous different enzymatic subunits, which are held together into the complex via a separate, *nonhydrolytic* subunit, called *scaffoldin* (Bayer et al., 1994). The scaffoldin subunit is structured into a series of functional domains, one of which is a cellulose-binding domain (CBD). This CBD is very similar to the CBDs derived from simple microbial cellulases and xylanases. The latter CBDs usually deliver a single type of catalytic domain to the substrate (Gilkes et al., 1991). In contrast, the CBD from the scaffoldin subunit of the cellulosome delivers its varied array of complementary enzymes—together with the entire cell—to the cellulosic substrate.

The CBD from the cellulosomal scaffoldin subunit of *C. thermocellum* has been cloned and expressed (Morag et al., 1995), the recombinant protein has been crystallized (Lamed et al., 1994), and its structure has been elucidated (Tormo et al., 1996). By comparing the complementary surface topologies of the CBD and of cellulose, a plausible mode of interaction at the molecular level has been postulated. Although the interaction between the CBD and cellulose is essentially a surface interaction, specific types and combinations of amino acids appear to interact selectively with glucose residues positioned on three adjacent chains of the cellulose surface. Thus, selective interaction of the CBD with three separate glucose chains on cellulose is required for strong and stable binding of the bacterium to the insoluble substrate.

6.2. CELLULOSE AND THE ENVIRONMENT

Cellulose makes up the major portion of plant matter and, as such, represents the most abundant renewable source of organic raw material on the planet. The content of cellulose in plant tissues ranges from 20% in certain grasses to over 90% in cotton fiber. It is also the major component of wood in trees, making up nearly half of the biomass.

Cellulose is a very stable, linear polymer of β-1,4–linked glucose units (Hon, 1994). From the chemical point of view, the repeating unit is glucose, but, structurally, the repeating unit is the disaccharide cellobiose, since each glucose residue is rotated 180° relative to its neighbor (Fig. 6.1). This chemically simple primary structure belies a very complicated three-dimensional arrangement of the cellulose fibers, both *in situ* and in purified form. The

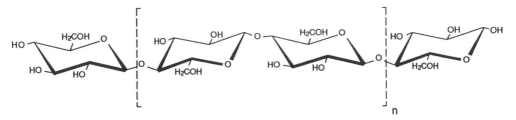

Fig. 6.1. The primary structure of cellulose. Cellulose is a polymer of glucose, consisting of repeating units of cellobiose [4-O-(β-D-glucopyranosyl)-D-glucopyranose], where n is the degree of polymerization of the cellulose chain.

degree of polymerization of the cellulose chains ranges from about 100 to more than 10,000. The resultant chains form parallelly arranged, hydrogen-bonded fibrils, which are collectively coalesced by strong associative forces. For this reason, cellulose is a relatively difficult substrate for microbes to degrade. Nevertheless, a wide variety of bacteria and fungi have evolved that can efficiently hydrolyze cellulose-containing plant material, thus counteracting its limitless accumulation.

The cellulolytic microbes occupy a broad range of habitats. Some are free living and rid the environment of cellulosic materials by converting them in nature to the simple sugars that they assimilate. Others occur associated with cellulolytic animals, such as in the gut of wood-eating termites and worms or in the digestive tracts of ruminants and other grazers (Haigler and Weimer, 1991).

In any event, in any given niche, a cellulolytic microbe is not in pure culture, but exists in a mixture with other bacterial and/or fungal species (Bayer et al., 1994). One or more strains serve as the central polymer degrader(s), which convert the cellulose and associated polymers (mainly lignin and xylan) to their respective simple sugars and other degradation products (Fig. 6.2). Due to its great abundance in nature, the difficulty in its enzymatic breakdown, and the reward (pure cellobiose and glucose) in accomplishing the latter process, the major cellulose-degrading strain plays a major and critical role in this ecosystem. The polymer degraders are assisted by other satellite microbes that contribute to the purging of the microenvironment of the polymer breakdown products (i.e., oligosaccharides and simple sugars), end-products (organic acids, molecular hydrogen), and toxic by-products.

Prior to the accumulated excesses that accompanied the spread of human civilization, the collection of cellulolytic bacteria and fungi were fully capable of handling the amounts of cellulosic waste in the environment (Fig. 6.3). This situation changed drastically when the accumulated agricultural, residential, commercial, and industrial wastes were discarded mainly into landfills and have thus eventually become environmental pollution. The prominence of cellulose in our society is clear: Paper-related products occupy almost half of the space in landfills; newspapers are the largest single item.

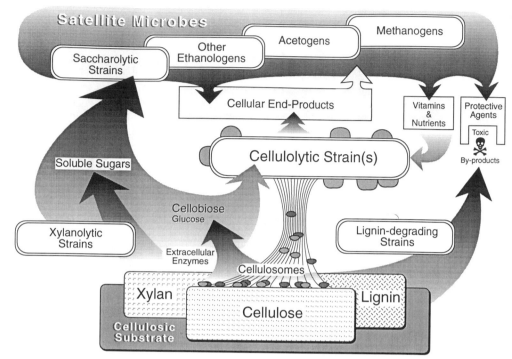

Fig. 6.2. Schematic description of a characteristic cellulosic ecosystem. Cellulolytic strains, aided by xylan- and lignin-degrading microbes, play a central role in solubilizing plant matter, primarily to low-molecular-weight sugars. The excess sugars and other cellular end-products are further assimilated by a host of satellite microorganisms, thus contributing to a well-balanced microenvironment. (Reproduced from Bayer et al., 1994, with permission of the publisher.)

Over and above the environmental factor, cellulose is a potential source of biomass for conversion to soluble sugars and other useful products, leading to fuels and chemicals (Bayer and Lamed, 1992). The microbial or enzymatic degradation of cellulosics for combined waste management and by-product and energy resource has been a stated goal for over half a century.

6.3. BACTERIAL ADHESION TO CELLULOSE

It is logical to propose that a bacterium that attacks an insoluble substrate does so by maintaining a physical association with the substrate. This can be achieved either by specific mechanisms (i.e., an adhesin) or via less specific forces (e.g., charge-mediated or hydrophobic interactions). Sometimes, the initial adhesion is superceded by more intensive secondary interactions that lead to a firmer colonization of the organism.

The primary event in the degradation of cellulose is the tight adhesion of the cellulolytic bacterium to its substrate (Fig. 6.4). Adhesion of the bacterium

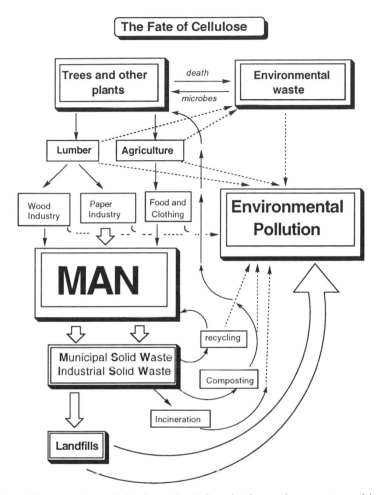

Fig. 6.3. Schematic view of the fate of cellulose in the environment, resulting from both microbial- and human-induced activities. The death of plant life conveys cellulosic biomass to the soil, where cellulolytic microbes, which occupy virtually every niche and clime, prevent its accumulation as environmental pollution by virtue of their natural process of recycling. In contrast, human civilization and its utilization of cellulosic reserves have produced vast quantities of cellulosic waste, most of which is collected into landfills, representing the largest single component, by mass, of environmental pollution. (Reproduced from Bayer and Lamed, 1992, with permission of the publishers.)

brings the cell into close proximity to the substrate and concentrates the hydrolytic enzymes on the cellulose surface. Such bacteria have been shown to display a new type of surface organelle that appears to harbor the relevant cellulose-adhesion factor(s) (Bayer and Lamed, 1986; Bayer et al., 1985). Ultrastructural studies of various cellulolytic bacteria have revealed extensive structural elements on the cell surface that have been described alternatively

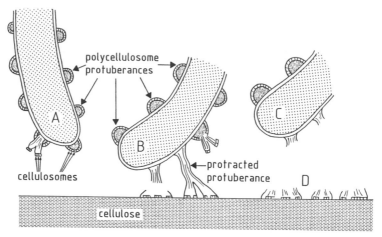

Fig. 6.4. Schematic diagram of the major events in the interaction of a cellulolytic bacterium with cellulose. This model was described initially for the highly efficient, cellulose-degrading bacterium *C. thermocellum.* Cell **A** represents a cell immediately prior to adhesion. The cell surface bears protuberance-like surface organelles that contain multiple copies of the cellulosome, which functions, at this stage, as the major cellular agent responsible for adhesion. Upon contact with the substrate (cell **B**), some of the protuberances protract dramatically, depositing their cellulosomes along the cellulose surface. The enzymes on the cellulosome commence their degradation of the substrate, introducing large quantities of cellobiose into the immediate environment. After an extended period of growth (cell **C**), the cell detaches from the cellulose, leaving cell-free cellulosome clusters (**D**), that coat the entire surface of the residual cellulose matrix and presumably continue the degradation process. (Reproduced from Lamed and Bayer, 1988a, with permission of the publisher.)

as glycocalyces, protuberances, projections, and so forth (Lamed et al., 1987a,b; Mayer et al., 1987; Salamitou et al., 1994a). Concomitant with the faculty of adhesion, cellulolytic enzymes are a common feature of the cell surface of cellulolytic bacteria.

The study of the adhesion process in bacteria is aided invaluably by the acquisition of defined adhesion-defective mutants (Bayer et al., 1983). By comparing the surface characteristics and binding properties of the mutant and those of the substrate-adhering parent forms, the molecular basis for adhesion can be elucidated (Lamed et al., 1983a). Thus far, however, only limited work in this direction has been performed, but the results of this line of research have been very revealing.

The molecular mechanism of adhesion of only a few cellulolytic bacterial species has been studied extensively. These include the precursory work on *C. thermocellum* and the more recent findings on the adhesion of *Fibrobacter succinogenes* and *Ruminococcus albus* to cellulose (Forsberg et al., 1993; Gong and Forsberg, 1989, 1993; Miron, 1991; Morris, 1988; Morris and Cole, 1987; Roger et al., 1990). In many other pure and mixed bacterial cultures, particularly from the rumen, the binding characteristics and surface ultrastructure have been examined in some detail (Bhat et al., 1990; Cheng et al., 1983; Kauri and Kushner, 1985; Minato et al., 1993; Miron et al., 1989).

In the following analysis, we focus on our work that led to the molecular description of the adhesion to and degradation of cellulose by *C. thermocellum*. This bacterium produces a multisubunit, protein complex termed the *cellulosome,* which is exquisitely designed for efficient binding and hydrolysis of the substrate. In this case, the quality of adhesion appears to be a function of a CBD, which is contained on the crucial cellulosomal subunit, scaffoldin. In itself, scaffoldin lacks a hydrolytic function but includes additional domains that integrate the various catalytic components into the complex (Salamitou et al., 1994b; Tokatlidis et al., 1991, 1993; Yaron et al., 1995). Some of the catalytic subunits, thus incorporated into the cellulosome complex of *C. thermocellum,* have also been shown to contain their own CBD (Durrant et al., 1991; Hazlewood et al., 1988; Navarro et al., 1991), which may or may not participate in the initial adhesion event.

These findings do not necessarily imply that all cellulolytic bacteria will produce such intricate CBD-containing cellulosomes, which exhibit multiple functions including both adhesion and hydrolysis of the substrate. Nor do our findings imply that all cellulosomes will closely resemble, in sequence and structure, the cellulosome of *C. thermocellum*. The nature of the cellulase system of *C. thermocellum* may represent but one theme of which innumerable variations or alternatives may exist. In fact, an uncomplexed, nonhydrolytic cellulose-binding protein has been described in *F. succinogenes* (Gong and Forsberg, 1993) that exhibits characteristics reminiscent of those of the cellulosomal scaffoldin in *C. thermocellum*. One can conceive that the independent binding of such a protein to the substrate surface might serve as a source of subsequent complexation for hydrolytic components. One can also conceive of CBDs, attached individually to the cell surface, that would directly effect the adhesion of the cell to cellulose without defined molecular linkage to hydrolytic components. Moreover, additional secondary adhesive events, which consolidate cell–substrate contact, may follow the initial binding of the cell. Further work on different cellulolytic bacteria will certainly benefit our overall understanding of the adhesion processes that characterize this broad range of bacteria.

6.3.1. *Clostridium thermocellum*

C. thermocellum is an anaerobic, thermophilic, cellulolytic, Gram-positive bacterium. Since its initial isolation in 1926 (Viljoen et al., 1926), the great majority of anaerobic, thermophilic, cellulolytic bacteria isolated from nature turned out to be members of this particular species. *C. thermocellum* is extremely selective in its preference for substrates. Growth of the bacterium is normally restricted to cellulose and its degradation products (Lamed and Bayer, 1991). Although, the cellulase system in *C. thermocellum* includes additional enzyme types, notably xylanases, the corresponding degradation products (e.g., xylose and xylobiose) cannot serve as substrates for growth (Morag et al., 1990). It is difficult to reconcile this apparent paradox. Nevertheless, the hydrolytic potential for such a polymer, which fails to be consumed by the bacterium, may simply serve the purpose of getting to the good stuff (cellulose) while enriching the environment with soluble sugars that are utilized by satellite bacteria (Bayer et al., 1994).

One of the important properties of *C. thermocellum* is the production of very high levels of extracellular and cell-associated cellulases. The cellulase system in this bacterium is capable of complete solubilization of highly crystalline forms of cellulose—an accomplishment restricted to relatively few examples of cellulolytic microbes (Johnson et al., 1982). Cellulose solubilization by *C. thermocellum* is considered to be exceptionally high, and interest in its mode of action has persisted for decades, fueled by the belief that either the bacterium or its enzymes would eventually find commercial application (Bayer and Lamed, 1992). In reality, this belief has yet to be fulfilled.

6.3.2. Adhesion-Defective Mutant

One of the major breakthroughs in the initial stages of the work with *C. thermocellum* was the isolation of an adhesion-defective mutant (Bayer et al., 1983). Isolation of this mutant enabled the isolation of adhesion-specific antibodies, which in turn led to the isolation of a "cellulose-binding factor" and consequent description of the cellulosome concept (Lamed et al., 1983b).

The mutant was isolated by an enrichment procedure that included repetitive cycles of growth on cellobiose and selective removal of adhering bacteria by introducing cellulose to the suspension. A single colony was eventually isolated. The mutant cells, grown on cellobiose, failed to adhere to cellulose, in contrast to the wild-type cells (Fig. 6.5). To prepare adhesion-specific antibodies, a general antibody preparation (elicited against intact wild-type cells) was produced, and the mutant cells were used to absorb species of immunoglobulin that were common to both the mutant and wild type. Their removal afforded a residual fraction of immunoglobulins that failed to bind to the adherence-defective mutant. The fact that the resultant antibody preparation was indeed selective only for the wild-type cells corroborated the instinctive supposition that molecule(s) responsible for cell adhesion were associated with the cell surface.

The major consequence of the mutation was to delay the binding of the cells to cellulose (Fig. 6.6). The growth of the wild-type cells on cellulose was characterized by immediate attachment of the cells to the substrate and a relatively slow, almost linear growth on the surface of the substrate (Fig. 6.6A). After an extended period (about 5 to 7 hours), during which cells were not observed in the unbound state (Fig. 6.6B), cells began to leave the cellulose surface, and the exponential phase ensued.

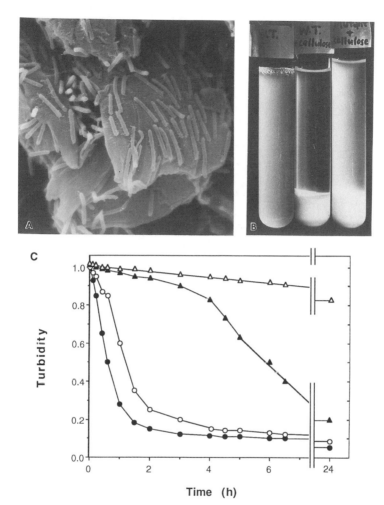

Fig. 6.5. Adhesion of *C. thermocellum* to cellulose. **A:** Scanning electron micrograph of cells attached to cellulose particles. **B:** Test tubes containing a suspension of wild-type cells (left), a similar suspension with added cellulose (center), and a suspension of adhesion-defective mutant cells with added cellulose. Note that the wild-type cells have bound strongly to cellulose, but the mutant fails to do so. **C:** Agglutinability of wild-type cells versus mutant cells by a general and an adhesion-specific antibody preparation. A species-specific antibody preparation (circles) causes the agglutination of both the wild-type (filled symbols) and mutant (open symbols), whereas the adhesion-specific antibody (triangles) agglutinates only the wild-type cells. (Reproduced from Bayer et al., 1983, with permission of the publisher.)

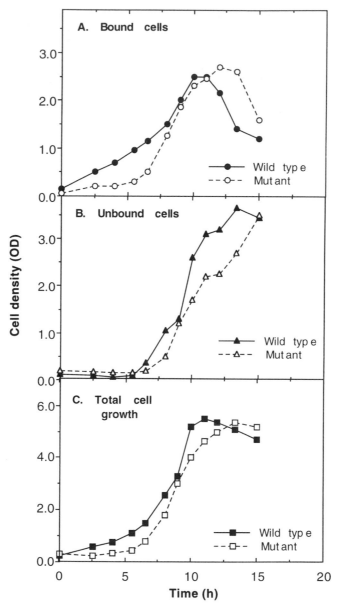

Fig. 6.6. Growth of *C. thermocellum* on cellulose. Wild-type (filled symbols) and mutant (open symbols) cells were grown on cellulose, and the amount of cells bound to the substrate (**A**) or in the substrate-free state (**B**) was determined. Note that the wild-type cells bind relatively quickly to the substrate, whereas the binding (and growth) of the mutant is delayed for several hours. Total cell growth (**C**) is also shown.

The number of cells bound to cellulose reached a peak after about 10 hours of growth, signifying the onset of the stationary phase of growth as seen in the total growth curve (Fig. 6.6C).

Growth of the mutant on cellulose was delayed for about 6 hours, during which very little adhesion of mutant cells to substrate could be observed. After this lag period, the property of adhesion appeared to have been gained in the mutant, thus indicating that the elements responsible for adhesion of the bacterium were now present on the cell surface. This was corroborated by the interaction of the cellulose-grown mutant cells with the adhesion-specific antiserum, in contrast to the lack of interaction of these antibodies with cellobiose-grown mutant cells. The mutant thus appears to be regulatory in nature.

6.4. THE CELLULOSOME OF *C. THERMOCELLUM*

6.4.1. Isolation of the Cellulosome

The availability of an adherence-defective mutant and adhesion-specific antibodies provided us the means with which to isolate and purify a cellulose-binding factor from the bacterium (Lamed et al., 1983a). This was accomplished by exploiting the quality of its inherent affinity for the insoluble substrate. Thus, extracellular or cell surface extracts were simply adsorbed onto a cellulose matrix (e.g., microcrystalline cellulose, amorphous cellulose, or even tissue paper), and the nonadsorbing material was washed away. The cellulose-bound protein was released either by distilled water or by mild base treatment. In later works, once it became clear that the cellulose-binding factor (i.e., the adhesion factor) included a highly efficient collection of cellulose-degrading enzymes arranged into the cellulosome, its catalytic potential was incorporated directly into the purification scheme (Morag et al., 1992a). In this context, cell-derived material was adsorbed onto the cellulose as before, but, instead of incomplete elution by the abovementioned treatments, the cellulosic carrier was allowed to undergo digestion by the adsorbed enzyme complex. In this manner, near-complete yields of the cellulose-binding factor, the cellulosome, were attained.

Indeed, one of the initial surprises in the original isolation of the cellulose-binding factor from *C. thermocellum* was the unexpected discovery that the purified protein was not a simple polypeptide, but consisted of a discrete, multifunctional, multienzyme complex capable of complete solubilization of cellulosic substrates. The complex was termed the *cellulosome* (Lamed et al., 1983b), and the cellulosome concept is currently considered to be a major paradigm of bacterial cellulolysis (Felix and Ljungdahl, 1993; Lamed and Bayer, 1988b, 1993).

6.4.2. Cellulosomal Subunits

The cellulosome in *C. thermocellum* is composed of a number of subunits that are packed into polycellulosomal protuberance-like organelles (Fig. 6.7). The cellulosomes, which are contained in these structures, mediate the adhesion of the cells to cellulose. Upon binding, the polycellulosomes undergo a dramatic conformational change (Bayer and Lamed, 1986), forming protracted contact corridors between the cell and the substrate. In addition to the cellulosomal enzymes, free, uncomplexed, noncellulosomal cellulases also contribute to the degradation process (Morag et al., 1990). The ultrastructure of the surface organelles of *C. thermocellum* (e.g., Fig. 6.7B,C; see Bayer and Lamed, 1986) substantiate the model described in Figure 6.3.

The multiple subunit arrangement of the isolated cellulosomes is revealed by high-resolution electron micrographs and by SDS-PAGE (Fig. 6.7D,F) (Lamed et al., 1983a; Mayer et al., 1987). The cellulosome is a 2 megadalton conglomerate that consists of a dozen or more different subunits (Hon-nami et al., 1986; Kohring et al., 1990; Lamed et al., 1983a). In some strains of *C. thermocellum* the cellulosome appears to be even larger (Coughlan et al., 1985; Wu and Demain, 1988). Most of the cellulosomal subunits are relatively large enzymes, i.e., cellulases and xylanases, which range in mass from about 50 to about 170 kDa (Lamed et al., 1983a). The cohesive association of these enzymes into the complex is accomplished by *scaffoldin,* the multifunctional,

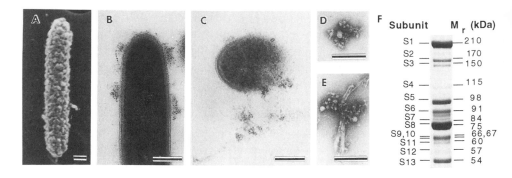

Fig. 6.7. Disposition and subunit composition of the cellulosome of *C. thermocellum* at increasing levels of resolution. **A:** SEM of a single, protuberance-adorned bacterial cell. **B:** TEM of anticellulosome-labeled cells. **C:** TEM of labeled bacterium bound to cellulose, showing protracted protuberance that connects cell to substrate. **D:** Negatively stained preparation of purified cellulosome. **E:** Purified cellulosomes adsorbed to cellulose fibers. **F:** Subunit profile of the isolated cellulosome separated by SDS-PAGE. Bar markers: A–C, 200 nm; D and E, 50 nm. (Reproduced from Bayer et al., 1994, with permission of the publisher.)

nonhydrolytic glycopolypeptide subunit (originally identified in *C. thermocellum* as the 210 kDa S1 subunit), which plays a central role in the cellulosome structure.

The structure of the cellulosome in *C. thermocellum* is exceptionally stable (Lamed and Bayer, 1988b). Nevertheless, it seems to be flexible, and, upon interaction with the substrate, the cellulosome is considered to undergo conformational rearrangement, which is presumably critical to its subsequent interaction with, and efficient degradation of, the cellulosic substrate (Morag [Morgenstern] et al., 1991; Morag et al., 1992b). In any event, the original hypothesis still stands: that the organization of the various complementary enzymes (notably the cellulases [the endo- and exo-glucanases]) into a defined cellulosome complex serves to promote their synergistic action (Lamed et al., 1983a,b).

6.4.3. Cellulosomal Domains and Their Functions

Whereas the multisubunit arrangement of the cellulosome was originally characterized by biochemical means (Lamed et al., 1983a), the multidomain structure of the subunits has been elucidated by recombinant DNA technology (Béguin, 1990, Béguin et al., 1985; Cornet et al., 1983; Gerngross et al., 1993; Hall et al., 1988). On the basis of both biochemical and molecular biological findings, a simplistic schematic view of the structure of the cellulosome has been constructed (Fig. 6.8). In this assessment, the scaffoldin subunit contains the major cellulose-binding function in a separate domain, the CBD (Gerngross et al., 1993; Poole et al., 1992). In addition, the scaffoldin subunit is also responsible for organizing the cellulolytic components into the multienzyme complex. This is accomplished by a unique affinity interaction that includes two complementary classes of domain that are located on the two separate interacting subunits (Salamitou et al., 1992, 1994b; Tokatlidis et al., 1991, 1993) i.e., a *cohesin* domain on scaffoldin and a *dockerin* domain on the enzymatic subunit (Bayer et al., 1994).

The cohesin domain consists of about 150 amino acids, and the dockerin domain contains about 50 to 60 amino acid residues, of which a characteristic duplicated sequence of about 23 amino acids is present (Béguin and Aubert, 1994). The cohesins of *C. thermocellum* are all very similar, exhibiting between 60% and almost complete identity among the nine cohesin domains of the scaffoldin subunit. Six of the nine cohesin domains show greater than 90% identity. Likewise, the sequences of the dockerin domains of the various cellulosomal enzyme subunits are very similar (Béguin and Aubert, 1994), indicating that their general structural fold is also very similar. It is not clear how selective a given cohesin domain is for a given dockerin domain, although the evidence currently favors a low level of selectivity (Yaron et al., 1995).

The scaffoldin subunit may also serve to anchor the cellulosome into the cell surface (Bayer et al., 1985; Fujino et al., 1993a). In this regard, the scaffoldin subunit possesses a C-terminal dockerin domain of its own. The dockerin domain of scaffoldin fails to interact with its own cohesin

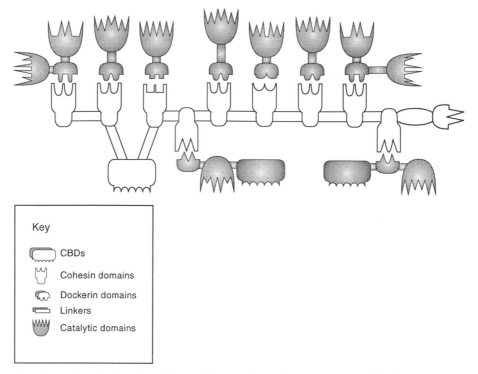

Fig. 6.8.   Simplified model of the cellulosome from *C. thermocellum*. All of the subunits are composed of multiple domains. The nonhydrolytic *scaffoldin* subunit (shown as the central structure in white) integrates the cellulolytic and xylanolytic subunits (shaded structures) into the complex via a novel, intersubunit interaction. The association between scaffoldin and the hydrolytic subunits is induced by reciprocal interacting domains: the *cohesin* domain on scaffoldin and the *dockerin* domain on the respective enzymatic subunit. The scaffoldin subunit itself bears a dockerin domain, which is believed to mediate its implantation into the cell surface. The enzymatic nature of the hydrolytic subunit is dictated by one or more catalytic domains. The domains of the various subunits are usually separated by distinct linker sequences. Both scaffoldin and the enzymatic subunits contain other domains (not shown), the functions of which are still unknown (Reproduced from Bayer et al., 1994, with permission of the publisher.)

domains but could conceivably interact with a distinct cohesin-bearing protein that is implanted in the cell surface for that purpose. Several novel cell surface proteins have recently been described (Fujino et al., 1993a; Lemaire et al., 1995; Salamitou et al., 1994b), one or more of which may prove to be components of the protuberance-like surface organelles shown in Figure 6.7 and/or to play a role in anchoring therein the scaffoldin subunit (and, thus, the intact cellulosome).

6.4.4. Cellulosomes in Other Cellulolytic Microbes

Although many species of cellulolytic bacteria are known to bind to cellulose, little is known about the exact molecular mechanism(s) for adhesion in each case. Thus far, the only well-defined mode of cell adhesion for a cellulolytic microorganism is that described above for *C. thermocellum,* whereby the bacterium produces a cellulosome (an intricate, multienzyme complex), one of the components of which contains a CBD for that purpose. In view of this paucity of information, it is certainly not clear how widespread are cellulosomes among the cellulolytic bacteria and whether their role in bacterial adhesion is a general phenomenon.

At the molecular level, a telltale sign of whether a cellulosome exists in a given bacterium is the presence of sequence(s) in the genome that can be classified as either a cohesin domain (indicative of a scaffoldin subunit) or a dockerin domain (indicative of a putative catalytic subunit that may bind to a scaffoldin subunit) (Bayer et al., 1994; Lamed and Bayer, 1993). On the basis of sequence comparison alone, thus far cellulosomes have been demonstrated unequivocally to exist in four species of bacteria—three mesophiles (Fierobe et al., 1993; Fujino et al., 1993b; Shoseyov et al., 1992) and one thermophile (Tokatlidis et al., 1991), all of the genus *Clostridium*—and one or two anaerobic fungal species (Wilson and Wood, 1992). In these microbial species, the adhesion phenomenon has been demonstrated only for *C. thermocellum* (Bayer et al., 1983) and *C. cellulolyticum* (Gelhaye et al., 1993).

6.4.5. The Scaffoldin Gene

In two of the cellulolytic bacteria, *C. thermocellum* and *C. cellulovorans,* the entire genes encoding for the scaffoldin subunit have been sequenced (Gerngross et al., 1993; Shoseyov et al., 1992). In both cases, the corresponding gene encodes for a very large protein of approximately 1,850 amino acid residues. Both of the scaffoldins from these two bacteria contain a single CBD and nine cohesin domains. Both contain additional domains that the other lacks and for which defined functions are as yet unclear. The sequencing of such large genes is exacerbated by the presence of multiple copies of the cohesin domains; their near-but-not-quite-perfect repetition within the gene ensures that the accuracy in the sequencing of the scaffoldins will remain a formidable task in the future. Consequently, the availability of complete primary sequences of such genes from cellulolytic bacteria will probably remain sparse. Nevertheless, various components of cellulosomes from other bacteria, notably from another clostridial species (Fierobe et al., 1993; Fujino et al., 1993b; Pagès et al., 1996) are currently being sequenced, which should eventually enrich our future understanding of their structural diversity.

In lieu of full sequences for such genes, there are other signs that may indicate the presence of a cellulosome. These include the presence of multicellulase complexes in a given bacterium, the cross reaction of a bacterium or bacterial extract with cellulosome-specific antibody, and, in view of the glycopeptidic nature of the scaffoldin subunit, the interaction of a bacterium or bacterial extract with cellulosome-specific lectin (Lamed et al., 1987b). In fact, two oligosaccharides from two different cellulolytic organisms have been sequenced. One of these sugars was derived from the cellulosome from *C. thermocellum* (Gerwig et al., 1991), and the other from a 230 kDa scaffoldin-like glycopolypeptide from the rumen bacterium *Bacteroides cellulosolvens* (Gerwig et al., 1992). The striking conclusion from comparison of the two oligosaccharides is that they both contain an identical backbone structure, consisting of an unusual trisaccharide but they are distinguished by what appear to be species-specific substitutions. These similarities in sugar structure imply that the relevant glycosyltransferases and other enzymes responsible for their synthesis are similar in these two very different types of cellulolytic bacteria. Their function in the cellulosome, albeit currently unknown, must also be similar.

6.5.   THE CELLULOSE-BINDING DOMAIN

The concept of a cellulosome refers to the organization of the cellulolytic enzymes into a multicomponent complex, and the presence of a CBD in a cellulosome may prove to be only incidental to its structure. Thus, in the future, cellulose-adherent cellulolytic microorganisms may eventually be described that contain a definitive cellulosome wherein the enzymatic subunits are indeed incorporated into its structure via a scaffoldin-like subunit that *lacks* a CBD. Such an organism may contain a separate CBD on the cell surface that mediates cell adhesion to cellulose, which, instead of being associated with the cellulosome, could either be associated with another type of subunit or implanted into the cell surface via its linkage to a relevant domain or to another type of component.

On the other hand, it seems that, in the final analysis, the phenomenon of bacterial adhesion to cellulose will indeed be traced to the presence of a special type of CBD distributed in some manner on the bacterial cell surface. Whether or not such CBDs are associated with a cellulosome and/or protuberance-like surface organelles, such as those observed for *C. thermocellum,* remains to be determined in future analyses of adhesion in other cellulolytic bacteria.

However, surely not every CBD is involved in cell adhesion to cellulose. As mentioned above, the great majority of CBDs are simply involved in the binding of a given cellulolytic enzyme, in the free state, to the substrate. In fact, quite a large number of CBDs from cellulases, xylanases, and chitinases have been described in detail. Over 100 CBDs have been sequenced, and, on the basis of sequence alignment, they have thus far been categorized into nearly a dozen different families (Béguin and Aubert, 1994; Gilkes et al., 1991; Tomme et al., 1995).

CBDs from cellulolytic fungi are generally quite small, consisting of about 33 to 36 amino acids. Fungal CBDs have been collectively grouped into a distinct family, designated type I CBD (Coutinho et al., 1992; Reinikainen et al., 1992). The three-dimensional structure of one of these has been elucidated by NMR spectroscopy (Kraulis et al., 1989). It has a wedge-shaped, β-sheet structure, consisting of three antiparallel β-strands. On one of its faces, three exposed aromatic amino acids have been implicated in the binding to cellulose on the basis of mutation and chemical modification experiments (Linder et al., 1995; Reinikainen et al., 1995). More recently, the involvement of these and other amino acids in cellulose binding has been supported by simulation analysis (Hoffrén et al., 1995).

The most extensively studied bacterial CBDs have been derived from free, uncomplexed cellulases of *Cellulomonas fimi* (Gilkes et al., 1988) and *Pseudomonas fluorescens* (Kellett et al., 1990; Poole et al., 1991). These CBDs were very similar in their primary structure and consisted of about 100 amino acid residues. On the basis of sequence homology, other CBDs from other bacteria have been shown to be very similar to those of *C. fimi* and *P. fluorescens,* and this type has been grouped into a second family, designated type II CBD (Coutinho et al., 1992). As in the type I fungal CBD, solvent-exposed aromatic residues have been implicated in the binding to cellulose (Din et al., 1994b; Poole et al., 1993). Recently, an NMR structure for this type of CBD was determined (Xu et al., 1995). The protein is organized into a half-closed β-barrel of nine β-strands, contained in two β-sheets, packed face to face in a jellyroll topology.

The CBDs have been proposed to enhance the enzymatic degradation of insoluble forms of cellulose in a manner that cannot be explained simply on the basis of delivering its associated catalytic (i.e., hydrolytic) domain to the substrate (Din et al., 1991). A similar activity has been suggested for the scaffoldin subunit of the cellulosome, and its CBD may eventually prove to be responsible for this activity. A catalytic, but not hydrolytic, function for the CBD itself has thus been suggested (Din et al., 1994a) in which it seems to disrupt *noncovalent* bonds between adjacent cellulose chains of the cellulose fibrils, thereby increasing their accessibility to subsequent attack by the hydrolytic domain.

6.6. THE RECOMBINANT CBD FROM THE CELLULOSOME OF *C. THERMOCELLUM*

6.6.1. Cloning and Expression

The CBD of the scaffoldin subunit from the cellulosome of *C. thermocellum* has been cloned and overexpressed in an appropriate bacterial host, and the expressed protein was purified (Morag et al., 1995). The CBD consists of about 155 amino acid residues and is thus significantly larger than the type

II CBD. In addition, it exhibits no sequence homology to CBDs of type II. On the other hand, its sequence is similar to a different family of CBD, collectively grouped and designated type III CBD. Interestingly, the only other verified cellulosomal CBD (i.e., the CBD of the scaffoldin subunit from *C. cellulovorans*) also belongs to the same family. The two sequences show approximately 50% identity in their amino acid residues (Fig. 6.9).

The CBD from *C. cellulovorans* has also been expressed and purified (Goldstein et al., 1993). Despite the sequence similarity, polyclonal antibodies raised against the CBD from *C. thermocellum* failed to interact with the CBD from *C. cellulovorans* (Morag et al., 1995), indicating that their exposed epitopes are different, although their overall fold would be expected to be very similar.

The cellulose-binding properties of the two recombinant type III CBDs were shown to be very similar. Both proteins recognize cellulose with an affinity constant of about 10^6 M^{-1}. Their substrate specificities are also very similar; both bind to insoluble cellulosic substrates and to chitin, but not to xylan. Cellobiose and other simple oligosaccharides fail to inhibit the binding or to release the bound CBD from the insoluble substrate. It is clear that the binding of the CBD to cellulose requires a highly structured arrangement of the cellulosic substrate.

6.6.2. Three-Dimensional Crystal Structure

The crystal structure of the recombinant CBD from the cellulosomal scaffoldin subunit of *C. thermocellum* has been determined to 1.75 Å resolution (Lamed et al., 1994; Tormo et al., 1996). The CBD folds into a compact, prisma-shaped

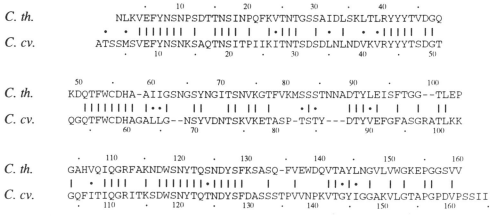

Fig. 6.9. Similarity between the aligned, type III CBDs of *C. thermocellum (C.th.)* and *C. cellulovorans (C.cv.)*. Identical residues are indicated by vertical lines. Similar residues (A, G; S, T; and V, L, I, M) are denoted by dots. (Reproduced from Morag et al., 1995, with permission of the publisher.)

structure with approximate dimensions of 30 × 30 × 45 Å. The structure belongs to the all-β family of proteins and is arranged in two antiparallel β-sheets that stack face to face to form a β-sandwich. The general structure of the CBD and its putative interaction with cellulose is shown schematically in Figure 6.10. This structure for a type III CBD enables comparison with those of the type I fungal CBD and type II bacterial CBD, both of which were determined by NMR spectroscopy.

As mentioned above, the cellulosomal, type III CBD from *C. thermocellum* is much larger than the type II CBD from *Cellulomonas*. In addition, the two families of CBD exhibit no recognizable amino acid sequence homology. Nonetheless, the two proteins share a similar nine-stranded jellyroll topology. One striking difference is the presence of a shallow groove in the former CBD that has no counterpart in the type II CBD from *Cellulomonas*. This groove

A   B

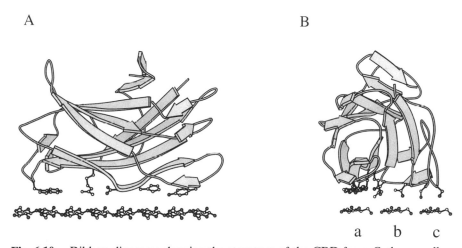

                                                        a   b   c

Fig. 6.10. Ribbon diagrams showing the structure of the CBD from *C. thermocellum* and its interaction with cellulose chains. **A:** The CBD is positioned along the length of a single glucose chain. Shown, protruding from the lower surface of the CBD molecule, are the amino acid residues that form the planar strip (from left to right: Trp-118, Arg-112, Asp-56, His-57, Tyr-67). The charged amino acids in this strip form a salt bridge and, together with the aromatic residues, stack upon four of six successive glucose pyranose rings in the cellulose chain. Note the shallow groove at the top of the molecule. This groove is filled with amino acid side chains, notably four tyrosine residues, and does not appear to constitute a binding pocket *per se.* It may serve to form secondary protein–protein interactions with other components of the scaffoldin subunit or with other cellulosomal subunits. **B:** The CBD molecule has been rotated 90° such that the view shows its interaction with three adjacent glucose chains of the cellulose surface. Chain a represents the glucose chain shown in A, which interacts with the planar strip of amino acids. Polar amino acids are proposed to form hydrogen bonds that anchor the CBD with the other two glucose chains. Asn-10, Asn-16, and Gln-110 form hydrogen bonds with glucose residues on chain b, and Ser-12 and Ser-133 form hydrogen bonds with residues on chain c.

is laden with a cluster of four tyrosines and three other charged or polar residues, which are conserved in the other CBDs of the type III family, but the function of this groove is not yet known.

The opposite side of the CBD molecule is composed of relatively long β-strands, which form a smoothly bent surface. This is the surface that appears to interact with the cellulose surface. Mapping of the residues conserved in other members of the type III CBD family has revealed a planar linear array of aromatic and charged residues located on one edge of this surface, which can be aligned with glucose moieties on one chain of the cellulose (Fig. 6.10A). This planar strip is analogous to those of the types I and II CBD, which are also characterized by aromatic amino acids. In addition, other amino acids, polar in nature, are available for additional interaction with adjacent cellulose chains (Fig. 6.10B). In this case, these polar residues appear to form hydrogen bonds with the cellulose chains, thereby anchoring the molecule to the substrate. The implications are that a structural element of three adjacent cellulose chains is required for strong binding of the CBD to the substrate.

This assessment is supported by the NMR structures of the types I and II CBD (Table 6.1). A planar strip of aromatic amino acids has also been noted in both of these fungal and bacterial CBDs, although in both cases an Arg-Asp salt bridge is absent. Moreover, both types I and II CBD contain additional polar residues on this particular surface, which could form hydrogen bonds and interact as anchoring points similar to those proposed in the structure of the cellulosomal CBD from *C. thermocellum*. The fact that such surface features are conserved among the three separate families of CBDs for which three-dimensional structures are known reinforces the assumption that they provide the cellulose-binding surface in all of them. It would be anticipated that similar surface features would exist in other, as yet uncharacterized CBD families as well.

However, the distribution of these amino acids on the face of the CBDs appears to be contributed by different structural elements on the individual types of protein. This is especially significant regarding the two bacterial proteins that exhibit a similar fold but show no sequence homology. Thus, most of the above-described amino acids on the two CBDs, which are purported to interact with cellulose as described above, are contributed by different loops and β-strands that traverse this particular surface of the CBD molecule. In fact, only one of these related amino acids (His-57) in *C. thermocellum* and Trp-54 in the CBD *Cellulomonas* CBD) is located in a topologically equivalent position.

This structurally contrasting distribution of amino acid residues between the two families of CBD is reflected in their proposed interaction with cellulose (Table 6.1). Interestingly, the character of the interaction with cellulose of the bacterial type II CBD from *Cellulomonas* more closely resembles that of the fungal type I CBD. In both, the planar strip interacts with a central chain on the cellulose polymer, and the presumed polar anchoring residues bind to glucose moieties on the two chains that occur on both sides of this central

TABLE 6.1. Proposed Interactions of CBD Residues With Successive Chains of the Crystalline Cellulose Lattice

Chain a (anchor)	Chain b (Planar strip)	Chain c (anchor)
Type I fungal CBD from cellobiohydrolase of *Trichoderma reesei*		
Gln-7	Tyr-5	Asn-29
	Tyr-31	
	Tyr-32	
	(Gln-34)[a]	

Chain a (anchor)	Chain b (Planar strip)	Chain c (anchor)
Type II bacterial CBD from exoglucanase of *Cellulomonas fimi*		
Gln-52	Trp-17	Asn-15
	Trp-54	Asn-24
	Trp-72	Gln-83
	(Asn-87)[a]	

Chain a (Planar strip)	Chain b (anchor)	Chain c (anchor)
Type III bacterial CBD from scaffoldin of *C. thermocellum*		
Asp-56[b]	Asn-10	Ser-12
His-57	Asn-16	Ser-133
Tyr-67	Gln-110	
Arg-112[b]		
Trp-118		

[a] In the *T. reesei* CBD, Gln-34 stabilizes the position of Tyr-32 by hydrogen bond formation, and in the *Cellulomonas* CBD, Asn-87 stabilizes Trp-54 similarly.
[b] Within the planar strip of the *C. thermocellum* CBD, Asp-56 and Arg-112 form a salt bridge, which is proposed to align across one of the glucose pyranose rings in the cellulose chain.

chain. This is in contrast to the interaction proposed for the type III CBD from *C. thermocellum*, wherein the anchoring residues interact with two vicinal glucose chains that occur on one side of the chain that purportedly interacts with the planar strip.

This would indicate that the two families of bacterial CBD did not evolve from a common molecular prototype; rather, their common structural architecture and the selection of their functional elements appear to have been dictated by evolutionary convergence. It is still perplexing why such a fold and functional residues may be important while the arrangement of the participating residues appears on different elements of the CBD structure. When three-dimensional structures of CBDs from other families become available, it should become clear whether these features constitute a general phenomenon of this unusual type of binding protein.

6.7. CONCLUDING COMMENTS

For want of more information on the mechanism of adhesion of cellulolytic bacteria, one should certainly hesitate to proffer sweeping generalizations of a speculative nature. Unlike earlier studies on cellulolytic bacteria and their cellulases, our contributions in the area were initially prompted by our simple observations on the adhesion of the anaerobic thermophilic bacterium *C. thermocellum* to cellulose. This alternative avenue of research led to the discovery of an adhesion factor and its close association with the cellulases in this bacterium. This unconventional approach allowed us to define the cellulosome concept.

What is clear is that in nature bacteria and fungi are faced with a mammoth task when it comes to degradation of all the cellulosic waste on the planet. Microbial adhesion, mediated by the CBD of the cellulosome, plays a vital primary role in this process. The inherent difficulties in the degradation of cellulose are twofold: (1) the massive volume and continuous supply of cellulose and (2) the innate recalcitrance of cellulosic substrates. To tackle global cellulose accumulation, the cellulolytic bacteria have evolved to produce intricate multienzyme complexes (the cellulosomes), specialized cell surface organelles that bear them, and a complicated, dynamic mode of interaction.

All of this complexity provides the bacterium with an effective mechanism to deliver itself and its mixed enzyme system to the surface of its very specified, but very abundant, insoluble substrate. There, it efficiently degrades the cellulose. Thus, the degradation process takes place *extracellularly,* but in close proximity to the cell, which assimilates the soluble sugars (i.e., cellobiose and glucose) that have enriched its immediate vicinity. Once inside the cell, the metabolism of the primary by-products is carried out in a relatively routine fashion by conventional enzymes and pathways. In the final analysis, this specialized machinery—where the definitive adhesion component is intimately connected to the enzymes necessary for the degradation of such a specialized substrate—provides the cellulose-degrading bacterium with an elegant means to play its central and decisive role in its natural habitat.

ACKNOWLEDGMENTS

The authors acknowledge support by a grant from the Israel Science Foundation, administered by the Israel Academy of Sciences and Humanities, Jerusalem.

REFERENCES

Bayer EA, Kenig R, Lamed R (1983): Studies on the adherence of *Clostridium thermocellum* to cellulose. J Bacteriol 156:818–827.

Bayer EA, Lamed R (1986): Ultrastructure of the cell surface cellulosome of *Clostridium thermocellum* and its interaction with cellulose. J Bacteriol 167:828–836.

Bayer EA, Lamed R (1992): The cellulose paradox: Pollutant *par excellence* and/or a reclaimable natural resource? Biodegradation 3:171–188.

Bayer EA, Morag E, Lamed R (1994): The cellulosome—A treasure-trove for biotechnology. Trends Biotechnol 12:378–386.

Bayer EA, Setter E, Lamed R (1985): Organization and distribution of the cellulosome in *Clostridium thermocellum*. J Bacteriol 163:552–559.

Béguin P (1990): Molecular biology of cellulose degradation. Annu Rev Microbiol 44:219–248.

Béguin P, Aubert J-P (1994): The biological degradation of cellulose. FEMS Microbiol Lett 13:25–58.

Béguin P, Cornet P, Aubert J-P (1985): Sequence of a cellulase gene of the thermophilic bacterium *Clostridium thermocellum*. J Bacteriol 162:102–105.

Bhat S, Wallace RJ, Ørskov ER (1990): Adhesion of cellulolytic ruminal bacteria to barley straw. Appl Environ Microbiol 56:2698–2703.

Cheng KJ, Stewart CS, Dinsdale D, Costerton JW (1983): Electron microscopy of bacteria involved in the digestion of plant cell walls. Anim Feed Sci Technol 10:93–120.

Cornet P, Millet J, Béguin P, Aubert J-P (1983): Characterization of two *cel* (cellulose degradation) genes of *Clostridium thermocellum* coding for endoglucanases. Bio/Technology 1:589–594.

Coughlan MP, Hon-nami K, Hon-nami H, Ljungdahl LG, Paulin JJ, Rigsby WE (1985): The cellulolytic enzyme complex of *Clostridium thermocellum* is very large. Biochem Biophys Res Commun 130:904–909.

Coutinho JB, Gilkes NR, Warren RAJ, Kilburn DG, Miller RCJ (1992): The binding of *Cellulomonas fimi* endoglucanase C (CenC) to cellulose and Sephadex is mediated by the N-terminal repeats. Mol Microbiol 6:1243–1252.

Din N, Damude HG, Gilkes NR, Miller RCJ, Warren RAJ, Kilburn DG (1994a): $C_1 - C_X$ revisited: Intramolecular synergism in a cellulase. Proc Natl Acad Sci USA 91:11383–11387.

Din N, Forsythe IJ, Burtnick LD, Gilkes NR, Miller RC Jr, Warren RAJ, Kilburn DG (1994b): The cellulose-binding domain of endoglucanase A (CenA) from *Cellulomonas fimi:* Evidence for the involvement of tryptophan residues in binding. Mol Microbiol 11:747–755.

Din N, Gilkes NR, Tekant B, Miller RCJ, Warren RAJ, Kilburn DG (1991): Non-hydrolytic disruption of cellulose fibres by the binding domain of a bacterial cellulase. Bio/Technology 9:1096–1099.

Durrant AJ, Hall J, Hazlewood GP, Gilbert HJ (1991): The noncatalytic C-terminal region of endoglucanase E from *Clostridium thermocellum* contains a cellulose-binding domain. Biochem J 273:289–293.

Felix CR, Ljungdahl LG (1993): The cellulosome—the exocellular organelle of Clostridium. Annu Rev Microbiol 47:791–819.

Fierobe H-P, Bagnara-Tardif C, Gaudin C, Guerlesquin F, Sauve P, Belaich A, Belaich J-P (1993): Purification and characterization of endoglucanase C from *Clostridium cellulolyticum*—Catalytic comparison with endoglucanase A. Eur J Biochem 217:557–565.

Forsberg CW, Gong J, Malburg LMJ, Zhu H, Iyo A, Cheng K-J, Krell PJ, Phillips JP (1993): Cellulases and hemicellulases of *Fibrobacter succinogenes* and their roles in fibre digestion. In Shimada K, Hoshino S, Ohmiya K, Sakka K, Kobayashi Y, Karita S (eds): Genetics, Biochemistry and Ecology of Lignocellulose Degradation. Tokyo, Japan: Uni Publishers Co., pp 125–136.

Fujino T, Béguin P, Aubert J-P (1993a): Organization of a *Clostridium thermocellum* gene cluster encoding the cellulosomal scaffolding protein CipA and a protein possibly involved in attachment of the cellulosome to the cell surface. J Bacteriol 175:1891–1899.

Fujino T, Karita S, Ohmiya K (1993b): Nucleotide sequences of the *celB* gene encoding endo-1,4-β-glucanase-2, ORF1 and ORF2 forming a putative cellulase gene cluster of *Clostridium josui*. J Ferment Bioeng 76:243–250.

Gelhaye E, Petitdemange H, Gay R (1993): Adhesion and growth rate of *Clostridium cellulolyticum* ATCC 35319 on crystalline cellulose. J Bacteriol 175:3452–3458.

Gerngross UT, Romaniec MPM, Kobayashi T, Huskisson NS, Demain AL (1993): Sequencing of a *Clostridium thermocellum* gene (cipA) encoding the cellulosomal S$_L$-protein reveals an unusual degree of internal homology. Mol Microbiol 8:325–334.

Gerwig G, Kamerling JP, Vliegenthart JFG, Morag (Morgenstern) E, Lamed R, Bayer EA (1991): Primary structure of *O*-linked carbohydrate chains in the cellulosome of different *Clostridium thermocellum* strains. Eur J Biochem 196:115–122.

Gerwig G, Kamerling JP, Vliegenthart JFG, Morag (Morgenstern) E, Lamed R, Bayer EA (1992): Novel oligosaccharide constituents of the cellulase complex of *Bacteroides cellulosolvens*. Eur J Biochem 205:799–808.

Gilkes NR, Henrissat B, Kilburn DG, Miller RCJ, Warren RAJ (1991): Domains in microbial β-1,4-glycanases: Sequence conservation, function, and enzyme families. Microbiol Rev 55:303–315.

Gilkes NR, Warren RAJ, Miller RCJ, Kilburn DG (1988): Precise excision of the cellulose-binding domains from two *Cellulomonas fimi* cellulases by a homologous protease and the effect on catalysis. J Biol Chem 263:10401–10407.

Goldstein MA, Takagi M, Hashida S, Shoseyov O, Doi RH, Segel IH (1993): Characterization of the cellulose binding domain of the *Clostridium cellulovorans* cellulose-binding protein A. J Bacteriol 175:5762–5768.

Gong J, Forsberg CW (1989): Factors affecting adhesion of *Fibrobacter succinogenes* subsp. *succinogenes* S85 and adherence-defective mutants to cellulose. Appl Environ Microbiol 55:3039–3044.

Gong J, Forsberg CW (1993): Cellulose binding proteins and their potential roles in adhesion of *Fibrobacter succinogenes* to cellulose. In Shimada K, Hoshino S, Ohmiya K, Sakka K, Kobayashi Y, Karita S (eds): Genetics, Biochemistry and Ecology of Lignocellulose Degradation. Tokyo, Japan: Uni Publishers Co., p 138.

Haigler CH, Weimer PJ (eds) (1991): Biosynthesis and Biodegradation of Cellulose." New York: Marcel Dekker, 694 pp.

Hall J, Hazlewood GP, Barker PJ, Gilbert HJ (1988): Conserved reiterated domains in *Clostridium thermocellum* endoglucanases are not essential for activity. Gene 69:29–38.

Hazlewood GP, Romaniec MPM, Davidson K, Grépinet O, Béguin P, Millet J, Raynaud O, Aubert J-P (1988): A catalogue of *Clostridium thermocellum* endoglucanase, β-glucosidase and xylanase genes cloned in *Escherichia coli*. FEMS Microbiol Lett 51:231–236.

Hoffrén A-M, Teeri TT, Teleman O (1995): Molecular dynamics simulation of fungal cellulose-binding domains: Differences in molecular rigidity but a preserved cellulose binding surface. Protein Eng 8:443–450.

Hon DN-S (1994): Cellulose: A random walk along its historical path. Cellulose 1:1–25.

Hon-nami K, Coughlan MP, Hon-nami H, Ljungdahl LG (1986): Separation and characterization of the complexes constituting the cellulolytic enzyme system of *Clostridium thermocellum*. Arch Biochem Biophys 145:13–19.

Johnson EA, Sakojoh M, Halliwell G, Madia A, Demain AL (1982): Saccharification of complex cellulolosic substrates by the cellullase system from *Clostridium thermocellum*. Appl Environ Microbiol 43:1125–1132.

Kauri T, Kushner DJ (1985): Role of contact in bacterial degradation of cellulose. FEMS Microbiol Ecol 31:301–306.

Kellett LF, Poole DM, Ferreira LMA, Durrant AJ, Hazlewood GP, Gilbert HJ (1990): Xylanase B and an arabinofuranosidase from *Pseudomonas fluorescens* subsp. *cellulosa* contain identical cellulose-binding domains and are coded by adjacent genes. Biochem J 272:369–376.

Kohring S, Weigel J, Mayer F (1990): Subunit composition and glycosidic activities of the cellulase complex from *Clostridium thermocellum* JW20. Appl Environ Microbiol 56:3798–3804.

Kraulis PJ, Clore GM, Nilges M, Jones TA, Pettersson G, Knowles J, Gronenborn AM (1989): Determination of the three-dimensional solution structure of the C-terminal domain of cellobiohydrolase I from *Trichoderma reesei*. A study using nuclear magnetic resonance and hybrid distance geometry-dynamical simulated annealing. Biochemistry 28:7241–7457.

Lamed E, Naimark J, Morgenstern E, Bayer EA (1987a): Scanning electron microscopic delineation of bacterial surface topology using cationized ferritin. J Microbiol Methods 7:233–240.

Lamed E, Naimark J, Morgenstern E, Bayer EA (1987b): Specialized cell surface structures in celluloylytic bacteria. J Bacteriol 169:3792–3800.

Lamed R, Bayer EA (1988a): The cellulosome concept: Exocellular/extracellular enzyme reactor centers for efficient binding and cellulolysis. In Aubert J-P, Béguin P, Millet J (eds): Biochemistry and Genetics of Cellulose Degradation. London: Academic Press, pp 101–116.

Lamed R, Bayer EA (1988b): The cellulosome of *Clostridium thermocellum*. Adv Appl Microbiol 33:1–46.

Lamed R, Bayer EA (1991): Cellulose degradation by thermophilic anaerobic bacteria. In Haigler CH, Weimer PJ (eds): Biosynthesis and Biodegradation of Cellulose and Cellulose Materials. New York: Marcel Dekker, pp 377–410.

Lamed R, Bayer EA (1993): The cellulosome concept—A decade later! In Shimada K, Hoshino S, Ohmiya K, Sakka K, Kobayashi Y, Karita S (eds): Genetics, Biochemistry and Ecology of Lignocellulose Degradation. Tokyo, Japan: Uni Publishers Co., pp 1–12.

Lamed R, Setter E, Bayer EA (1983a): Characterization of a cellulose-binding, cellulase-containing complex in *Clostridium thermocellum.* J Bacteriol 156:828–836.

Lamed R, Setter E, Kenig R, Bayer EA (1983b): The cellulosome—A discrete cell surface organelle of *Clostridium thermocellum* which exhibits separate antigenic, cellulose-binding and various cellulolytic activities. Biotechnol Bioeng Symp 13:163–181.

Lamed R, Tormo J, Chirino AJ, Morag E, Bayer EA (1994): Crystallization and preliminary x-ray analysis of the major cellulose-binding domain of the cellulosome from *Clostridium thermocellum.* J Mol Biol 244:236–237.

Lemaire M, Ohayon H, Gounon P, Fujino T, Béguin P (1995): OlpB, a new outer layer protein of *Clostridium thermocellum,* and binding of its S-layer-like domains to components of the cell envelope. J Bacteriol 177:2451–2459.

Linder M, Mattinen ML, Kontteli M, Lindeberg G, Ståhlberg J, Drakenberg T, Reini-kainen T, Pettersson G, Annila A (1995): Identification of functionally important aminoacids in the cellulose-binding domain of *Trichoderma reesei* cellobiohydrolase I. Prot Sci 4:1056–1064.

Mayer F, Coughlan MP, Mori Y, Ljungdahl LG (1987): Macromolecular organization of the cellulolytic enzyme complex of *Clostridium thermocellum* as revealed by electron microscopy. Appl Environ Microbiol 53:2785–2792.

Minato H, Mitsumori M, Cheng K-J (1993): Attachment of microorganisms to solid substrates in the rumen. In Shimada K, Hoshino S, Ohmiya K, Sakka K, Kobayashi Y, Karita S (eds): Genetics, Biochemistry and Ecology of Lignocellulose Degrada-tion. Tokyo, Japan: Uni Publishers Co., p 138.

Miron J (1991): The hydrolysis of lucerne cell-wall monosaccharide components by monocultures or pair combinations of defined ruminal bacteria. J Appl Bacteriol 70:245–252.

Miron J, Yokohama MT, Lamed R (1989): Bacterial cell surface structures involved in lucerne cell walls degradation by pure cultures of cellulolytic rumen bacteria. Appl Microbiol Biotechnol 32:218–222.

Morag (Morgenstern) E, Bayer EA, Lamed R (1991): Anomalous dissociative behavior of the major glycosylated component of the cellulosome of *Clostridium thermocel-lum.* Appl Biochem Biotechnol 30:129–136.

Morag E, Bayer EA, Lamed R (1990): Relationship of cellulosomal and noncelluloso-mal xylanases of *Clostridium thermocellum* to cellulose-degrading enzymes. J Bacte-riol 172:6098–6105.

Morag E, Bayer EA, Lamed R (1992a): Affinity digestion for the near-total recovery of purified cellulosome from *Clostridium thermocellum.* Enzyme Microb Technol 14:289–292.

Morag E, Bayer EA, Lamed R (1992b): Unorthodox intra-subunit interactions in the cellulosome of *Clostridium thermocellum.* Appl Biochem Biotechnol 33:205–217.

Morag E, Lapidot A, Govorko D, Lamed R, Wilchek M, Bayer EA, Shoham Y (1995): Expression, purification and characterization of the cellulose-binding domain of the scaffoldin subunit from the cellulosome of *Clostridium thermocellum.* Appl Environ Microbiol 61:1980–1986.

Morris EJ (1988): Characteristics of the adhesion of *Ruminococcus albus* to cellulose. FEMS Microbiol Lett 51:113–118.

Morris EJ, Cole OJ (1987): Relationship between cellulolytic activity and adhesion to cellulose in *Ruminococcus albus*. J Gen Microbiol 133:1023–1032.

Navarro A, Chebrou M-C, Béguin P, Aubert J-P (1991): Nucleotide sequence of the cellulase gene *celF* of *Clostridium thermocellum*. Res Microbiol 142:927–936.

Pagès S, Belaich A, Tardif C, Reverbel-Leroy C, Gaudin C, Belaich J-P (1996): Interaction between the endoglucanase CelA and the scaffolding CelC of the *Clostridium cellulolyticum* cellulosome. J Bacteriol (in press).

Poole DM, Durrant AJ, Hazelwood GP, Gilbert HJ (1991): Characterization of hybrid proteins consisting of the catalytic domains of *Clostridium* and *Ruminococcus* endoglucanases, fused to *Pseudomonas* non-catalytic cellulose-binding domains. Biochem J 279:787–792.

Poole DM, Hazelwood GP, Huskisson NS, Virden R, Gilbert HJ (1993): The role of conserved tryptophan residues in the interaction of a bacterial cellulose binding domain with its ligand. FEMS Microbiol Lett 106:77–84.

Poole DM, Morag E, Lamed R, Bayer EA, Hazlewood GP, Gilbert HJ (1992): Identification of the cellulose binding domain of the cellulosome subunit S1 from *Clostridium thermocellum*. FEMS Microbiol Lett 99:181–186.

Reinikainen T, Ruohonen L, Nevanen T, Laaksonen L, Kraulis P, Jones TA, Knowles JKC, Teeri TT (1992): Investigation of the function of mutated cellulose-binding domains of *Trichoderma reesei* cellobiohydrolase I. Proteins 14:475–482.

Reinikainen T, Teleman O, Teeri TT (1995): Effects of pH and high ionic strength on the adsorption and activity of native and mutated cellobiohydrolase I from *Trichoderma reesei*. Proteins 22:392–403.

Roger V, Fonty G, Komisarczuk-Bony S, Gouet P (1990): Effects of physicochemical factors on the adhesion of cellulose Avicel of the ruminal bacteria *Ruminococcus flavefaciens* and *Fibrobacter succinogenes* subsp. *succinogenes*. Appl Environ Microbiol 56:3081–3087.

Salamitou S, Lemaire M, Fujino T, Ohayon H, Gounon P, Béguin P, Aubert J-P (1994a): Subcellular localization of *Clostridium thermocellum* ORF3p, a protein carrying a receptor for the docking sequence borne by the catalytic components of the cellulosome. J Bacteriol 176:2828–2834.

Salamitou S, Raynaud O, Lemaire M, Coughlan M, Béguin P, Aubert J-P (1994b): Recognition specificity of the duplicated segments present in *Clostridium thermocellum* endoglucanase CelD and in the cellulosome-integrating protein cipA. J Bacteriol 176:2822–2827.

Salamitou S, Tokatlidis K, Béguin P, Aubert J-P (1992): Involvement of separate domains of the cellulosomal protein S1 of *Clostridium thermocellum* in binding to cellulose and in anchoring of catalytic subunits to the cellulosome. FEBS Lett 304:89–92.

Shoseyov O, Takagi M, Goldstein MA, Doi RH (1992): Primary sequence analysis of *Clostridium cellulovorans* cellulose binding protein A. Proc Natl Acad Sci USA 89:3483–3487.

Strobel HJ, Caldwell FC, Dawson KA (1995): Carbohydrate transport by the anaerobic thermophile *Clostridium thermocellum* LQRI. Appl Environ Microbiol 61:4012–4015.

Tokatlidis K, Dhurjati P, Béguin P (1993): Properties conferred on *Clostridium thermocellum* endoglucanase CelC by grafting the duplicated segment of endoglucanase CelD. Prot Eng 6:947–952.

Tokatlidis K, Salamitou S, Béguin P, Dhurjati P, Aubert J-P (1991): Interaction of the duplicated segment carried by *Clostridium thermocellum* cellulases with cellulosome components. FEBS Lett 291:185–188.

Tomme P, Warren RAJ, Gilkes NR (1995): Cellulose hydrolysis by bacteria and fungi. Adv Microb Physiol 37:1–81.

Tormo J, Lamed R, Chirino AJ, Morag E, Bayer EA, Shoham Y, Steitz TA (1996): Crystal structure of a bacterial family-III cellulose-binding domain: A general mechanism for attachment to cellulose. Submitted.

Viljoen JA, Fred EB, Peterson WH (1926): The fermentation of cellulose by thermophilic bacteria. J Agric Sci 16:1–17.

Wilson CA, Wood TM (1992): The anaerobic fungus *Neocallimastix frontalis:* Isolation and properties of a cellulosome-type enzyme fraction with the capacity to solubilize hydrogen-bond-ordered cellulose. Appl Microbiol Biotechnol 37:125–129.

Wu JHD, Demain AL (1988): Proteins of the *Clostridium thermocellum* cellulase complex responsible for degradation of crystalline cellulose. In Aubert J-P, Béguin P, Millet J (eds): Biochemistry and Genetics of Cellulose Degradation. London: Academic Press, pp 117–131.

Xu G-Y, Ong E, Gilkes NR, Kilburn DG, Muhandiram DR, Harris-Brandts M, Carver JP, Kay LE, Harvey TS (1995): Solution structure of a cellulose-binding domain from *Cellulomonas fimi* by nuclear magnetic resonance spectroscopy. Biochemistry 34:6993–7009.

Yaron S, Morag E, Baker EA, Lamed R, Shoham Y (1995): Expression, purification and subunit-binding properties of cohesins 2 and 3 of the *Clostridium thermocellum* cellulosome. FEBS Lett 360:121–124.

PSEUDOMONAS AERUGINOSA: VERSATILE ATTACHMENT MECHANISMS

ALICE PRINCE

Department of Pediatrics, College of Physicians & Surgeons, Columbia University, New York, New York 10032

7.1. INTRODUCTION

Pseudomonas aeruginosa are versatile organisms that flourish under diverse conditions. As a human pathogen, *P. aeruginosa* may be associated with infections in immunocompromised patients, superinfection of burn wounds, and corneal trauma and with the chronic lung disease characteristic of cystic fibrosis. In each of these settings, it is necessary for environmental strains of bacteria to attach to a host receptor and rapidly adapt to the selective pressures imposed by the host immune response. Even within a single anatomic site, it may be necessary for the organism to alter its surface properties to adapt to a given environmental niche as the local conditions are modified by the presence of phagocytes, antibodies, neutrophil products such as oxygen radicals and superoxide, as well as antimicrobial agents.

   P. aeruginosa have a large repertoire of virulence genes that undoubtedly contribute to its success as an opportunistic pathogen. The selective expression of specific adhesins in response to given environmental demands is an important mechanism that allows the organism to proliferate, particularly in the lung under conditions that facilitate clearance of most other pathogens. Within the airway there are chiefly three types of receptors for *P. aeruginosa* important in the pathogenesis of human infection: the epithelial cell, respiratory mucins,

Bacterial Adhesion: Molecular and Ecological Diversity, pages 183–199
© *1996 Wiley-Liss, Inc.*

and phagocytic cells. In this review, the data describing the *P. aeruginosa* ligands for each of these classes of receptors are discussed.

7.2. ADHESINS ASSOCIATED WITH INFECTION OF MUCOSAL SURFACES

Much of the data describing the ligand–receptor interactions between *P. aeruginosa* and the airway epithelium have been derived from studies of cystic fibrosis patients. There has been a great deal of interest in the pathogenesis of *P. aeruginosa* infection in the lungs of patients with cystic fibrosis (CF). This is the most common lethal genetic disease of Caucasians, and virtually all CF patients eventually succumb to *P. aeruginosa*–related pulmonary failure. The relationship between this specific pathogen and patients who have a mutation in an epithelial chloride channel, the CFTR (Welsh and Smith, 1993), provides an example of the specificity of the ligand–receptor relationship and the diverse adhesins that may be associated with different stages of clinical pulmonary disease.

Histologic studies have demonstrated that *P. aeruginosa* associate with ciliated epithelial cells via polar ligands (Irvin et al., 1989). However, in early studies of animal models of *Pseudomonas* infection, few bacteria were found to be adherent unless the respiratory epithelium was injured by acid or viral infection (Ramphal and Pyle, 1983). Studies of human pathology, looking at the distribution of organisms in the infected lung, demonstrated that large numbers of bacteria were enmeshed in mucin within secretions of the airway lumen. Few, if any, organisms were found in contact with the intact epithelial surface but were aggregated at areas of injured epithelium (Baltimore et al., 1989).

To account for these observations, it has been postulated that transiently inspired organisms infrequently find epithelial receptors and most often are removed by normal mucociliary clearance mechanisms, as occurs in the normal host. These organisms are usually motile, piliated, and express smooth lipopolysaccharide. However, under specific conditions bacteria do occasionally find epithelial receptors and establish a nidus of infection. The organisms grow in microcolonies, forming a biofilm within the airway surrounded by their own secretions, as well as the accumulated inflammatory products from the host airways. In CF, *P. aeruginosa* with a particular phenotype overexpressing alginate, a polymer of guluronic and mannuronic acid, are often selected in response to the environmental conditions within the lung, particularly osmolarity and general stress, and may predominate in chronic infection. Organisms associated with these chronic infections are found to have flagellar mutations and rough lipopolysaccharide, and do not express pili (Hancock et al., 1983, Mahenthiringham et al., 1994). Thus, the pathogenesis of *P. aeruginosa* infection within the lung includes several stages of infection: ligands that recognize mucin components, which serve to clear inspired organisms as would occur in normal hosts; epithelial receptors that trap organisms, and mucin components that surround and bind organisms adapted for chronic infection within the lung.

7.2.1. Pilin Binding to Asialylated Glycolipids

P. aeruginosa pili (fimbriae) were first recognized by Woods and coworkers (1985) to mediate attachment to buccal epithelial cells. Pili are polar filaments consisting of repeated protein subunits of 14,000 to 18,000 molecular weight with a characteristic *N*-methyl-phenylalanine (NMetPhe) residue at the N terminus as well as conserved hydrophobic N-terminal sequences. Several loci are involved in pilus biosynthesis: *pilA,* the structural gene, and associated loci *pilBCD,* thought to be involved in pilus assembly (Nunn et al., 1990); *pilT,* which is associated with a phenomenon called "twitching motility" (Whitchurch et al., 1991); a σ_{54} RNA polymerase initiation factor, RpoN (Totten et al., 1990), and a two-component regulatory system, PilR and PilS, predicted to function as a sensor activated by environmental signals that in turn activates PilR via kinase activity (Hobbs et al., 1993). PilR then activates transcription of *pilA.* PilD (*xcpA*) functions in the export of pilin subunits (Bally et al., 1992). Mapping studies with PAO1538 suggest that these genes are in the 72 to 75 minute region of the chromosome along with other housekeeping genes (Hobbs et al., 1993). A recently described gene, *pilO,* may be involved in the glycosylation of certain strains of *P. aeruginosa* pili (Castric, 1995).

The antigenically dominant portion of pilin is in the central region, while the C terminus contains a disulfide loop encompassing an epithelial-binding domain that varies significantly among strains (Irvin et al., 1989). These NMetPhe or type 4 pili are common to *Vibrio cholera, Neisseria gonorrhoea, Neisseria meningitidis* (all human mucosal pathogens), *Bacteroides nodosus,* and *Moraxella bovis.* There is significant antigenic diversity among different *P. aeruginosa* pili. Based on the epitopes recognized by antipilin antibodies, *P. aeruginosa* are roughly divided into two pilin types: group I, which includes common laboratory strains that have been genetically well characterized such as PAO1, PAK, and PA103; and group II, which contains numerous other strains as well as PA1244, originally isolated from a burn wound (Castric, 1995). However, the adhesin domain is relatively heterogeneous with a few critical shared components. Antibody specific to one pilus type does not necessarily recognize the binding epitope of pili from unrelated strains (Saiman et al., 1989). Biochemical and genetic data suggest that PA1244 pilin is glycosylated, sharing the carbohydrate designated by the O side chain of lipopolysaccharide (Castric, 1995). However, the site of glycosylation on the pilin polypeptide and its biological significance are not known.

Pili recognize carbohydrate sequences available on asialylated glycolipids, such as asialoGM binding to a GalNAcβ1–4Gal moiety, which is normally sialylated in gangliosides (Saiman and Prince 1993; Sheth et al., 1994). Pili are thus responsible for the binding of *P. aeruginosa* to various asialylated glycolipids, as was initially demonstrated by solid-phase binding studies to glycolipids separated by thin layer chromatography (Krivan et al., 1988). The pilin adhesin domain in strains PAO and PAK has been mapped to residues 128–144 of the pilin subunit (Sheth et al., 1994). Interruption of this region by the insertion of a chloramphenicol determinant, while maintaining the structural integrity of the pilus, abrogates binding to the GalNAcβ1–4 receptor (Farinha et al., 1994). This carbohydrate receptor is not commonly available on mucosal surfaces of the human airway (Saiman et al., 1992; Saiman and Prince, 1993; Imundo et al., 1995), as the galactose moiety is usually sialylated. *P. aeruginosa* adherence *in vitro* to tracheal explants or to human epithelial cells, both primary or transformed cell lines, is minimal. However, patients with cystic fibrosis have more asialylated receptors available for *P. aeruginosa* adherence (Saiman and Prince, 1993; Imundo et al., 1995). Respiratory epithelial cells from patients who are homozygous for the most common and severe ΔF508 mutation have significantly more receptors for *P. aeruginosa* binding than cells from patients with normal CFTR (Zar et al., 1995) (Fig. 7.1).

The biological relevance of pilin-mediated binding to respiratory epithelial cells was suggested in recent studies that demonstrate that *P. aeruginosa* attachment can initiate an inflammatory response by respiratory epithelial cells. Isolated pili and piliated *Pseudomonas* elicit the production of interleukin-8 (IL-8) by respiratory epithelial cells (DiMango et al., 1995) (Fig. 7.2). IL-8 is a major neutrophil chemokine, attracting neutrophils to the site of infection on the airway epithelium and initiating the inflammatory response with elaboration of elastase and superoxide, which cause symptomatic pulmo-

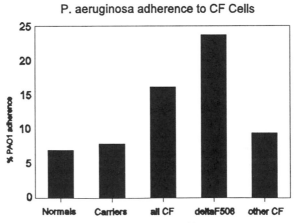

Fig. 7.1. The percent adherence of an inoculum of ^{35}S-labeled PAO1 to nasal epithelial cells in primary culture from normal control subjects, parents of children with CF (carriers), CF patients homozygous for the most common CFTR mutation (ΔF508), and CF patients with other CFTR mutations.

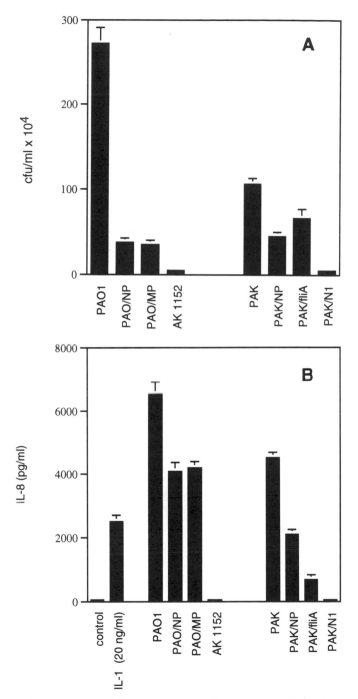

Fig. 7.2. **A:** A comparison of the adherence of *P. aeruginosa* strains to a monolayer of 1HAEo cells, an SV40-transformed epithelial cell line. PAO1, wild type; PAO/NP, *pilA;* PAO/MP, modified pilin adhesin; AK1152, PAO Pil⁻FLa⁻. PAK, wild type; PAK/NP, *PilA;* PAKfliA, Fla⁻; PAKN1, *rpoN,* Pil Fla⁻. **B:** IL-8 produced by the 1HAEo cells, measured 18 h after a 60 min incubation with the bacteria. (Reproduced from DiMango et al., 1995, with permission of the publisher.)

nary disease in CF patients. Thus, the earliest interaction of piliated *P. aeruginosa* and specific asialylated epithelial glycoconjugates is sufficient to trigger inflammation, even prior to interactions between the bacteria and professional phagocytic cells, such as alveolar macrophages or polymorphonuclear leukocytes. It is of note that clinical studies of infants with CF demonstrate significant amounts of IL-8 and inflammatory cells very early in the course of *P. aeruginosa* pulmonary infection (Khan et al., 1995).

The contribution of pili to the initial stages of infection was tested in a mouse model of acute *Pseudomonas* pneumonia (Tang et al., 1995). The most virulent strains, that is, those that readily caused acute pneumonia and septicemia, were piliated. The inoculation of purified pili alone was found to elicit a significant inflammatory response *in vivo,* consistent with the *in vitro* evidence that pili trigger an IL-8 response. However, *pil* mutants were also capable of causing both pneumonia and bacteremia. Although *pil* mutants were less virulent than their respective parental strains, some *pil* mutants were more virulent than unrelated piliated *P. aeruginosa* strains lacking expression of other virulence factors that apparently facilitate colonization of the lungs. The virulence of *pilA* mutants with interrupted adhesin domains (PAO-MP) was tested in a mouse model of intraperitoneal sepsis, demonstrating an increase in the LD_{50} by almost two orders of magnitude (Farinha et al., 1994). These animal experiments suggest that pilin-mediated adherence is important in pathogenesis but that other *P. aeruginosa* virulence factors must also be capable of mediating attachment and contributing to the virulence of the organisms at different anatomic sites.

7.2.2. Nonpilus Adhesins

Several lines of evidence support a role for nonpilus adhesins of *P. aeruginosa* in binding to epithelial receptors. Studies using monoclonal antipilin antibody (Saiman et al., 1989) or *pil* mutants suggest that approximately 50% of the total binding to epithelial cells is pilin independent. *Pseudomonas* lipopolysaccharide may account for some of this binding, as experiments performed with radiolabeled lipopolysaccharide demonstrated binding to asialoGM1 (Gupta et al., 1994). Outer membrane proteins and flagellar genes have been suggested to account for nonpilus-mediated adherence, although specific receptors have not been defined. A *fliO* mutant is significantly impaired in attachment to A549 cells (as well as to mucins) (Simpson, 1995a). Since mutants lacking either the flagellar structural gene *fliC* or *fliA,* the σ_{28} necessary for *fliC* expression, were able to adhere to A549 cells under these conditions, it is unlikely that flagellin itself is the ligand responsible for binding (Simpson et al., 1992). The function of FliO is currently unknown. It is possible that flagellin may contribute to epithelial binding in addition to the role of functional flagella in motility.

7.2.3. Additional Epithelial Receptors for Nonpilus Adhesins

In addition to pilus recognition of a disaccharide, there is also evidence for cholesterol and cholesterol esters functioning as receptors for *P. aeruginosa* (Rostand and Esko, 1993). In these studies Chinese hamster ovary (CHO) cells and CHO mutants defective in the synthesis of *O*-linked and *N*-linked carbohydrates were compared in binding assays with several clinical *P. aeruginosa* strains. *Pseudomonas* adherence to these cells was diminished by ethanol, and the associated receptor was identified to be cholesterol and cholesterol esters. These findings suggest that *P. aeruginosa* can use different ligands to attach to various substrates depending on the composition of the cell membranes. As cholesterol is found in regenerating membranes, as well as in purulent airway secretions, attachment to lipid receptors may also be involved in certain stages of pathogenesis. Epithelial regeneration is a well-described component of pulmonary disease (Baltimore et al., 1989), and these additional receptors may be important after the initial infecting organisms establish a nidus of infection and secrete damaging exoproducts.

The accumulated data support the hypothesis that *P. aeruginosa* binds to epithelial glycolipids primarily by pilus-mediated mechanisms and that additional ligands and receptors are involved secondarily. The interactions between *P. aeruginosa* and respiratory mucins are less well defined. As pilin does not appear to recognize mucin-derived receptors, there has been a substantial effort to identify both the *P. aeruginosa* ligand and the mucin glycoprotein(s) that mediate mucin binding.

7.2.4. *P. aeruginosa* Ligands for Mucin

Transiently inhaled *P. aeruginosa* are efficiently removed from the airways of normal hosts by mucociliary action. This process implies binding of the organisms to human respiratory mucins that are then cleared by the concerted action of ciliated airway cells. Components of human mucin, *n*-acetylglucosamine and *n*-acetylneuraminic acid, were suggested to be constituents of the mucin receptor (Vishwanath and Ramphal, 1985). More recent studies identified carbohydrate chains containing Galβ1–3GlcNAc or Galβ1–4GlcNAc disaccharides as receptors for nonpiliated mutants of *P. aeruginosa*, PA1244 and PAK, and sialylation (α2,6 linkage)–inhibited binding (Ramphal et al., 1991). These carbohydrate chains are particularly abundant in mucins and thus may represent biologically important ligands.

However, other investigators have not found compelling evidence for mucin receptors for *P. aeruginosa*. In a study of the binding kinetics of *P. aeruginosa* to highly purified human intestinal and respiratory mucin there was no evidence of specific binding to mucin, nor did the data support the concept that mucins compete for bacterial ligands on the respiratory epithelial cells (Sajjan et al., 1992a). Similar conclusions were reached from studies comparing

Pseudomonas attachment to matched human epithelial cell lines in which adherence to a mucin-producing epithelial cell line was less than to the nonmucin-producing line (Rostand and Esko, 1993). These observations contrast with the demonstration of mucin adherence associated with *B. cepacia*, which bind with high affinity and specificity for mucin components (Sajjan et al., 1992b). In studies comparing the adherence of *P. aeruginosa* to CF and normal mucins, adherence to CF glycoconjugates was actually less than to the secretions from normal subjects (Ramphal et al., 1989). Failure to bind mucin and subsequent failure of mucociliary clearance mechanisms may allow persistence of bacteria in the airway lumen. However, this conjecture requires direct testing in animal model systems as well as studies to define the *Pseudomonas* ligand for mucin and composition of the corresponding mucin glycoconjugate receptor.

7.2.4.1. A Genetic Analysis of P. aeruginosa Ligands for Mucin

A genetic approach was used to identify *P. aeruginosa* loci associated with adherence to mucin and/or epithelial receptors. Strains with mutations in either *pilA* or *rpoN*, which is required for the transcription of several genes, including *pilA* and *fliC*, were mutagenized with a miniTn5 derivative selecting for mutants entirely unable to adhere to either epithelial cells or mucins. Two classes of mutants were identified, one group that was defective in attachment to mucin, and a second group in which adherence to both mucin and epithelial cells appeared to be mediated by the same nonpilus adhesin (Simpson et al., 1992). Complementation experiments revealed several loci within flagellar genes that are involved in nonpilus-mediated mechanisms of attachment (Simpson et al., 1995a; Ritchings et al., 1995). *fliO*, a gene purported to be involved in flagellar synthesis, was demonstrated to be necessary for motility and adherence to both mucin and A549 cells (Simpson et al., 1995a). Neither the structural gene for flagella, *fliC*, nor the flagellar σ factor, FliA, was necessary for adherence to mucin, although other flagellar proteins that are regulated by FleS and FleR, flagellar response regulators, are essential for motility and adhesion to mucin (Ritchings et al., 1995). It remains unclear exactly which flagellar component is actually mediating attachment, whether it is a flagellar component on the cell surface or an unrelated and unidentified gene product that uses the flagellar machinery for transport to the cell surface. As mucins present a diverse array of carbohydrate structures, it has not yet been demonstrated exactly what mucin component acts as the receptor for the flagellar-associated ligand or if the receptor is one of the disaccharides described above.

The biological role of these flagellar genes in virulence has not been tested directly. Although the flagellar structural gene *fliC* is not involved in mediating adherence to mucin, Pil⁻Fla⁻ mutants are significantly less virulent in a neonatal mouse model of pneumonia (Tang et al., 1995). As shown in Figure 7.3, loss of the flagella renders the organism avirulent, as Fla⁻ mutants were not associated with any mortality and significantly less pneumonia than attributed to the parental strain. The *fliC* mutant and *pilAfliC* were virtually identical in virulence, and both caused significantly less pneumonia than even a *pilA*

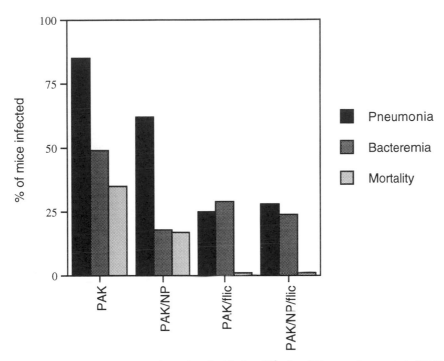

Fig. 7.3. Neonatal mice were inoculated with 5×10^8 cfu of *P. aeruginosa* strain PAK (wild type), PAK/NP, (*pilA*), PAK/fliC, (Fla⁻), or PAK/NP/*fliC* (Pil⁻Fla⁻), and relative rates of pneumonia, bacteremia, and mortality were compared.

mutant. Thus, motility and/or the presence of flagellin appears to contribute to the virulence of the organism, although the mechanism involved is not entirely clear.

7.3. ADHESINS THAT MEDIATE ATTACHMENT TO PHAGOCYTIC CELLS

An additional component in virulence is the ability of the host to clear potential pathogens from mucosal surfaces. This process usually requires opsonization of the organisms, either with specific antibody or complement and professional phagocytic cells. In the initial stages of airway infection in an immunologically naive host, *P. aeruginosa* may be present in the airway in the absence of opsonic antibody. Pilus-deficient mutants of *P. aeruginosa* strain PAO1 were less susceptible to nonopsonic phagocytosis by macrophages, stimulated by fibronectin, than wild-type piliated strains, suggesting that the pilus can act as a ligand for macrophage uptake (Kelly et al., 1989). Besides pili, flagella are necessary to expedite ingestion and killing of the organism by either alveolar macrophages or polymorphonuclear leucocytes (Speert, 1988; Mahenthiralingam and Speert, 1995). Phagocytosis of unopsonized *P. aeruginosa* by macrophages appears to be mediated by mannose receptors (Speert et al., 1988), and the flagella functions as a ligand required for ingestion by phagocytes.

The expression of flagellar adhesins and/or motility itself is a variable phenotype. Although infecting strains of *P. aeruginosa* are usually motile, clinical isolates of *P. aeruginosa* from CF patients with chronic infection often are not motile and do not express functional flagella (Mahenthiralingam et al., 1994). Longitudinal genetic analysis of such isolates cultured from individual patients has demonstrated the deletion of segments of the *P. aeruginosa* genome resulting in an *rpoN*-type phenotype (Simpson, 1995b). The selection of such Fla⁻ mutants *in vivo* may represent a mechanism to avoid phagocytic clearance, mediated by the flagellar ligand. It is not known if the Fla⁻ mutants selected *in vivo* are also less adherent to mucin, but their ability to escape phagocytic clearance is probably a significant benefit leading to their selection and persistence. Organisms that adhere to mucin via these flagellar-associated ligands may be more efficiently removed by normal mucociliary clearance mechanisms; thus these Fla⁻ mutants may be selected *in vivo* since they are less efficiently cleared.

7.4. ALGINATE AS AN ADHESIN

Phenotypic mutants of *P. aeruginosa* isolated from the airways of CF patients with long-standing infection are often visibly different than the isolates from early stages of disease in that they express copious amounts of an exopolysaccharide, alginate. This is a heterogeneous polymer composed of D-mannuronic acid and its 5′ epimer L-guluronic acid. Mucoid strains have undergone mutations in the *muc* loci (Fyfe and Goran, 1980). These include *algU* (Yu, 1995) (*algT*), an analog of *Escherichia coli* σ^E, as well as a number of other regulatory genes involved in the expression of *algD,* which encodes GDP mannose dehydrogenase and initiates the enzymatic reactions that result in the production of alginate. Transcription of *algD* and associated loci are modulated by environmental factors such as osmolarity and stress (Deretic et al., 1989; Yu et al., 1995). Although initial reports suggested that alginate from clinical isolates of *P. aeruginosa* mediated adherence to buccal cells, significant diversity was observed among the adherence attributed to the alginate produced by specific strains (Doig et al., 1987). Scanning electron micrographs characterizing the interactions of mucoid organisms and tracheal cell monolayers revealed clusters of microcolonies (Hata and Fick, 1991) similar to what has been described in human autopsy studies. Alginate may serve a more important function as an interbacterial adhesin than as a ligand for host tissues. It may facilitate the formation of microcolonies and biofilms, which protect at least part of the bacterial population from exposure to the immune surveillance system, and act as a barrier to phagocytosis.

The respiratory tract represents a complex environment providing receptors that allow for colonization, bacterial replication, and the eventual selection of organisms proliferating in microcolonies and biofilms. Bacterial adaptation to this milieu is a dynamic process with the stepwise selection of both phenotypic and genotypic mutants that express specific adhesins to facilitate persistence and avoid physical and immune-mediated clearance mechanisms. This flexibility undoubtedly contributes to the success of *P. aeruginosa* as an opportunistic pathogen at additional sites.

7.5. *P. AERUGINOSA* ADHESINS AND RECEPTORS IN THE CORNEA

Extrapolation of the adherence data characterizing *P. aeruginosa* infection in the respiratory tract to the eye again demonstrates the versatility of the organism. *Pseudomonas* corneal infections occur as a complication of damage to the cornea usually from foreign bodies, including contact lenses, injury, and surgical procedures. AsialoGm1 has been postulated to be the relevant receptor in the damaged cornea, as determined by immunofluorescence techniques and binding studies to scarified corneas (Hazlett et al., 1993). *P. aeruginosa* pili and lipopolysaccharide were shown to bind to purified asialoGM1, as well as to neutral glycolipids isolated from bovine corneas that migrated to a position similar to that of an asialoGM1 standard (Gupta et al., 1994). However, other investigators have been unable to demonstrate asialoGM1 in either rabbit or human corneas (Zhao and Panjwani, 1995). Several neolactoglycosphingolipids from regenerating corneal epithelial surfaces have been identified, and such neutral lipids from the cornea may serve as receptors (Panjwani et al., 1995). As *P. aeruginosa* do not adhere to the undamaged cornea, it is plausible that the relevant receptors are only available at sites of regenerating tissue. The pathogenesis of *P. aeruginosa* corneal infections may be somewhat distinct from that of the mucosal infections discussed above. Recent studies suggest that *P. aeruginosa* actually invade corneal cells and can replicate intracellularly (Fleiszig et al., 1995). These organisms may not interact with superficial receptors in the same manner as has been postulated for the respiratory epithelium.

7.6. STRATEGIES TO BLOCK *P. AERUGINOSA* ADHERENCE

Particularly in CF, the inflammation associated with even low numbers of adherent *P. aeruginosa* initiates clinically significant symptoms and is associated with pathological changes in the lung (Khan et al., 1995). Thus, it would be useful to develop strategies to prevent or at least delay the acquisition of the organism in the respiratory tract. As pilin-mediated adherence appears to be important in the early stages of infection and elicits significant amounts of inflammation, agents to block pilin binding to GalNAcβ1–4Gal receptors may be clinically beneficial. Dextran, a polymer of D-glucose, has been shown to block effectively *in vitro* adherence of *P. aeruginosa* to epithelial cells. At clinically achievable levels (5 to 10 mM), dextran-blocked *P. aeruginosa* attachment to transformed epithelial cell lines, demonstrating an effect that was specific for piliated organisms (Fig. 7.4). In addition, dextran blocked PAO adherence to primary cultures of nasal polyp cells and blocked the epithelial IL-8 response associated with *Pseudomonas* binding. In a neonatal mouse model of pulmonary infection, 10-day-old mice exposed to an aerosol of dextran for 15 min failed to develop pneumonia, whereas control littermates became infected (Fig. 7.5). These types of experiments may lead to the development of clinically useful therapies to block acquisition of *P. aeruginosa* from the environment based on the identification of the specific ligand–receptor interactions that initiate infection.

The binding of Pil⁻ and Fla⁻ mutants in the presence of dextran

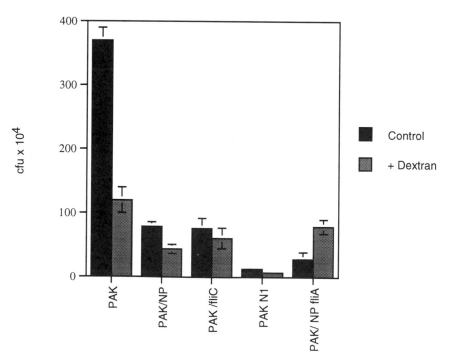

Fig. 7.4. A comparison of the binding of *P. aeruginosa* strains under control conditions or in the presence of 5 mM dextran. PAK, wild type; PAK/NP, Pil⁻; PAK/*fliC*, Fla⁻; PAKN1 (*rpoN*), Fla⁻Pil⁻; PAK/NP/*fliA*, Fla⁻Pil⁻.

The protective effect of dextran in a mouse model of pulmonary infection

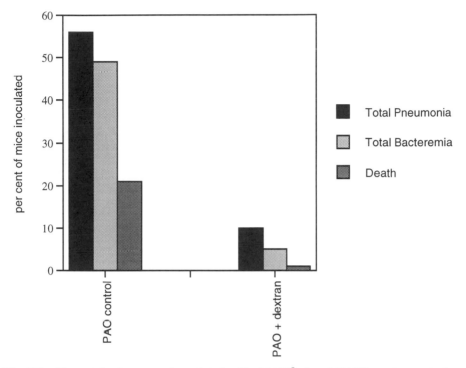

Fig. 7.5. Neonatal mice were inoculated with 5×10^8 cfu of PAO1 under control conditions (n = 33) or after the mice were exposed for 15 min to an aerosol of 5 mM dextran (n = 21). The animals were sacrificed 24 h later, and rates of pneumonia (defined both histologically and by culture of PAO1 from the lung), bacteremia (defined by positive splenic cultures), or death was compared, demonstrating a statistically significant decrease protective effect ($P < 0.01$, Mann-Whitney).

7.7. CONCLUSIONS

P. aeruginosa flourish as opportunistic pathogens in a variety of clinical settings. This is due to their large repertoire of metabolic and virulence-related genes, which allow them to proliferate in environments that are hostile to many other bacteria. The ability to express numerous different adhesins and the heterogeneity in the structure and antigenicity within individual classes of adhesins not only allow infection to occur on different substrates, but enable *P. aeruginosa* to avoid several levels of immune recognition. Unlike other less flexible organisms, *P. aeruginosa* can readily adapt to environmental stresses: lack of nutrients or iron; the presence of complement, antibody, or phagocytes undergoing both phenotypic and genotypic selection. Thus, from even a single anatomic site, *P. aeruginosa* isolates may express flagellar adhesins, pili, and lipopolysaccharide that mediate attachment to epithelial receptors or may be mucoid, nonpiliated, and Fla⁻ enmeshed within a biofilm. This level of diversity will ensure the presence of *P. aeruginosa* as significant human pathogens for the foreseeable future.

REFERENCES

Bally M, Filloux A, Akrim M, Ball G, Lazdunski A, Thomassen J (1992): Protein secretion in *Pseudomonas aeruginosa:* Characterization of seven *xcp* genes and processing of secretory apparatus components by prepilin peptidase. Mol Microbiol 6:1121–1131.

Baltimore RS, Christie CDC, Walker-Smith GJ (1989): Immunohistopathologic localization of *Pseudomonas aeruginosa* in lungs from patients with cystic fibrosis. Am Rev Respir Dis 140:1650–1661.

Castric P (1995): *pilO,* a gene required for glycosylation of *Pseudomonas aeruginosa* 1244 pilin. Microbiology 141:1247–1254.

Deretic V, Dikshit R, Konyecsni M, Misra T, Chakrabarty AM, Misra KT (1989): The algR gene which regulates mucoidy in *Pseudomonas aeruginosa* belongs to a class of environmentally responsive genes. J Bacteriol 171:1278–1283.

DiMango ED, Zar H, Bryan R, Prince A (1995): Diverse *Pseudomonas aeruginosa* gene products stimulate respiratory epithelial cells to produce interleukin-8. J Clin Invest 96:2204–2210.

Doig P, Smith NR, Todd T, Irvin RT (1987): Characterization of the binding of *Pseudomonas aeruginosa* alginate to human epithelial cells. Infect Immun 55:1517–1522.

Farinha MA, Conway BD, Glasier LMG, Ellert NW, Irvin RT, Sherburne R, Paranchych W (1994): Alteration of the pilin adhesin of *Pseudomonas aeruginosa* PAO results in normal pilus biogenesis but loss of adherence to human pneumocyte cells and decreased virulence in mice. Infect Immun 62:4118–4123.

Fleiszig SMJ, Zaidi TS, Pier GB (1995): *Pseudomonas aeruginosa* invasion of and multiplication within corneal epithelial cells in vitro. Infect Immun 63:4072–4077.

Fyfe JAM, Govan J (1980): Alginate synthesis in mucoid *Pseudomonas aeruginosa:* Achromosomal locus involved in control. J Gen Microbiol 119:443–450.

Gupta SK, Berk RS, Masinick S, Hazlett LD (1994): Pili and lipopolysaccharide of *Pseudomonas aeruginosa* bind to the glycolipid asialo GM1. Infect Immun 62:4572–4579.

Hancock REW, Mutharia L, Chan L, Darveau RP, Speert DP, Pier GB (1983): *Pseudomonas aeruginosa* isolates from patients with cystic fibrosis: A class of serum-sensitive, nontypable strains deficient in lipopolysaccharide O side chains. Infect Immun 42:170–177.

Hata JS, Fick RB (1991): Airway adherence of *Pseudomonas aeruginosa:* Mucoexopolysaccharide binding to human and bovine airway proteins. J Lab Clin Med 117:410–422.

Hazlett LD, Masinick S, Barrett R, Rosol K (1993): Evidence for asialo GM1 as a corneal glycolipid receptor for *Pseudomonas aeruginosa* adhesion. Infect Immun 61:5164–5173.

Hobbs M, Collie ESR, Free PD, Livingston SP, Mattick JS (1993): PilS and PilR, a two component transcriptional regulatory system controlling expression of type 4 fimbriae in *Pseudomonas aeruginosa.* Mol Microbiol 7:669–682.

Imundo L, Barasch J, Prince A, Al-awqati Q (1995): CF epithelial cells have a receptor for pathogenic bacteria on their apical surface. Proc Natl Acad Sci USA 92:3019–3023.

Irvin RT, Doig P, Lee KK, Sastry PA, Paranchych W, Todd T, Hodges RS (1989): Characterization of the *Pseudomonas aeruginosa* pilus adhesin: Confirmation that the pilin structural protein subunit contains a human epithelial cell-binding domain. Infect Immun 57:3720–3726.

Kelly NM, Kluftinger JL, Pasloske B, Paranchych W, Hancock REW (1989): *Pseudomonas aeruginosa* pili as lignads for nonopsonic phagocytosis by fibronectin-stimulated macrophages. Infect Immun 57:3841–3845.

Khan TZ, Wagener JS, Bost T, Martinez J, Accurso FJ, Riches DWH (1995): Early pulmonary inflammation in infants with cystic fibrosis. Am J Respir Crit Care Med 151:1075–1082.

Krivan HC, Roberts DD, Ginsburg V (1988): Many pathogenic bacteria bind specifically to the carbohydrate sequences GalNAcβ1–4Gal found in some glycolipids. Proc Natl Acad Sci USA 85:6157–6161.

Mahenthiralingham E, Campbell ME, Speert DP (1994): Nonmotility and phagocytic resistance of *Pseudomonas aeruginosa* isolates from chronically colonized patients with cystic fibrosis. Infect Immun 62:596–605.

Mahenthiralingham F, Speert DP (1995): Nonopsonic phagocytosis of *Pseudomonas aeruginosa* by macrophages and polymorphonuclear leukocytes requires the presence of the bacterial flagellum. Infect Immun 63:4519–4523.

Nunn D, Bergman S, Lory S (1990): Products of three accessory genes, *pilB, pilC,* and *pilD,* are required for biogenesis of *P. aeruginosa* pili. J Bacteriol 172:2911–2919.

Panjwani N, Zhao Z, Ahmad S, Yang Z, Jungalwala F, Baum J (1995): Neolactoglycosphingolipids, potential mediators of corneal epithelial cell migration. J Biol Chem 270:14015–14023.

Ramphal R, Carnoy C, Fievre S, Michalski J-C, Houdret N, Lamblin G, Strecker G, Roussel P (1991): *Pseudomonas aeruginosa* recognizes carbohydrate chains containing type 1(Galβ1–3GlcNAc) or type 2 (Galβ1–4GlcNAc) disaccharide units. Infect Immun 59:700–704.

Ramphal R, Houdret N, Koo L, Lamblin G, Roussel P (1989): Differences in adhesion of *Pseudomonas aeruginosa* to mucin glycopeptides from sputa of patients with cystic fibrosis and chronic bronchitis. Infect Immun 57:3066–3071.

Ramphal R, Pyle M (1983): Adherence of mucoid and non-mucoid *Pseudomonas aeruginosa* tp acid injured tracheal epithelium. Infect Immun 41:345–351.

Ritchings BW, Almira EC, Lory S, Ramphal R (1995): Cloning and phenotypic characterization of *fleS* and *fleR,* new response regulators of *Pseudomonas aeruginosa* which regulate motility and adhesion to mucin. Infect Immun 63:4868–4876.

Rostand KS, Esko JD (1993): Cholesterol and cholesterol esters: Host receptors for *Pseudomonas aeruginosa* adherence. J Biol Chem 268:24053–24059.

Saiman L, Cacalano G, Gruenert D, Prince A (1992): Comparison of the adherence of *Pseudomonas aeruginosa* to cystic fibrosis and normal respiratory epithelial cells. Infect Immun 60:2808–2814.

Saiman L, Prince A (1993): *Pseudomonas aeruginosa* pili bind to asialo GM1 which is increased on the surface of cystic fibrosis epithelial cells. J Clin Invest 92:1875–1890.

Saiman L, Sadoff J, Prince A (1989): Cross reactivity of *Pseudomonas aeruginosa* anti-pilin monoclonal antibodies with heterogeneous strains of *P. cepacia* and *P. aeruginosa*. Infect Immun 57:2764–2770.

Sajjan US, Corey M, Karmali MA, Forstner JF (1992a): Binding of *Pseudomonas cepacia* to normal human intestinal mucin and respiratory mucin from patients with cystic fibrosis. J Clin Invest 89:548–656.

Sajjan U, Reisman J, Doig P, Irvin RT, Forstner G, Forstner J (1992b): Binding of nonmucoid *Pseudomonas aeruginosa* to normal human intestinal mucin and respiratory mucin from patients with cystic fibrosis. J Clin Invest 89:657–665.

Sheth HB, Lee KK, Wong WY, Srivastava G, Hindsgaul O, Hodges RS, Paranchych W, Irvin RT (1994): The pili of *Pseudomonas aeruginosa* strains PAK and PAO bind specifically to the carbohydrate sequence βGalNAc(1–4βGal) found in glycosphingolipids asialo-GM_1 and asialo-GM_2. Mol Microbiol 114:715–723.

Simpson DA, Ramphal R, Lory S (1992): Genetic analysis of *Pseudomonas aeruginosa* adherence: Distinct genetic loci control attachment to epithelial cells and mucins. Infect Immun 60:3771–3779.

Simpson DA, Ramphal R, Lory S (1995a): Characterization of *Pseudomonas aeruginosa fliO*, a gene involved in flagellar biosynthesis and adherence. Infect Immun 63:2950–2957.

Simpson DA, Mahenthirlingham E, Speert DP (1995b): Characterization of non-motile *Pseudomonas aeruginosa* isolated from cystic fibrosis patients. Pediatr Pulmonol S12:238.

Speert DP, Wright SD, Silverstein SD, Mah B (1988): Functional characterization of macrophage receptors for *in vitro* phagocytosis of unopsonized *Pseudomonas aeruginosa*. J Clin Invest 82:872–879.

Tang H, Kays M, Prince A (1995): Role of *Pseudomonas aeruginosa* pili in acute pulmonary infection. Infect Immun 63:1278–1285.

Totten PA, Lara JC, Lory S (1990): The *rpoN* gene product of *Pseudomonas aeruginosa* is required for expression of diverse genes including the flagellin gene. J Bacteriol 172:389–396.

Vishwanath S, Ramphal R (1985): Tracheobronchial mucin receptor for *Pseudomonas aeruginosa:* Predominance of amino sugars in binding sites. Infect Immun 48:331–335.

Welsh MJ, Smith AE (1993): Molecular mechanisms of CFTR chloride channel dysfunction in cystic fibrosis. Cell 73:1251–1254.

Whitchurch CB, Hobbs M, Livingston SP, Krishnapillai V, Mattick JS (1991): Characterisation of a *Pseudomonas aeruginosa* twitching motility gene and evidence for a specialised protein export system widespread in eubacteria. Gene 101:33–44.

Woods DE, Straus DC, Johanson WG, Berry VK, Bass JA (1980): Role of pili in adherence of *Pseudomonas aeruginosa* to mammalian buccal epithelial cells. Infect Immun 29:1146–1151.

Yu H, Schurr MJ, Deretic V (1995): Functional equivalence of *E. coli* σ^E and *Pseudomonas aeruginosa* AlgU: *rpoE* restores mucoidy and reduces sensitivity to reactive oxygen intermediates in *algU* mutants of *P. aeruginosa*. J Bacteriol 177:3259–3268.

Zar HB, Saiman L, Quittell L, Prince A (1995): The binding of *Pseudomonas aeruginosa* to respiratory epithelial cells from cystic fibrosis patients with various mutations in CFTR. J Pediatr 126:230–233.

Zhao Z, Panjwani N (1995): *Pseudomonas aeruginosa* infection of the cornea and asialo GM_1. Infect Immun 63:353–355.

8

CONCEPTUAL ADVANCES IN RESEARCH ON THE ADHESION OF BACTERIA TO ORAL SURFACES[1]

RICHARD P. ELLEN

Department of Periodontics, University of Toronto, Faculty of Dentistry, Toronto, Ontario M5Q 1G6, Canada

ROBERT A. BURNE

Department of Dental Research, University of Rochester, Rochester, New York 14642

8.1. INTRODUCTION

Oral microbial ecologists have been credited with many of the conceptual advances that have been made during the past three decades of intensive research into microbial adhesion to surfaces. By the mid-1960s, pragmatic incentives led to the recognition that prevention of dental caries and periodontal diseases might be achieved by disrupting or preventing the establishment of microbial biofilms on teeth (dental plaque). Direct access for sampling and experimental manipulation of the indigenous microbiota facilitated rapid advancements, because many findings established by laboratory research could be tested in humans without resorting to invasive techniques. In all fields, research progress is limited by available technology. While need often leads

---

[1] In memory of William B. Clark IV, whose adhesion assay based on saliva-coated hydroxyapatite became the standard in the field and whose research into fimbria-mediated adhesion of *Actinomyces* species always seemed ahead of mine.—R.P.E.

Bacterial Adhesion: Molecular and Ecological Diversity, pages 201–247
© *1996 Wiley-Liss, Inc.*

to fresh ideas that improve technology, most applications of ideas are framed within the limits of existing methodology. Advances in research on the adhesion of oral bacteria have stayed close to the leading edge of the applied technology of the times.

This chapter concentrates on the oral bacterial adhesion interactions that have had the greatest conceptual impact on the field of microbial adhesion in general—contributions of ideas and findings that have had sufficient staying power to keep pace with current advances in biotechnology (Table 8.1). For example, the notion that selectivity in bacterial adherence is a determinant

TABLE 8.1. Some Novel Concepts Advanced Through Research on Oral Microbial Adhesion

1960s	Dental diseases are transmissible infections caused by the accumulation of biofilms (dental plaque) on teeth
	Agglutination and adhesion of (specifically) mutans streptococci are mediated by glucans in plaque matrix
	GTFs and FTFs and their catalysis of polysaccharide synthesis are discovered
1970s	Bacterial adhesion is a principal determinant of colonization of host surfaces exposed to a fluid flow
	Specificity of adhesion determines tissue tropisms and host range of indigenous bacteria and exogenous pathogens
	The function of S-IgA is to prevent bacterial adhesion
	Gram-positive bacteria bear fimbriae with adhesive (including lectin) functions
1980s	Bacteria enzymatically prime host cell oligosaccharides to enhance their adhesion
	Kinetics of bacterial adhesion can be analyzed using models based on kinetics of other molecular interactions
	Bacterial adhesion is mediated by multiple classes of adhesins, both stereospecific and nonspecific, and has some characteristics of cooperativity
	S. mutans GTFs and its other adhesins are useful immunogens for caries-protective vaccines
	Bacteria colonizing in a salivary milieu adhere to cryptic receptors (cryptitopes)
	Bacterial surface proteases expose cryptitopes and/or serve as bacterial adhesins themselves
	Protective secretory immunity can be stimulated by manipulating adjuvants, route of delivery, and recombinant immunogens based on adhesin molecules
1990s	Polar adhesion of spirochetes is associated with "capping" of adhesins toward their tips
	Expression of adhesins and plaque polysaccharide-synthesizing enzymes are regulated at the genetic level in response to environmental conditions

of host-specific and tissue-specific colonization of surfaces exposed to a fluid flow was a major conceptual advancement that followed from the first series of studies that Gibbons and van Houte (1971, 1975) conducted with oral streptococci. This notion became a hypothesis, then a theory, then a fact, and it is still one of the principal tenets guiding today's research on the molecular regulation of specific adhesins and their ligands and receptors. Specificity will be a dominant theme both in this chapter on bacterial adhesion to oral surfaces and in London and Kolenbrander's chapter on intergeneric bacterial coaggregation (Chapter 9). Like the bacteria that colonize all host surfaces bathed by secretions, the few hundred species of oral bacteria have evolved a great variety of specific adhesive mechanisms that allow them to avoid physical removal by a myriad of innate and specific host defenses that may interfere with or inhibit the functions of their surface adhesins. In the mouth, these include physiological desquamation of mucosal epithelium and the rinsing actions of saliva with its mucinous, blood-group reactive, and other glycoproteins, immunoglobulins, lysozyme, and a variety of other molecules that bind to oral bacterial adhesins. Unique to the oral cavity is the opportunity for bacteria to accumulate as complex biofilms on the teeth.

8.2. BACTERIAL POLYSACCHARIDES AND CARBOHYDRATE-BINDING PROTEINS OF ORAL STREPTOCOCCI

Dental plaques, exemplified by biofilms located on supragingival tooth surfaces, are composed of densely arranged bacterial microcolonies within an extracellular matrix of salivary proteins and polysaccharides derived from bacteria. Therefore, it seems natural to suspect that some of the polysaccharides would function in some way to foster the adherence of bacteria to the teeth or to each other. Yet it took some fortuitous observations and experimentation in the mid-1960s to determine that adhesive interactions of streptococci with extracellular polysaccharides (EPS) are specific and that they are significant for plaque formation and colonization of a limited range of species (Gibbons and van Houte, 1975). Whereas several species synthesize EPS, and thus contribute to the array of homopolymers in a dental plaque matrix largely comprised of glucans and fructans, it is primarily the species belonging to the group known today as the mutans streptococci that actually exploit EPS for adhesion. These bacteria have apparently evolved to exploit adhesive interactions with glucans for accumulation as significant populations on the teeth. Research into the synthesis, adhesion specificity, and genetic regulation of expression of the glucan-producing glucosyltransferase (GTF) enzymes, fructan-producing fructosyltransferase (FTF) enzymes, and glucan-binding proteins (GBP) has been promoted because the mutans streptococci are significant in the etiology of dental caries (Hamada and Slade, 1980).

Cariogenic microorganisms and, in particular, the mutans streptococci have a variety of enzyme systems that are devoted to the hydrolysis of sucrose (Fig.

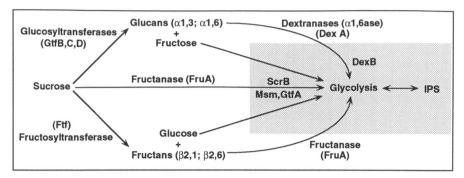

Fig. 8.1. Schematic flow diagram for the pathways involved in sucrose metabolism by *S. mutans*. Sucrose can be metabolized directly, or the glucose or fructose moieties can be shunted into a polymeric form. Direct metabolism can involve extracellular cleavage by a fructanase (FruA) to produce glucose and fructose, which can be transported intracellularly by one of a few high-affinity active transport systems (Russell et al., 1991) and metabolized to organic acids. Alternatively, sucrose can be transported by a phosphoenolpyruvate-dependent sugar phosphotransferase system (EII^{scr}, ScrA), and the internalized sucrose-6-phosphate cleaved by sucrose-6-phosphate hydrolase (ScrB) to produce glucose-6-phosphate and fructose. Sucrose can also enter the cell through a binding-dependent transport system (encoded in the Msm operon) and be cleaved by the GtfA enzyme (sucrose phosphorylase). Extracellularly, three glucosyltransferases (GtfBCD) and a dextranase (DexA) cooperate to produce a glucan polymer rich in α1,3 linkages (see text). A single fructosyltransferase cleaves sucrose to produce polymers of fructose. The α1,6-linked dextrans and the β2,1- and β2,6-linked fructans are susceptible to attack by an extracellular dextranase (DexA) and fructanase (FruA), respectively. The released carbohydrate can then be transported and metabolized. Under certain conditions, mutans streptococci can also shunt carbohydrate inside the cells into a glycogen-like material termed IPS, for intracellular polysaccharide or iodophilic polysaccharide. The stippled area represents events that occur within the cell.

8.1). Most germane to the topic presented here is the conversion of sucrose into high molecular mass, extracellular polysaccharides composed entirely of fructose or glucose. The synthesis of fructans and glucans by the FTFs and GTFs of dental plaque streptococci is an extracellular event that depends on the disaccharide sucrose as the principal substrate (Hamada and Slade, 1980; Kuramitsu, 1993). The enzymes catalyze the transfer of either the fructose or glucose moiety of sucrose onto a growing polysaccharide chain with the concomitant liberation of free monosaccharides, which can be transported intracellularly and catabolized for the generation of energy. Fructans are more labile than glucans in dental plaque and are degraded relatively rapidly by fructanase enzymes produced by a variety of oral bacteria (DaCosta and Gibbons, 1968; Burne et al., 1987). There is no direct evidence that fructans are significant for adhesion or coadhesion of bacteria in dental plaque, although the synthesis and metabolism of these polysaccharides may impact plaque ecology by other means (Burne, 1991). In contrast, much of the glucan synthesized in dental plaque is predominantly water insoluble and is largely refractile to enzymatic hydrolysis by oral bacteria (Birkhed et al., 1979). These high-molecular-weight polymers function in multiple capacities: binding specific bacteria together, mostly the mutans streptococci, and serving as receptors to allow binding of some of the mutans streptococci to the salivary proteins coating the teeth (enamel pellicle). The length, solubility, degree of branching, and proportion of $\alpha1,3$- to $\alpha1,6$-glucosidic linkages determine the solubility of the polymers and presumably the degree to which the extracellular glucans mediate adhesion and accumulation. The physical characteristics of the polysaccharides are primarily dictated by the production and activities of various GTFs. The mutans streptococci also produce an extracellular endo-dextranase (Lawman and Bleiweis, 1991; Barrett et al., 1987) that may modify the physicochemical properties of glucans by attacking $\alpha1,6$ linkages, thus increasing the proportion of $\alpha1,3$-linked glucan with a concomitant decrease in the water solubility of the polymer.

8.2.1. Genetics of Polysaccharide Synthesis and Glucan Binding

The advent and application of biotechnology to the study of mutans streptococci has contributed to the rapid advancement in knowledge about GTFs, FTFs, and GBPs. Three true *gtf* genes have been cloned from *Streptococcus mutans* GS-5 and characterized: *gtfB, gtfC,* and *gftD* (Hananda and Kuramitsu, 1988, 1989; Honda et al., 1990). Ablation of these genes by directed mutagenesis eliminates all detectable Gtf activity from *S. mutans* (Munro et al., 1991; Kuramitsu, 1993). GtfB (GTF-I) and GftC (GTF-SI) synthesize a branched, insoluble glucan rich in $\alpha1,3$ linkages, and GtfD (GTF-S) produces a less branched soluble glucan. The *gtfB* and *gtfC* genes are located in tandem on the GS-5 chromosome, quite distant from the *gtfD* gene (Hananda and Kuramitsu, 1989), and share extensive nucleotide sequence homology. The GtfD protein is roughly 50% identical to the other GTFs of *S. mutans* (Honda et al., 1990; Perry and Kuramitsu, 1990). Overall, the GTFs of *S. mutans* and other oral streptococci are structurally similar (Giffard et al., 1993; Giffard and Jacques, 1994; Russell et al., 1988), but there is less conservation at the sequence level, and polymorphisms among *gtf* genes in oral streptococci are common (Chia et al., 1991). This heterogeneity may account for the functional variability and levels of enzyme expression noted among GTFs from different strains. The enzymes also differ in their requirements for short $\alpha1,6$-linked dextran primers for catalytic functions (Kuramitsu, 1993; Hananda and Kuramitsu, 1989).

From genetic and functional analyses, GTFs are apparently composed of at least two domains involved in glucan synthesis and binding, respectively. The sucrose-binding portion of the GTFs of *S. mutans* and *Streptococcus sobrinus*—which have been functionally mapped (Mooser et al., 1989) and, in the case of the former, identified by site-directed mutagenesis (Kato et al., 1992)—is virtually identical among species of mutans streptococci (Cope and

Mooser, 1993). In particular, site-directed mutagenesis of an Asp residue (Asp-451) of the putative sucrose binding/cleavage site destroys all catalytic function of the *S. mutans* GS-5 GTF-I (GtfB), an enzyme that synthesizes insoluble glucan (Kato et al., 1992). More recently, construction of hybrid enzymes (Nakano and Kuramitsu, 1992) and "semirandom" mutagenesis of GTFs (Shimamura et al., 1994) have begun to identify residues that may confer substrate–product binding, product binding–linkage specificity, and primer dependence. For example, six different amino acid residues on GTF-I have been mutated to match the corresponding residues in the GTF-S enzyme to produce recombinant GTF enzymes that synthesize a greater proportion of water-soluble glucans (Shimamura et al., 1994). Similarly, mutagenesis of the GTF-S enzyme at position 589 from Thr to Asp or Glu produced an enzyme that produced insoluble glucans in the absence of a dextran acceptor (Shimamura et al., 1994). These studies have begun to provide knowledge about the functional elements of streptococcal GTFs, although a thorough picture of the functional characteristics of the active catalytic sites of GTFs and how the glucosyl units are actually linked to an acceptor is lacking. Likewise, a fundamental grasp of the molecular basis for the catalysis of different linkages is still enigmatic. Such detailed information should ultimately come with extensive genetic and biochemical analyses coupled with crystal structure information on the various enzymes.

The deduced amino acid sequences of the known *gtf* genes of oral streptococci are characterized by a group of directly repeated domains composed of approximately 23 amino acids arranged in tandem in the C-terminal one-third of the molecule (summarized by Giffard and Jacques, 1994). These repeats share strong structural and compositional homology among the GTFs of the oral streptococci and are similarly conserved with catalytically inactive glucan-binding proteins expressed on the surface of mutans streptococci (Banas et al., 1990). The repeated domains also share extensive structural homology with a family of ligand-binding proteins, including *Clostridium difficile* toxin, lysins of pneumococcal phages, and other surface-associated adhesins (Wren, 1991; von Eichel-Streiber et al., 1992; Giffard and Jacques, 1994). Interestingly, a recent report implicated GTF(s) of *Streptococcus gordonii* in its adhesion to endothelial cells *in vitro* (Vacca-Smith et al., 1994), a phenomenon that may be correlated with the observed structural homology between the glucan-binding domains of GTFs and mammalian cell-binding molecules. In *S. mutans,* deletion analysis has shown that the repeats in the C-terminal portion of GTFs are actually required for glucan binding, GTF polymerase activity (Kato and Kuramitsu, 1989), and, either directly or indirectly, surface localization of GTF (Kato and Kuramitsu, 1991).

As with the GTFs, the gene encoding a GBP of *S. mutans* (strain Ingbritt) has been cloned and sequenced (Banas et al., 1990). GBP is a 59 kDa, surface-localized adhesin with functional analogs present in other mutans streptococci (Drake et al., 1988; Singh et al., 1993; Smith et al., 1994a). The GBPs and the glucan-binding domains of the GTFs have been observed to share significant primary sequence homology within the presumed glucan-binding domains. Conclusive identification of the active sites for glucan binding in any of the glucan-binding lectins of oral streptococci has not yet been accomplished, but chemical modification of the *S. sobrinus* glucan-binding lectin implicated aspartic acid, glutamic acid, histidine, lysine, and tyrosine residues as significant for bacterial aggregating activity (Singh et al., 1993), perhaps not surprisingly because of the composition of most repeats in these proteins. Data also support that structural rather than sequence-specific constraints are key elements in carbohydrate binding by these repeated elements (Wren, 1991; von Eichel-Streiber et al., 1992; Giffard and Jacques, 1994). Examination of the activity of a number of glucan-binding domains by site-directed mutagenesis may be necessary to resolve the basis for substrate binding, recognition specificity, and the various affinities these lectins have for their cognate ligands.

8.2.2. Specificity and Significance of Glucan Binding

Glucan-mediated adhesion is a function shared by all mutans streptococci but by few, if any, other oral species. This phenomenon is mediated by surface-localized adhesins capable of specific stereochemical interactions with the glucan products of the GTFs. The candidate adhesins include the GTFs themselves (Fukushima et al., 1992), as well as the distinct cell surface GBPs. GTFs are synthesized with a classical signal sequence and in sucrose-free medium are secreted almost exlusively in a cell-free form. However, in the presence of even trace amounts of sucrose, tight associations between these enzymes, in complexes with glucan polysaccharides, and the cell wall of mutans streptococci are formed (Kuramitsu and Ingersoll, 1978). Kuramitsu and coworkers have demonstrated that association of the GTFs with the cell surface is obligately linked to possession of both glucan synthetic capabilities of the enzyme and intact glucan-binding domains in the protein (Kato and Kuramitsu, 1991).

One theory (Fig. 8.2) is that cell-association of GTFs may occur because a small amount of secreted GTF remains bound at the cell wall in some as yet undefined manner and, in the presence of sucrose, initiates bacterial surface-associated glucan synthesis. The remaining extracellular GTFs can then become associated with the growing polysaccharides or solution phase glucan–GTF complexes after binding to this growing matrix. Alternatively, cell-associated GBP may bind newly synthesized and exogenous glucans to the cell surface, thereby recruiting GTF to the cell surface and promoting bacterial aggregation. Recent unpublished results (J. A. Banas) indicate that *gbp* mutants of *S. mutans* do not bind dextran, but bind as avidly as wild-type cells to *in situ* synthesized glucans (R. A. Burne and W. H. Bowen, unpublished results) as measured by the *in vitro* model developed by Schilling et al. (1989), suggesting that the GTFs may be major contributors to the glucan-dependent adhesion and aggregation of mutans streptococci. This is also consistent with results from Kuramitsu's laboratory that demonstrated that expression of a functional GTF-I enzyme in *Streptococcus milleri* allowed these recombinant

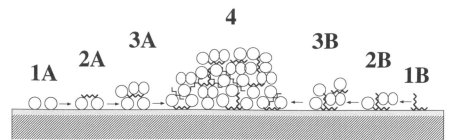

Fig. 8.2. Illustration representing adhesion and accumulation of streptococci (biofilm formation) on a pellicle-coated tooth surface. There are two putative mechanisms for primary colonization by mutans streptococci, both of which involve multiple adhesin–receptor interactions. **1A:** Mutans streptococci can bind via their adhesins to various proteins found in salivary pellicle (stippled) formed on teeth (hatched). When bound to the pellicles, the organisms can produce glucans **(2A)**, which help to stabilize the interactions with the tooth and serve as adhesins for recruitment of additional glucan-binding bacteria into the biofilm **(3A,B)**. The enzymes producing glucans can also be found in pellicles *in vivo* in an enzymatically active form, occurring there presumably as a result of secretion of these proteins by bacteria in plaque and saliva. In the presence of sucrose, glucans can be synthesized within the pellicle **(1B)**, which can serve as receptors for mutans streptococci **(2B)**. Through growth of the bacteria on the surface and recruitment of bacteria from other sites and saliva **(3A,B)**, microcolonies of irreversibly bound mutans streptococci become established, contributing to the formation of dental plaques **(4)**. In reality, dental plaque is composed of complex populations of bacteria, but, for the purpose of detailing binding mechanisms, other organisms have been omitted. For a more detailed treatment of interbacterial interactions, see Chapter 8 (this volume).

strains to aggregate and to adhere to glass in the presence of sucrose (Fukushima et al., 1992). Still, one cannot eliminate the possibility that it is a combination of GBPs and GTFs binding to glucans that leads to the thick biofilm formation and extensive aggregation of mutans streptococci that is observed when these organisms are cultured in the presence of sucrose (Fig. 8.2).

Despite the fact that numerous oral streptococci produce GTFs, glucan-mediated adhesion to teeth appears crucial only for the true mutans streptococci. It is also of interest that the GBPs produced by mutans streptococci appear to have the highest affinity for the α1,6-linked glucans, or dextrans, rather than α1,3-glucan, which is the predominant linkage produced by the GTFs of these organisms (Banas et al., 1990; Smith et al., 1994a). Likewise, the GBPs (lectin) of *S. sobrinus* (Wu-Yuan and Gill, 1992) and *Streptococcus cricetus* (Drake et al., 1988) demonstrate apparently high specificity for α1,6-glucan, their conformation and glucan-binding functions being dependent on their complexing with manganous ions during synthesis (Lü-Lü et al., 1992). Although it has not been proven, one can propose that glucans rich in α1,3 linkages may provide a supporting framework for the dextran-type linkages, this framework being resistant to any known enzymes produced by oral bacteria. In doing so, the α1,6 linkages may be protected sterically from attack by dextranases produced by numerous plaque microorganisms. Another unsupported but intriguing hypothesis is that the α1,6-linked glucans produced by bacteria in plaque, such as *Streptococcus sanguis,* could be exploited by *S. mutans* to gain a firm foothold on the tooth surface. Regardless, it seems that glucan binding has evolved as a major mechanism for adherence that is exploited principally by mutans streptococci. Nevertheless, many of these considerations need to be examined more closely in the context of dental plaque ecology, in the context of the temporal characteristics of plaque biofilm formation, and in terms of the succession of species, which clearly occurs in the dental plaque community.

Whether glucan-mediated adhesion to salivary pellicle is essential for the colonization of mutans streptococci has been a source of controversy since the original *S. mutans* was divided into at least five species. More recently, the controversy has been distilled to whether GTFs and glucan synthesis really are essential for the initial adhesion to salivary pellicle or whether they are more significant in the accumulation phase of mutans biofilm formation (Fig. 8.2) (Gibbons et al., 1986; Schilling et al., 1989; Schilling and Bowen, 1992). With the definitive discovery of glucan-binding domains in the *S. mutans* GTFs (Mooser and Wong, 1988) and subsequent genetic and functional characterization of GBPs on the cell surface of mutans streptococci (Banas et al., 1990), it is now reasonable to conclude that most strains of mutans streptococci have the potential to utilize glucans *in situ* as receptors for colonizing the human tooth surface if these polysaccharides were available. Moreover, a recent study found no decrease in cariogenicity of a strain of *S. mutans* that lacked the major saliva-binding adhesin P1 (see below) but retained its native GTF and FTF activities when the animals were fed a diet containing sucrose (Bowen et al., 1991). Although this strain could not bind well to saliva-coated hydroxyapatite *in vitro,* the P1-deficient mutant was in fact as cariogenic as the wild type independent of the site of the lesions examined. This implies that glucans may serve as its adhesins *in vivo* and that *S. mutans* has other mechanisms to adhere to a saliva-coated tooth, some of which are likely dependent on glucan synthesis and binding. Studies with strains defective in both P1 and glucan synthesis and binding should provide further insights on this issue.

8.2.3. Regulation of Genes Encoding Polysaccharide Synthesizing and Binding Molecules

Since the activity of GTFs, GBPs, and FTFs are crucial for adhesive interactions of mutans streptococci and for the synthesis of the plaque matrix polysaccharides, a thorough knowledge of how genes encoding them are regulated, particularly in response to environmental factors that are likely to be encountered *in vivo,* will be essential for a true understanding of polysaccharide metabolism in dental plaque. It had long been thought that GTFs and FTFs were constitutive enzymes, since expression of these proteins did not require supplementation with specific sugars, nor was expression markedly influenced by the carbon source in batch-grown organisms. This view began to change with early work in which continuous culture was used to modulate precisely the growth conditions of mutans streptococci (Hardy et al., 1981; Keevil et al., 1983; Walker et al., 1982, 1984). These studies dramatically demonstrated the phenotypic plasticity of oral streptococci and showed that changes in the measurable GTF and FTF activity occurred in response to variations in carbon source and availability, pH, and growth rate.

More recently, advances in biotechnology have offered the opportunity to construct gene and operon fusions to the cognate regulatory elements of the glucosyltransferase genes of *S. mutans* to facilitate the monitoring of these genes' expression under vaired environmental conditions. Operon and gene fusions with chloramphenicol acetyltransferase (CAT) to either *gtfBC* or *ftf* gene promoters have been used to study regulation of the transferase operons. Studies by Hudson and Curtiss (1990) provided evidence that suggested that the regulation of *gtf* gene expression could be stimulated by adhesion of the bacteria to saliva-coated hydroxyapatite. Subsequently, the same gene fusion strains were examined in continuous chemostat culture to study their differential expression in response to environmental stimuli (Wexler et al., 1993). This work established the feasibility of combining long-term continuous culture, which has traditionally proven to be a powerful tool in examining the physiology of oral streptococci, with gene fusion technology to study gene regulation in oral streptococci. These studies provided conclusive evidence that expression of GTF and FTF were influenced by growth rates, carbon source, and environmental pH. Likewise, sucrose was shown to stimulate GTF and FTF gene expression under certain conditions. Using a different reporter gene technology, a subsequent study confirmed the effects of sucrose on FTF expression and tentatively identified *cis*-acting sites for regulation of sucrose induction of *ftf* (Kiska and Macrina, 1994). Studies are now progressing in several laboratories to examine more closely the environmental effects on gene expression in the mutans streptococci, combining continous culture and other physiologic and genetic tools. Substantial advances in reporter gene technology, vector design, gene transfer systems, and transposon mutagenesis will facilitate even finer analyses in a wide variety of important oral bacteria.

S. mutans, other viridans streptococci, and many oral microorganisms elicit diseases in the oral cavity while colonizing surfaces as biofilms. In fact, it is noteworthy that the dependence on surface growth for mutans streptococci seems absolute, since these bacteria are almost never found in other environments. Importantly, there is a growing recognition that biofilm bacteria can have radically different phenotypic properties than their planktonic counterparts (Costerton et al., 1987; Davies et al., 1993). Recognizing this, Burne et al. (1995) have begun to examine *gtf* and *ftf* expression using gene fusion technology coupled with growth in biofilm reactors and scanning confocal laser microscopy to define more precisely how environmental factors and growth of oral streptococci on surfaces influence gene expression and polysaccharide synthesis. These studies are starting to provide insights into the phenotypic capacities of these organisms under conditions that presumably mimic those encountered by bacteria colonizing surfaces in the human oral cavity. The results obtained thus far support that certain aspects of the expression of these genes is conserved, e.g., the inducibility of *ftf* and *gtfBC* by sucrose, when the cells are cultivated in biofilms. However, surface growth, biofilm age and thickness, and substrate availability markedly affected the expression of sucrolytic gene products of *S. mutans* in a gene-dependent fashion (Burne et al., 1995). The implications of these results in the context of oral disease are clear; knowledge of the phenotypic characteristics and capacities of adherent oral pathogens needs to be gained if novel and effective strategies for the inhibition of bacterial colonization, accumulation, or other pathogenic strategies are to be realized.

There have been substantive advances in our understanding of the molecular basis for the adhesion of mutans streptococci to GTF-derived products in the oral environment. Yet as with most efforts to couple molecular biology to the "actual biology" of the organism, the insights gained have elicited many more questions and have highlighted complexities. In the case of polysaccharides and adhesion as they pertain to microbial ecology of the mouth, these studies have emphasized how little is actually understood about dental plaque formation and the interrelationships of enzyme systems and bacteria in microenvironments. Many challenges remain before fundamental aspects of polysaccharide synthesis, degradation, and binding in dental plaque are understood. Much of this stems from a lack of knowledge of the contribution of individual GTFs to the final glucan product. Recently, inroads have been made into dissecting enzymatic aspects of glucan synthesis and control of end-product structure and composition using active site reagents and molecular genetic techniques (Cope and Mooser, 1993; Kato et al., 1992; Shimamura et al., 1994). Yet these studies represent the beginnings to a full appreciation of how these complex enzymes split sucrose, make glucans, and catalyze the synthesis of polymers with different linkages and degrees of branching. An additional layer of complexity is superimposed on understanding glucan synthesis *in vivo* when one considers that the GTFs appear not to function independently but rather in cooperation with each other, with an endo-dextranase that can have its activity modified by a dextranase inhibitor protein (Dei) (Sun et al., 1994) and probably with other enzymes and plaque constituents to produce a final glucan product. Moreover, the kinetic properties of the GTFs are influenced by aggregation state, binding to surfaces, the presence of hydrophobic compounds, and perhaps through post-translational processing by proteolysis. When these variables are placed in the context of differential regulation of enzyme expression in response to environmental influences, surface growth, and the variety of bacteria commonly found in dental plaque, which could produce factors that might affect activity of the GTFs, the challenges become obvious. The use of molecular biological techniques, the ability to construct genetically engineered strains of oral bacteria, and the increasing sensitivity of physical and chemical measurement techniques should help to provide a much better picture of oral microenvironments in the coming years.

8.2.4. Other Significant Adhesins of Mutans Streptococci

While the concept of polysaccharide-mediated aggregation of mutans streptococci launched the whole era of intensive investigation into the specificity and
mechanisms for streptococcal adhesion, there has been considerable progress
in defining the structures and functions of mutans streptococcal adhesins other
than GTFs and GBPs. These proteins mediate adhesion not to bacterially
derived carbohydrate ligands but to host molecules, like high-molecular-
weight mucinous glycoproteins, that adsorb to teeth to form salivary pellicles
(Fig. 8.3). Mutans streptococci also express receptors for the serum-derived
molecules fibronectin and β_2-microglobulin (β_2m), which are found in oral
secretions (Babu and Dabbous, 1986; Ericson, 1984). β_2m has been used as
a ligand–gold probe to determine the distribution of specific adhesins on
mutans streptococci preserved by the freeze substitution method (Ericson et
al., 1987). If β_2m receptors are typical of adhesins for other salivary molecules,
they extend much further beyond the bacterial cell wall than previously expected.

   Several cell wall associated proteins have been scrutinized for their potential
use as caries vaccine candidates. One of these, toward which protective antibodies can be generated in primates, has been identified as an adhesin with
salivary-binding properties. This high-molecular-weight, 185 kDa protein has
been called protein B, protein (antigen) I/II, P1, Pac, or SR (genetic designation *spaP* or *pac*) in *S. mutans* and SpaA or Pag (genetic designation *spaA*)
in *S. sobrinus* (for review, see Russell, 1994). There is a great degree of
sequence similarity among all of these proteins. They also have considerable
sequence similarity to salivary-binding proteins of other streptococcal species
but not necessarily the same adhesion domain specificity (see below). Isogenic
mutants constructed by insertional inactivation of *spaP* in *S. mutans* lose the
ability to agglutinate with whole saliva and to adhere to hydroxyapatite coated
with a high-molecular-weight agglutinin prepared from whole saliva (Lee et
al., 1989). They are also less hydrophobic than the parent strain, a property
shown previously to foster greater adherence of wild-type *S. mutans* to saliva-
coated hydroxyapatite (McBride et al., 1985). Yet these mutants retain their
native GTF and FTF activities and are not diminished in virulence in rats fed
a high sucrose diet.

8.3. ADHESION OF "NONMUTANS" ORAL STREPTOCOCCI TO SALIVARY PELLICLE

Research into the adhesion of mutans streptococci has attracted a great deal
of attention because they are cariogenic and associated with the initiation and
distribution of carious lesions in humans. However, mutans streptococci are
generally slower to establish in the human mouth and to recolonize cleaned
tooth surfaces than are some of the other oral streptococci. *Streptococcus*

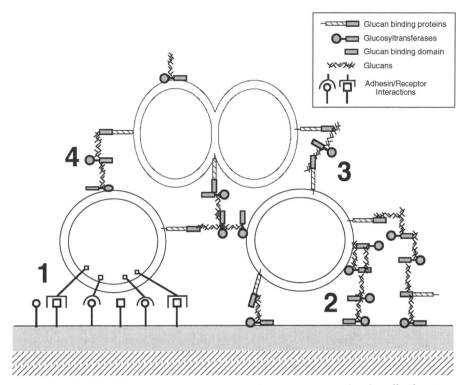

Fig. 8.3. Schematic diagram depicting molecular events occurring in adhesion to a pellicle-coated (stippled) dental substratum (hatched) as in Figure 8.2. **1:** Specific adhesin receptor interactions between cell surface adhesins on mutans streptococci, such as P1, with salivary macromolecules in pellicle, which include glycoproteins such as high-molecular-mass agglutinins like the mucins. **2:** Interaction of glucan-binding proteins and GTFs, which possess similar glucan-binding domains, with glucans synthesized in salivary pellicles. **3,4:** Interactions of cell-free and cell-associated GTFs and GBP with glucans to facilitate interbacterial aggregation. The actual valence of the various glucan-binding domains and the relative contributions of the various proteins to aggregation is undefined. Nevertheless, it is clear that both GBPs and GTFs are found attached at the cell surface and in a secreted but cell-free form. Importantly, these entities retain their enzymatic and binding activities, at least in part, in both forms. The exact molecular basis for anchoring of the GTFs and GBP are ill-defined (see text).

sanguis and strains of *S. sanguis,* later reclassified as *S. gordonii,* and *Streptococcus oralis* have received most attention because they colonize humans immediately upon tooth eruption and they have been shown experimentally to adhere rapidly and selectively to cleaned smooth tooth surfaces (Gibbons and van Houte, 1975). Early work in Gibbons' and van Houte's laboratories established that salivary coatings mimicking pellicle fostered the adhesion of *S. sanguis* to mineral sufaces *in vitro*. Then several others began to identify potential adhesins and mechanisms (Appelbaum et al., 1979; Busscher et al., 1986; Fives-Taylor and Thompson, 1985; Gibbons and van Houte, 1975; Liljemark and Bloomquist, 1981; Morris and McBride, 1984; Murray et al., 1982; Nesbitt et al., 1982a,b; Rosan et al., 1982).

Oral streptococci, mostly *S. sanguis* and *Steptococcus mitior,* were used to define the kinetics of bacterial adhesion to salivary pellicles by mathematical models that are usually reserved for analyzing the interfacial adsorption kinetics of molecules. Gibbons et al. (1976) introduced the application of Langmuir isotherms, Nesbitt et al. (1982b) the application of Scatchard and Hill plot analysis, and Cowan et al. (1986, 1987a) the application of Arrhenius plots to study the kinetics and energetics of the initial stage of *S. sanguis* adhesion to salivary pellicle. In general, conclusions drawn from these studies pointed to *S. sanguis* adhesion being mediated by multiple classes of adhesins that differ in affinity for salivary pellicles and/or its adhesion demonstrating positive cooperativity. These are not mutually exclusive. The key messages were that (1) The initial stage of bacterial adhesion to pellicle is reversible and possibly affected by nonspecific ionic and hydrophobic interactions. (2) Bacterial adhesion may reach a state of relative equilibrium, with binding curves suggesting saturation, but the adsorption/desorption kinetics do not necessarily equate to those seen for independent molecules when applying the molecular kinetics models used. (3) With time, bacteria adhere irreversibly to pellicle, apparently shifting from low specificity forces to high specificity, higher affinity attractive forces. (4) Bacteria may express multiple types of adhesins that interact with different salivary receptors by entirely different mechanisms (Fig. 8.4). One explanation for cooperative effects actually relates to these multiple adhesive interactions (Nesbitt et al., 1982b). Although mathematical molecular models applied to these nonideal microbiological systems could have provided erroneous information, many of the interpretations of the data have proved reasonable using isogenic mutants, biochemical analysis, and monospecific antibodies.

These studies were conducted during an era when lectin-like activities were discovered to impart selective binding properties to bacterial adhesins. One of the classes of streptococcal adhesins with high affinity for salivary glycoproteins in pellicles was identified as a lectin with sialic acid specificity (Duan et al., 1994; Gibbons et al., 1983; Levine et al., 1978; McBride and Gisslow, 1977; Morris and McBride, 1984; Murray et al., 1982, 1986, 1992) which, in *S. sanguis,* may be responsible for the time-dependent shift to high affinity binding (Cowan et al., 1987b). Sialic acid binding activity is apparently strain specific among the streptococci (Liljemark et al., 1989). Other proposed salivary receptors for *S. sanguis* and/or *S. gordonii* include a pH-sensitive receptor, salivary α-amylase, and acidic proline-rich proteins when absorbed on hydroxyapatite (Fig. 8.4) (Douglas, 1990; Gibbons et al., 1991; Scannapieco et al., 1989, 1990).

A number of proteins that may mediate the adhesion of these streptococci to, or aggregation by, salivary molecules have now been identified, cloned, and characterized. Two of them are evidently lipoproteins, a 34.7 kDa saliva-binding lipoprotein SsaB from *S. sanguis* 12 (Ganeshkumar et al., 1991, 1993) and a 76 kDa antigen SarA that may be required for the aggregation of *S. gordonii* Challis in saliva (Jenkinson, 1991). Others include an adhesin for salivary pellicle that has been cloned from *S. gordonii* G9B (Rosan et al., 1989), the FimA fimbrial protein associated with the adhesion of *S. sanguis* (or now *Streptococcus parasanguis*) FW213 to saliva-coated hydroxyapatite (Fenno et al., 1989; Fives-Taylor et al., 1991), and a 165 kDa sialic acid-specific lectin with calcium-binding activity, SSP-5, from *S. gordonii* M5, that interacts with high-molecular-weight mucinous glycoproteins in saliva (Duan et al., 1994). SsaB and FimA are highly homologous and may share some structural and functional relationship with other streptococcal proteins, like ScaA, that are implicated in streptococcal coaggregation with *Actinomyces naeslundii* (Kolenbrander et al., 1994). SSP-5 is structurally and immunologically related to the series of saliva-binding mutans streptococcal proteins, *S. mutans* SpaP (and synonyms), and *S. sobrinus* SpaA, described above.

8.4. HYDROPHOBICITY AND ITS RELATIONSHIP TO COOPERATIVITY

While much of the research focus has concentrated on the specificity of adhesin interactions with complementary salivary molecules, oral microbiologists have contributed a great deal to conceptual developments on the significance of electrostatic, hydrophobic, and other physicochemical interactions mediated at the interface between bacteria and substrate surfaces (Doyle et al., 1990; see also Chapters 1 and 2, this volume). Most bacteria isolated directly from the oral cavity, and many laboratory strains of oral bacteria are hydrophobic when tested by *in vitro* assays based on adhesion to hydrocarbons or hydrophobic interaction chromatography (Beighton, 1984; Clark et al., 1985; Gibbons and Etherden, 1983; Handley and Tipler, 1986; Rosenberg et al., 1983, 1991; Weiss et al., 1982). Further evidence for the significance of hydrophobicity in adhesion of oral streptococci includes the following: (1) Surface fibrils, fimbriae, and even specific surface proteins, like *S. mutans* P1, that are known to promote adhesion also confer hydrophobic characteristics to bacteria when expressed on their surfaces (Fives-Taylor and Thompson, 1985; Lee et al., 1989; Weerkamp et al., 1987). (2) Effects of medium composition and growth rate on steptococcal surface hydrophobicity and adhesion parallel each other (Hogg and Manning, 1987; Knox et al., 1985; Rogers et al., 1984). (3) Adhesion to salivary pellicles can be inhibited by chaotropic agents (Nesbitt et al., 1982a). (4) Hydrophobic strains of *S. mutans* adhere better *in vitro* and can be implanted experimentally more readily than less hydrophobic strains in the human mouth (McBride et al., 1985; Svanberg et al., 1984).

Stereospecific recognition by complementary adhesin–receptor pairs does not exclude hydrophobic or electrostatic effects from influencing initial adhesion or contributing to the affinity and irreversibility of adhesive interactions.

Eukaryotic cell

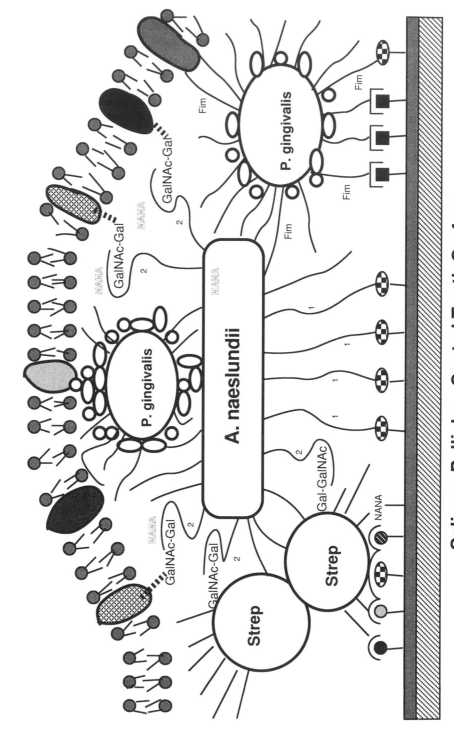

Salivary Pellicle - Coated Tooth Surface

Isolated adhesins and fimbriae are often composed of at least some hydrophobic domains among other domains that could account for polar interactions contributing to affinity (Irvin, 1990). Moreover, the carbohydrate-binding sites that confer sugar specificity to bacterial lectins are often in close proximity to hydrophobic domains, and at least some detergents disrupt bacterial interactions known to depend on lectin–carbohydrate recognition (Doyle et al., 1990; Ellen et al., 1994c; McIntire et al., 1982). Indeed the prevailing concept of how bacteria, including oral streptococci, adhere to protein-coated surfaces like saliva-coated hydroxyapatite is a multiple site model in which hydrophobic, cationic bridging, and electrostatic interactions may stabilize or in some other way affect the affinity of stereospecific interactions between adhesins and complementary receptors (Busscher et al., 1992; Doyle et al., 1990). Yet it is probably the specificity inherent in adhesin–receptor recognition that determines bacterial host range, host age of colonization, and tissue tropisms or perhaps even whether lasting adhesion will occur at all (Ellen, 1985; Ellen et al., 1994c; Gibbons and van Houte, 1971, 1975; Gibbons et al., 1985; Strömberg and Borén, 1992). In this context, it is also feasible that the interplay between recognition specificity and the various forces accounting for attraction or repulsion may lead to positive cooperativity, whereby the adhesion of one bacterium to a substratum enhances the probability for the adhesion of subsequent bacteria, thereby influencing the pattern in which oral bacteria accumulate as biofilms on exposed protein-coated surfaces.

Research into the physical association of oral bacteria with substrata has fostered the development of novel methodology for quantifying adhesion and cooperativity. Rather than restricting the study of adhesion to static conditions,

---

Fig. 8.4. Schematic diagram depicting some mechanisms accounting for adhesion of *A. naeslundii, P. gingivalis,* and some streptococcal species (Strep; e.g., *S. sanguis, S. gordonii,* or *S. oralis*) to salivary-coated dental surfaces, to eukaryotic cells, and to each other. *A. naeslundii* binds to conformationally receptive proline-rich proteins, PRPs (white circles enclosing smaller black boxes), in salivary pellicles via its type 1 fimbriae (1). It adheres to oral streptococci via stereospecific interactions of the lectin adhesin of its type 2 fimbriae (2) with Gal-β-1-3-GalNAc– or GalNAc-β-1-3-Gal– containing cell wall polysaccharides. It adheres to eukaryotic cell membrane glycoproteins and glycolipids via a similar lectin-specific mechanism after enzymatically cleaving sialic acid (NANA), thereby exposing the penultimate galactosides: some of the sialic acid is metabolized. *P. gingivalis* can adhere to salivary proteins (square symbols), including PRPs, to *A. viscosus,* to streptococci (not depicted), and to some eukaryotic membranes via its fimbriae (Fim). Fimbrial adhesion to erythrocytes is controversial (see text). Its abundant surface vesicles and vesicular and outer membrane proteinases also function as adhesins that bind to *A. naeslundii* and to host cell proteins, potentially partially digesting the proteins to expose cryptic receptors. Nonmutans oral streptococci (Strep) recognize several salivary pellicle receptors (circular symbols), including sialic acid (NANA) and PRPs.

real-time experiments using specially adapted flow cells allow for the investigation of cooperative effects both by calculating adsorption isotherms and by analyzing near-neighbor relationships using imaging systems. Using such an approach, Van der Mei and coworkers (1993) have found that positive cooperativity apparently depends on the nature of the protein coat that forms on hydrophobic or hydrophilic substrata but not necessarily on conformational changes in the proteins. Indeed, some cooperative effects are even seen when uncoated substrata are used. For bacteria incompletely immobilized, near-neighbor collection occurs in the direction of fluid flow, apparently more from the sliding of surface-adhering cells than from cooperative accumulation of planktonic cells. Such homotypic bacteria–bacteria interactions may depend in part on their hydrophobicity (Rosenberg et al., 1991). Although most of the novel methodology for quantifying adhesion kinetics has concentrated on oral streptococci as a model, the interplay between stereospecific and nonspecific hydrophobic interactions is evident in the adhesion of other species. The next section focuses on oral *Actinomyces* species, which have hydrophobic surfaces (Clark et al., 1985) but which adhere to oral surfaces and to other bacteria by dominant stereospecific mechanisms.

8.5. *ACTINOMYCES* LECTINS, *ACTINOMYCES* FIMBRIAE

One of the more elegant series of structure–function studies on the adhesion of oral bacteria has been directed toward understanding structural and molecular requirements of adhesin–receptor interactions for *Actinomyces naeslundii* and strains of *A. naeslundii* that were classified until recently as *Actinomyces viscosus* (see Chapter 9, Section 9.3.2.1, this volume). These are ubiquitous oral bacteria that colonize virtually all adults. By numerical taxonomy, they segregate into seven clusters (Fillery et al., 1978). Cluster-1 (*A. viscosus*), cluster-3 ("atypical" *A. naeslundii*), and cluster-5 (*A. naeslundii*), which contain most isolates, have been studied most extensively for their adhesion properties. Although each can be isolated from any surface in the mouth, cluster-5 *A. naeslundii* generally establishes earlier in the host's life, often being detected before tooth eruption, and it outnumbers cluster-1 *A. viscosus* on the tongue mucosa and in saliva. In contrast, *A. viscosus* colonizes the tooth surface more readily than cluster-5 *A. naeslundii*. Differences in their colonization patterns have been linked to differences in the expression of two types of fimbriae and in the specificity of lectin-like and protein-specific adhesins borne on the fimbriae (Figs. 8.4 and 8.5).

What are today called *Actinomyces* "fimbriae" were first described by Girard and Jacius (1974) as "fibrils" in *A. viscosus* and later by Ellen et al. (1978) as adherence-associated "long appendages" in *A. naeslundii*, because there was no accepted precedent then for fimbria-like or pilus-like structures in Gram-positive bacteria. The major thrust in understanding the structure and function of these distinct fimbriae was led by Cisar and coworkers (Cisar

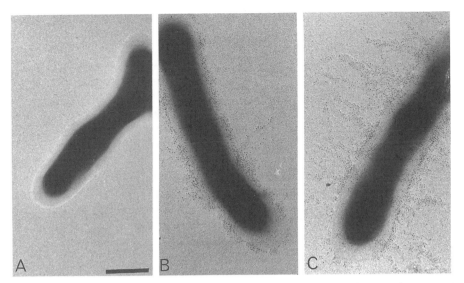

Fig. 8.5. Electron photomicrograph of *Actinomyces naeslundii* (*viscosus*) T14V-J1 labeled by indirect immunogold method with either KCl buffer **(A)**, anti-type 1 **(B)** or anti-type 2 **(C)** monospecific antifimbrial antibodies as the first step (antibodies provided by J. Cisar), followed by a goat antirabbit IgG gold probe, 5 nm diameter. Fimbriae bearing type 2 antigen are labeled along a greater length from the cell body. Bar = 500 nm. (Reproduced from Ellen ct al., 1989, with permission of the publisher.)

and Vatter, 1979; Cisar et al., 1989). They first used lactose-sensitive, intergeneric bacterial aggregation (coaggregation) as their adhesion model and then extended their work to eukaryotic cells and, through collaboration with Clark et al. (1984, 1989), to saliva-coated hydroxyapatite. They characterized types 1 and 2 fimbriae by their immunological specificity. Then, using antibodies prepared against the fimbriae, they demonstrated that the type 2 fimbriae carried the streptococcal aggregating activity, which was lactose inhibitable and reversible (Cisar and Vatter, 1979; Revis et al., 1982). From their work and from the biochemical and serological studies of fimbriae purified from representative strains of the taxonomic clusters by Masuda and coworkers (1981, 1983), it became evident that strains of *A. naeslundii* and strains then classified as *A. viscosus* expressed fimbriae that were to a degree antigenically distinct. Of ecological significance, strains representing the taxonomic groups were found to differ in both their adhesive properties and their expression of the two fimbrial types. *A viscosus* and strains from cluster-3 *A. naeslundii* adhered much better than strains from cluster-5 *A. naeslundii* to saliva-coated hydroxyapatite (Clark et al., 1985). Yet the latter strains hemagglutinated and coaggregated readily with oral streptococci, which are both lectin functions.

Cisar and coworkers (1984) subsequently demonstrated that a strain typical of cluster-5 *A. naeslundii* carried type 2 fimbriae exclusively, the fimbriae that bore the β-galactoside-specific lectin activity. Thus its relatively poor adherence to salivary pellicle could be explained by its lack of type 1 fimbriae, which had just been shown by Clark and coworkers (1981, 1982) to confer salivary pellicle adhesion properties to *A. viscosus*. The structural association of types 1 and 2 fimbriae with adhesion to salivary pellicle and lectin activities, respectively, was later confirmed using fimbria-deficient mutants selected on the basis of their failure to react with antibodies against the fimbrial antigens (Cisar et al., 1988; Clark et al., 1989).

The β-galactoside-specific adhesion of these *Actinomyces* species was among the first bacterial lectins reported. Although first associated by McIntire et al. (1978) with actinomycete–streptococcal coaggregation, β-galactoside specificity is also crucial for adhesive interactions with erythrocytes (Costello et al., 1979; Ellen et al., 1980), epithelial cells (Brennan et al., 1984), and polymorphonuclear leukocytes (PMNs) (Cisar et al., 1989; Kurashima et al., 1991; Sandberg et al., 1995), and it is associated with aggregation of *A. naeslundii* by submandibular/sublingual saliva (Ellen et al., 1983) (Fig. 8.4). Lactose has been the most frequently used inhibitor of such reactions, probably because of its availability. Coaggregation studies by McIntire and coworkers (1983) and biochemical analyses by Reddy et al. (1994) have characterized more definitively the receptor for the *Actinomyces* lectin in repeat oligosaccharide units of streptococcal polysaccharides. Streptococci currently classified as *S. oralis* (including the original *S. sanguis* 34 used in coaggregation experiments by McIntire et al. [1978, 1982]) and some strains of *S. sanguis, S. gordonii, Streptococcus mitis,* and *Streptococcus anginosus* bind the *Actinomyces* lectin via either Galβ1–3GalNAc- or GalNAcβ1–3Gal-containing polysaccharide receptors (Reddy et al., 1994). Studies in McIntire's laboratory and Ellen's laboratory have demonstrated that, while affected by nonspecific ionic or hydrophobic interactions, *Actinomyces* coaggregation with *S. oralis* 34 is far more sensitive to inhibitors of the lectin activity than to detergents or chaotropic agents (McIntire et al., 1982; Ellen et al., 1994c). These observations helped to define, in real terms, the relative importance of specific versus nonspecific adhesion.

Investigations of the receptors for the *Actinomyces* lectin on eukaryotic cells have implicated glycoproteins and glycolipids with glycosidic specificity resembling the streptococcal receptors. Using a solid-phase overlay assay in which adhesion could be blocked by receptor analogs and targeting glycoconjugates that often confer receptor specificity to mammalian membranes, Strömberg and Karlsson (1990) characterized the binding of *A. naeslundii* and *A. viscosus* strains to glycosphingolipids. Both strains bound well to lactosylceramide and to a series of other GalNAcβ1–3- and Galβ1–4-containing ceramides in a lactose-resistant, GalNAcβ1–3GalαOethyl-sensitive fashion. The *A. naeslundii* strain also bound to terminal or internal GalNAcβ units in a typically lactose-inhibitable manner. Their fine specificity in recognizing these glycoconjugates seemed to correspond to differences in their relative adherence to epithelial cells, streptococci, and saliva-coated hydroxyapatite and in their coaggregation with streptococci (Stömberg and Borén, 1992).

Studies to characterize receptor specificity for the adhesion of *Actinomyces* strains to eukaryotic cells themselves have used lectins with distinct glycoside specificity as probes to inhibit the interaction. Treatment of cultured KB oral epithelial cells with the Gal- and GalNAc-specific lectins from *Bauhinia purpurea* and *Arachis hypogaea* inhibits the adhesion of *Actinomyces* strains bearing type 2 fimbriae, as does treatment with lactose and other β-galactosides (Brennan et al., 1984). These lectins also bind to a putative KB cell glycoprotein receptor in nitrocellulose transblots. These and the Gal/GalNAc-specific lectins of *Ricinus communis* also bind specifically to PMNs and inhibit PMN killing of *Actinomyces* cells, suggesting that they share receptor specificity with the bacteria (Sandberg et al., 1995). Indeed, overlay methods have demonstrated a PMN glycoprotein and glycolipids with a common oligosaccharide requirement to that of KB cells (Brennan et al., 1987).

A prerequisite for *Actinomyces* strains to adhere to eukaryotic cells and glycoconjugates in overlays prepared from cell membrane extracts is pretreatment by sialidase, an enzyme that removes the terminal sialic acid that naturally masks the penultimate galactosidic units. For example, sialidase (i.e., neuraminidase) treatment of PMNs greatly enhances their galactoside-sensitive, phagocytic bacteriocidal activity toward *Actinomyces* cells bearing the fimbrial lectin. Discovery of this reaction was shared by Costello et al. (1979) and Ellen et al. (1980), who demonstrated that hemagglutination via the lactose-sensitive lectin of *Actinomyces* strains depended on pretreatment with a sialidase. Both groups found that the *Actinomyces* strains themselves could serve as a source of sialidase, thereby priming erythrocytes for subsequent agglutination, even by other strains. Ellen et al. (1980) showed that sialidase and hemagglutination priming activity varied among representative strains of the taxonomic clusters. In essence, these were among the first reports in oral microbiology that supported a subsequent hypothesis promulgated by Gibbons and coworkers that bacteria often adhere to cryptic, or hidden, receptors unmasked by enzymatic activities or by conformational changes in proteins adsorbed to surfaces (see below).

In a series of studies that used their standard assay for quantifying bacterial adherence to saliva-coated hydroxyapatite, Clark and coworkers (1981, 1984, 1989), established that the type 1 fimbriae mediated *Actinomyces* adhesion to salivary pellicle (Fig. 8.4). Monoclonal antibodies against the type 1 fimbriae but not the type 2 fimbriae blocked adhesion. Type 1 fimbrial preparations competitively inhibited the adhesion of strains bearing type 1 fimbriae. Moreover, only wild-type and Cisar's mutant strains bearing type 1 fimbriae adhered well to the salivary coatings, whereas mutants lacking type 1 or both types of fimbriae adhered in much lower numbers than wild-type strains.

The putative salivary pellicle receptors for type 1 fimbriae of *Actinomyces* species belong to the acidic proline-rich protein (PRP) family and the tyrosine-rich protein statherin. These salivary proteins are generally not glycosylated; they are known to bind readily to hydroxyapatite and to function in calcium sequestration in pellicle on the tooth surface (Bennick et al., 1979; Hay et al., 1988). Salivary fractions rich in these proteins form pellicles on hydroxyapatite *in vitro* that support the lactose-resistant adhesion of type 1 fimbria-bearing strains of *A. viscosus* (Clark et al., 1989; Gibbons and Hay, 1988; Gibbons et al., 1988). Furthermore, latex beads coated with purified PRPs and statherin are agglutinated by *A. viscosus* strains expressing type 1 fimbriae but not by mutants that are type 1 deficient. Similarly, PRP–gold probes adhere to fimbriae on only those *A. viscosus* strains that bear type 1 fimbriae (Leung et al., 1990). Gibbons and Hay (1988) have studied the adherence of *A. viscosus* to pellicles composed of single, purified PRPs and have found that PRP-1, PRP-2, and PIF-slow, which are 150 amino acid residue family members that differ at only one or two amino acid residues, support maximum *A. viscosus* adhesion. The 106 residue PRP-3 and PRP-4 proteins, which lack a 44 residue C-terminal post-translational cleavage fragment, require much higher concentrations to support equal bacterial adhesion, suggesting that the C-terminal domain enhances but is not essential for *A. viscosus* adsorption. A 30 residue N-terminal tryptic digest segment, which is known to bind to the enamel during pellicle formation, did not support *A. viscosus* adhesion when used as a pure *in vitro* mineral coating. Although PRPs are probably the major pellicle receptors that bind type 1 fimbriae via a protein–protein interaction, other pellicle proteins can theoretically serve as ligands for additional *Actinomyces* adhesins. Indeed, Strömberg et al. (1992) have recently reported very complex and variable interactions of *A. viscosus* and *A. naeslundii* strains with *in vitro* pellicles formed from different-sized fractions of parotid and submandibular salivas, including strain-specific adhesion to PRP-rich fractions, statherin-rich fractions, and high-molecular-weight fractions supporting GalNAcβ1–3GalαOethyl-sensitive adherence. They interpreted the specificity to depend on the differential glycosylation patterns of the salivary proteins.

Although the immunochemistry, functions, and receptor specificity of both types 1 and 2 *Actinomyces* fimbriae are well understood, the actual adhesins on the fimbriae have yet to be identified and isolated. The type 1 fimbrial subunit gene of *A. viscosus* T14V and the type 2 fimbrial subunit gene of *A. naeslundii* WVU45 have been cloned and expressed in *Escherichia coli* by Yeung and Cisar (1990) and characterized in Yeung's laboratory (Yeung, 1992). The encoded proteins have molecular weights approximating 55,000 to 60,000 kDa, which is very close to estimates based on biochemical determinations for the native fimbriae (Masuda et al., 1981, 1983). The amino acid sequences show no homology to other known proteins, but strains of other *Actinomyces* species hybridize with probes for type 1 fimbrial subunit sequences (Yeung, 1992). The cloned type 1 and type 2 fimbrial unit proteins share 34% amino acid sequence identity, including several conserved sequences near both the N and C termini, suggesting that they share common ancestry.

It is generally thought and supported by the immunochemical data of Cisar et al. (1991) that the antigenic activities that have been used to characterize the fimbriae relate to the major structural proteins of the fimbriae and do not define the adhesins. From immunogold labeling using the Cisar et al. fimbrial mutants and antibodies against type 1 and type 2 antigens, the distribution of antibody-reacting sites seems to differ on the two fimbrial types (Ellen et al., 1989). The labeling pattern on type 2 fimbriae extends for a greater distance along the length of the fimbriae, whereas the label on type 1 fimbriae is concentrated more closely to the cell body (Fig. 8.5). The type 2 fimbriae also bear epitopes that have been used by Firtel and Fillery (1988) to distinguish among *Actinomyces* taxonomy clusters. Progress in characterizing the structural basis of lectin activity of the type 2 fimbriae and the assembly of the fimbriae has been slow because the native fimbriae are difficult to solubilize and the cloned subunit does not self-assemble (J. Cisar, unpublished communication). Moreover, the lectin activity relies on multiple low affinity interactions for its avidity, which probably depends on maintaining the fimbrial unit intact. Like studies on the assembly of *E. coli* fimbriae, progress will probably come from directed mutagenesis of fimbrial genes once systems for experimental gene transfer in *Actinomyces* species become routine (Yeung and Kozelsky, 1994).

8.6. CRYPTITOPES: CRYPTIC RECEPTORS FOR BACTERIAL ADHESINS

Data obtained from a series of *A. viscosus* adhesion experiments led Gibbons and Hay to raise the hypothesis that bacteria that colonize in a salivary milieu have evolved adhesins that recognize cryptic receptors ("cryptitopes") that become exposed in a receptive conformation upon their adsorption to enamel to form salivary pellicle (Gibbons and Hay, 1988; Gibbons et al., 1990). Their supportive observations were that PRPs in solution do not bind to *A. viscosus,* nor do they inhibit the adhesion of *A. viscosus* to saliva-coated hydroxyapatite or to PRP-containing salivary pellicles. Yet *A. viscosus* adheres with high affinity to mineral-bound PRPs. It was known that PRPs changed conformation upon adsorption to mineral surfaces via their highly negatively charged N-terminal segment (Moreno et al., 1982), and thus a basis for explaining the plausibility of cryptitope exposure seemed evident. Experiments using *S. gordonii* essentially repeated and confirmed these observations, but the two species had some differences in their PRP adhesion specificity. With *S. gordonii,* Gibbons et al. (1991) went beyond experiments using native PRP fragments as pellicles to determine the minimum synthetic peptide sequence that would support bacterial aggregation of coated agarose beads. The minimum peptide that promoted adherence was the C-terminal dipeptide Pro-Gln that represents PRP residues 149 and 150. There is as yet no conclusive evidence that this peptide or any other that is specific for a known bacterial adhesin is the masked domain that becomes exposed due to a conformational change upon adsorption of PRPs to hydroxyapatite. Yet the cryptitope hypothesis is compelling, because it may explain how bacteria can colonize the teeth despite being suspended in high concentrations of salivary proteins that serve as their natural ligands when adsorbed to the teeth.

Unmasking of cryptic receptors may occur by other means than conformational change of a protein upon its adsorption to a surface. Receptors in pellicle, on mucosal surfaces in the mouth, and even on PMNs that migrate into the mouth via the gingival crevice offer a wide range of potential peptide and oligosaccharide sequences to which oral bacteria have probably adapted. Gibbons et al. (1990) also proposed that cryptic receptors for bacterial adhesins might be exposed by the action of degradative enzymes like the myriad of sialidases and proteases elaborated by oral bacteria. Many species of oral bacteria bear galactoside-specific lectins, and Childs and Gibbons (1990) found that the adherence of such species is enhanced when oral epithelial cells are treated with neuraminidase, as mentioned for the *Actinomyces* strains above. Similarly, the adherence of some of the Gram-negative anaerobes, which are associated with periodontal disease, to oral epithelial cells was enhanced by mild protease digestion or exposure of the cells to preparations of lysosomal enzymes from PMNs. The fact that some of these species, like *Porphyromonas gingivalis,* rely on catabolism of peptides for energy and thus have evolved a broad range of extracellular protease activities raises the likelihood that they have evolved adhesive mechanisms that recognize peptide sequences altered by proteolysis. For example, Naito and Gibbons (1988) have shown that tryptic digestion of fibronectin–collagen complexes promotes the adhesion of *P. gingivalis,* and this and a few other species of periodontal pathogens produce proteases with trypsin-like peptidase specificity. Therefore, this explanation for the unmasking of cryptic receptors probably has wide applicability in bacterial colonization throughout the gastrointestinal tract.

8.7. PROTEOLYTIC ENZYMES AS ADHESINS

In addition to unmasking receptors, there is emerging evidence that bacterial outer membrane proteases may function as adhesins themselves (see Chapter 9, Section 9.3.2.3, this volume). Since the kinetics of adhesin-ligand interactions shares many similarities with the kinetics of enzyme–substrate interactions, especially considering their specificity and avidity, proteases borne in a surface location should have the opportunity to mediate adhesive interactions through their recognition domains. Lantz and coworkers' demonstration of 150 and 120 kDa outer membrane proteases (porphypain-1 and -2) of *P. gingivalis* serving the dual function of adhesin and fibrinogenolytic enzyme is probably the best example of this concept (Lantz et al., 1991a,b). Grenier (1992) and Hoover et al. (1992) demonstrated a strong relationship between the trypsin-like activity of *P. gingivalis* and its ability to adhere to epithelial cells and erythrocytes, respectively. Nishikata and Yoshimaura (1991) isolated a *P. gingivalis* hemagglutinin that was proteolytic. Recently, Hoover and Yoshimura (1994) isolated *P. gingivalis* mutants by transposon mutagenesis that were both trypsin activity and hemagglutination deficient. Similarly, using inhibition assays and chemically induced trypsin-deficient mutants, Li et al. (1991) and Ellen et al. (1992) have published evidence implicating *P. gingivalis* trypsin-like proteases in its coadhesion with *A. viscosus*. Much of the protease activity of *P. gingivalis* is associated with its outer membrane and abundant surface vesicles (Figs. 8.4 and 8.6). Isolated vesicles bind to a variety of substrata to which whole *P. gingivalis* cells adhere. In the case of its adhesion

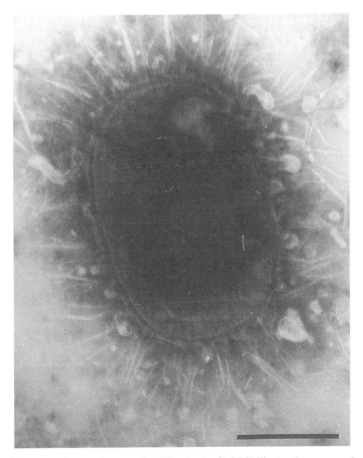

Fig. 8.6. Electron photomicrograph of *P. gingivalis* 2561 illustrating a complex surface with abundant outer membrane vesicles and fimbriae. This particular cell was a control exposed to preimmune rabbit serum and a goat antirabbit IgG–gold probe prior to negative staining with methylamine tungstate by the method of Handley and Tipler (1986). Bar = 500 nm. (Courtesy of I.A. Buivids and R. P. Ellen.)

to *A. viscosus*, pretreatment of *A. viscosus* monolayers with *P. gingivalis* vesicle preparations can inhibit the subsequent adhesion of *P. gingivalis* cells in a concentration-dependent fashion (Ellen and Grove, 1989).

A wide range of protease inhibitors can diminish *P. gingivalis* adherence to *A. viscosus*, collagen, and other proteins. Arginine, one of the amino acid residues exposed by trypsin activity, has also been shown to inhibit the adhesion of *P. gingivalis* and its adhesins to many protein-containing substrata, to other bacteria, and to eukaryotic cells (Bourgeau and Mayrand, 1990; Ellen et al., 1992; Inoshita et al., 1986; Okuda et al., 1986). Pike et al. (1994) have recently purified arginine-specific and lysine-specific cysteine proteinases (Arg- and Lys-gingipain) that complex noncovalently with hemagglutinating proteins. Their laboratory has provided evidence that *P. gingivalis*'s multiple forms of trypsin-like activity are due to the activities of these two proteinases (Potempa et al., 1995). They have also cloned a gene encoding the 50 kDa Arg-gingipain (RGP-1) and have proposed that RGP-1 is synthesized as a polyprotein complexed with multiple adhesin (hemagglutinin) molecules located at the C terminus (Pavloff et al., 1995). Thus, these new findings suggest that molecules important for both adhesion and acquisition of required peptides and hemin may have evolved in tandem and may be co-regulated.

8.8.   BACTERIAL ADHESION TO MATRIX PROTEINS

Like many bacteria that colonize human mucosal surfaces, *P. gingivalis* and several other oral bacteria can adhere to human matrix proteins (Winkler et al., 1988; Patti et al., 1994). Their adhesion seems to be a more dynamic process than previously appreciated, affected either by exposure of [cryptic] complementary receptor domains in the target substrate or even in the adhesin itself or by reorientation of bacterial adhesins upon substrate contact. For example, like its adhesion to hydroxyapatite-adsorbed salivary proteins, *S. sanguis* adheres readily to fibronectin (Fn)-coated surfaces and to vegetations that typically form on damaged heart valves. The Fn interaction is evidently affected by conformational changes in Fn when it adsorbs on inanimate surfaces (Lowrance et al., 1988, 1990). Other oral streptococci, like *S. salivarius* and *S. mitior,* can absorb fibronectin from saliva (Ericson and Tynelius-Bratthall, 1986), and mutans streptococci can bind to Fn when it adsorbs as a component of salivary pellicles *in vitro*(Babu and Dabbous, 1986).

S. sanguis also expresses a platelet-aggregating protein adhesin (PAAP) with a platelet interactive domain conforming to a predicted 7-mer that also confers platelet-activating activity to intercellular matrix collagens I and III (Erikson and Herzberg, 1993). PAAP[+] phenotypes aggregate platelets in the presence of plasma. Strains of *S. sanguis* with the PAAP phenotype have been shown to cause larger heart valve vegetations and more extensive endocarditis lesions than PAAP-deficient phenotypes in a rabbit model (Herzberg et al.,

1992). Some of the nonplatelet interacting strains of *S. sanguis* evidently express cryptic adhesins that can be detected by PAAP-specific monoclonal antibodies only upon mild tryptic digestion of the bacterial surface (Herzberg et al., 1990). The PAAP of *S. sanguis* is but one example of mimicry of common human proteins among the natural oral microbiota.

8.8.1. Polar Adhesion of Spirochetes to Matrix Proteins

There has been a great deal of recent attention to matrix protein interactions of the cultivable, small oral spirochete *Treponema denticola* (Ellen et al., 1994a). Like *Treponema pallidum* and many other spirochetes, *T. denticola* often adheres to host cells and to inanimate surfaces coated with matrix proteins in a polar orientation. *T. denticola* adhesion is selective; it adheres more readily to fibronectin and laminin than to collagen and serum albumin (Dawson and Ellen, 1990). Strains of *T. denticola* differ in their ability to bind to fibronectin, and those that adhere best apparently bind more frequently in a polar orientation. Similar to *T. pallidum, T. denticola* adheres to plastic surfaces coated with peptides containing the sequence RGDS, which is common to fibronectin, laminin, and some other extracellular matrix proteins. However, the adhesion of *T. denticola* to Fn cannot be inhibited by these peptides in solution. Therefore, it is possible that *T. denticola* interacts with more than one domain in the Fn molecule, or it may alter the Fn molecule. A key feature of *T. denticola* tip-oriented adhesion is that, upon contact with Fn-coated surfaces, its Fn-specific adhesins cluster toward one end of the bacterium (Fig. 8.7) (Dawson and Ellen, 1994). Reorientation of the Fn-specific adhesins may be a kind of primitive capping phenomenon resembling that seen in eukaryotic membrane receptors exposed to complementary ligands, and, like capping, it probably increases the avidity of the adhesive interaction. When adhesion assays are run in the cold, *T. denticola* still adheres to Fn but polar orientation is much less frequently observed, suggesting that the clustering of Fn adhesins toward the poles may depend on outer membrane fluidity. These observations support the concept that bacteria have evolved mechanisms to concentrate adhesins toward one region of the cell, thereby overcoming the limitations of their simple structure.

Like *P. gingivalis, T. denticola* is highly proteolytic and largely depends on the transport of peptides for energy. This spirochete elaborates endopeptidases and proteases with both trypsin-like and chymotrypsin-like activities. The chymotrypsin-like protease is the major protein degradative activity associated with intact cells, and Grenier and coworkers (1990) have demonstrated that it is located on the bacterial surface. This enzyme is probably involved in both the adhesion and degradation of Fn as well as other matrix proteins. The protease inhibitor PMSF inhibits the chymotrypsin-like but not the trypsin-like activities of *T. denticola*. PMSF pretreatment inhibits tip-oriented adhesion of *T. denticola* to Fn coated on plastic (Dawson and Ellen, 1994). PMSF also inhibits the degradation of endogenous Fn on human gingival fibroblasts by

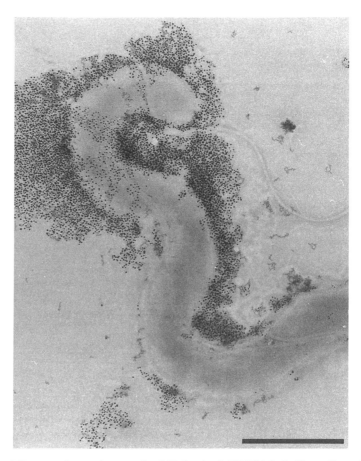

Fig. 8.7. Electron photomicrograph of *T. denticola* 33520 labeled by a direct fibronectin (Fn)–gold method after migrating through a methyl cellulose column into contact with Fn immobilized on a nitrocellulose membrane. In such experiments, the Fn probe clusters toward one end of the spirochete and does not redistribute upon incubation, presumably due to crosslinking of the Fn ligand. This particular cell, incubated several hours at room temperature after labeling, shows an extreme degree of gold label associated with the cell and with outer sheath material shedding from the cell. Bar = 500 nm. (Courtesy of J.R. Dawson and I.A. Buivids; based on Dawson and Ellen, 1994.)

T. denticola. However, it does not affect the adhesion of *T. denticola* to the fibroblasts, suggesting once again that the bacterium has alternative mechanisms for attaching to host cells (Ellen et al., 1994b). There are several outer membrane proteins of *T. denticola* to which soluble Fn binds *in vitro* (Haapasalo et al., 1991; Umemoto et al., 1993). Among these is a prominent 53 kDa antigen that has been cloned and characterized in McBride's laboratory (Haapasalo et al., 1992). The 53 kDa adhesin also functions like a large porin *in vitro,* and it is closely associated with chymotrypsin-like proteolytic activity

(Egli et al., 1993). Similarly, Weinberg and Holt (1991) have reported a major 64 kDa protein likely involved in the adhesion of some *T. denticola* strains to fibroblasts, and it too is part of a large protein complex with *in vitro* characteristics of a porin (A. Weinberg, personal communication). Thus, there seems to be a unique juxtaposition of adhesive, degradative, and pore-forming surface proteins that might optimize the adhesion and the uptake of peptides by *T. denticola*.

8.9. *PORPHYROMONAS GINGIVALIS* FIMBRIAE

In addition to outer membrane proteins and enzymes that may serve as adhesins, *P. gingivalis* has structurally distinct fimbriae that have been implicated in adhesion to a variety of target substrates (Fig. 8.6) (Yoshimura et al., 1984, 1985). In electron photomicrographs, the fimbriae appear wavy rather than straight and rigid like the fimbriae of many other Gram-negative bacteria (Handley and Tipler, 1986; Yoshimura et al., 1984). Investigations of the encoding gene sequence for the fimbrial subunit, fimbrillin, have also yielded a unique nucleotide sequence (Dickinson et al., 1988). The size of the purified native fimbrial protein was originally estimated at 43 kDa, but subsequent studies of several strains have yielded values from 41 to 48 kDa. Based on the nucleotide sequence of the cloned fimbrial subunit, the predicted size of fimbrillin is around 35,924.

Research into adhesive interactions of *P. gingivalis* fimbriae has followed two major paths: adhesion to saliva-coated hydroxyapatite, which may be significant for colonization; and adhesion to host cells, which may be significant for the stimulation of cytokine release as part of the complex pathogenesis of periodontitis. Although the significance of adhesion to salivary pellicles for the establishment of *P. gingivalis* in the human mouth may be questioned (Slots and Gibbons, 1978), there is evidence that fimbriae contribute significantly to its adhesion to saliva-coated hydroxyapatite *in vitro* and to its colonization in laboratory rodents. Both the native fimbriae and recombinant fimbrillin, expressed as amino acids 10 to 337 of the deduced sequence, have been shown to bind to saliva-coated hydroxyapatite and to inhibit the adhesion of *P. gingivalis* cells to saliva-coated hydroxyapatite (Lee et al., 1992). Synthetic peptides based on sequences mostly from the C-terminal end are also inhibitory. An insertionally inactivated mutant in *fimA,* the gene encoding fimbrillin, has a greatly reduced capacity to adhere to saliva-coated hydroxyapatite compared with the wild-type (Genco et al., 1994). Recent studies by Amano et al. (1994) have shown salivary binding interactions of recombinant fimbrillin that are reminiscent of some of the experiments on the adhesion of *A. viscosus* mentioned above. Recombinant fimbrillin bound to pellicles from whole, parotid, or submandibular saliva and to pellicles of PRP-1 and statherin adsorbed to saliva-coated hydroxyapatite. Yet the interaction was not inhibited by the pure salivary proteins in solution. It resembled a protein–protein interaction and was not inhibited by carbohydrates, EDTA, or charged amino acids. The fimbriae and synthetic peptides are immunogenic, stimulating both humoral and cellular responses (Ogawa et al., 1994). An initial study in gnotobiotic rats has found that *P. gingivalis* colonization may be impaired by immunization with a vaccine containing native fimbriae (Evans et al., 1992). Therefore, there is interest in the potential use of recombinant fimbrillin as a vaccine component.

Like fimbriae from many species, *P. gingivalis* fimbriae adhere to a variety of cell types, including other bacteria, fibroblasts, epithelial cells, and various inflammatory cells (Genco et al., 1994; Goulbourne and Ellen, 1991; Isogai et al., 1988; Lamont et al., 1993). There is some controversy over the function of native fimbriae in hemagglutination. *P. gingivalis* agglutinates sheep erythrocytes, and most strains are fimbriated. However, many investigations to identify the hemagglutinins have yielded proteins that are immunologically distinct from native fimbriae, that do not copurify with the fimbriae, but that might complex with the fimbriae on the bacterial surface (Mouton et al., 1991). Furthermore, cloning from a *P. gingivalis* library in *E. coli* by screening for expression of functional hemagglutinins has yielded genes distinct from *fimA*, (Lépine and Progulske-Fox, 1993). Recently, Ogawa and Hamada (1994) demonstrated that native fimbriae and a few synthetic peptides based on the sequence for fimbrillin agglutinated erythrocytes from a wide variety of animal species and that hemagglutination could be inhibited by the amino acids L-arginine and L-lysine but not by carbohydrates. Their data using eukaryotic cells as adhesion substrates support a common theme for *P. gingivalis* adhesion, that it is based most often on protein–protein interactions, like those of *A. viscosus* type 1 fimbriae, and therefore might be uncommon when considering fimbrial functions that often display characteristics of lectins. As mentioned above, some chemically induced and transposon-induced mutants of *P. gingivalis* are both hemagglutination- and trypsin deficient and less hydrophobic than their parent strains. Contrasting this work, experiments with mutants specifically inactivated in *fimA* have shown no change in hemagglutination or in hydrophobicity (Genco et al., 1994; Hamada et al., 1994).

The biological responses of a variety of human cells to stimulation by *P. gingivalis* fimbriae exemplify the multifunctional nature of structures bearing bacterial adhesins. The fimbriae are chemotactic for monocytes (Ogawa and Hamada, 1994). The native fimbriae and some peptides based on its sequence are mitogenic for murine spleen cells and cause the release of interleukin-6 and tumor necrosis factor-α from monocytes (Hanazawa et al., 1988, 1991). They stimulate thymocyte-activating factor from human gingival fibroblasts. They also stimulate macrophages to express interleukin-1 and granulocyte macrophage colony-stimulating factor associated with osteoclastic bone resorption in calvaria *in vitro* (Kawata et al., 1994). Therefore, *P. gingivalis* fimbriae have the capacity to perturb many of the pathways involved in inflammatory resorption of periodontal tissues and in natural connective tissue remodeling.

8.10. HOST CELL CYTOSKELETAL PERTURBATION AND INVASION

It is evident that *P. gingivalis* can have profound effects on the host cells to which it or its adhesins bind. In addition to stimulating cytokine expression and release, the adhesion of periodontal pathogens to host cells evidently sends transmembrane signals that perturb other critical cellular functions; among these are cytoskeletal reorganization and the uptake of some of the bacteria themselves. Following the fast pace of advances with enteric species, it is now becoming apparent that endogenous pathogens, including several oral species, have the capacity to penetrate host tissues *in vivo* and to invade human cells *in vitro*. Regardless of whether frank invasion of the periodontium is a key element of the pathogenesis of periodontal infections, the ability of bacteria or their products to adhere to and thereby to initiate a cascade of intracellular reactions that can subvert normal functions of periodontal cells is an important line of investigation that is paralleling studies on exogenous pathogens.

Much recent research on epithelial cell invasion has centered on *Actinobacillus actinomycetemcomitans,* one of the principal pathogens in early onset and some recurrent forms of periodontitis. *A. actinomycetemcomitans* is known to penetrate the gingival epithelium during natural periodontal infections (Christersson et al., 1987). Its adhesion and invasion have been studied mostly in Fives-Taylor's laboratory, using the KB epithelial cell line and assays that have been used to study the invasion of salmonellae. Morphologically, *A. actinomycetamcomitans* interacts with the surface of KB cells in an analogous fashion to invasive enteropathic bacteria. Upon contact, it stimulates cytoskeletal responses characterized by the accumulation of F-actin near the point of cell contact and the engulfment of the adherent bacteria via an endocytic pathway (Meyer et al., 1991; Sreenivasan et al., 1993). Although the precise mechanism of adhesion is yet to be identified, there is some evidence of strain specificity and the association of surface fimbriae, extracellular vesicles, and an extracellular amorphous material with the ability to adhere (Meyer and Fives-Taylor, 1993, 1994; Mintz and Fives-Taylor, 1994). Because methods for genetic transfer in *A. actinomycetamcomitans* were not available until only recently, there has been little investigation of the molecular basis of its adhesion and invasion. Meyer and coworkers have just presented an initial report indentifying a few cloned genetic fragments that permit the naturally noninvasive host *E. coli* strain to invade KB cells (Meyer et al., 1995). Strains of *P. gingivalis* have also been reported to invade KB cells in culture, and photomicrographs suggest that it too may be engulfed concomitant with rearrangement of the actin cytoskeleton (Duncan et al., 1993; Sandros et al., 1993).

The adhesion and resultant cytoskeletal perturbation by *T. denticola* has been studied using both human gingival fibroblasts (HGF) and oral epithelial cells. Upon adhesion of *T. denticola* to the dorsal surface of cultured HGF, the cells undergo membrane ruffling, shape changes, and gross rearrangement of the cytoskeleton that has been measured by monitoring changes in stress fiber integrity and fluorimetry of F-actin stained with phalloidin conjugated to fluorophores (Baehni et al., 1992; Weinberg and Holt, 1990; Ellen et al., 1994a). Within 1 hour of contact and concomitant with (but not necessarily a result of) degradation of endogenous cell surface Fn, the actin stress fibers become disorganized and much of the cortical F-actin in the ventral third of the cell is depolymerized (Yang and Ellen, 1995). There is condensation of F-actin in a perinuclear array (Baehni et al., 1992). Profiles of *T. denticola* have been detected inside some of the HGF, but not in endocytic vacuoles, by electron microscopy and by double-labeling fluorescence microscopy (Ellen et al., 1994a). Subsequently, the cells detach from the substratum. Since HGF locomotion through intercellular matrix and their phagocytosis of collagen depends on cytoskeletal integrity, changes of this magnitude initiated by contact with bacterial adhesins or degradation by adhesive surface proteases would probably have serious consequences for normal tissue remodeling and wound healing.

T. denticola also adheres to and causes cytoskeletal perturbation, surface blebbing, endogenous fibronectin degradation, and detachment of oral epithelial cells. Some cellular responses are cell-line specific. Adhesion of *T. denticola* to KB cells stimulates cellular shrinkage associated with reduced expression of F-actin and desmoplakin II, a protein that localizes in cell junctions (De Filippo et al., 1995). The KB cells' physiological volume regulation is also perturbed, but the viability of the cells remains unaffected until they detach from the substratum. As yet, there is no evidence for intracellular invasion of KB cells. In contrast, *T, denticola* can apparently invade some cultured porcine epithelial rest of Mallesez cells (PRM), especially those that are subconfluent and still migrating (Uitto et al., 1995). When the PRM cells are cultured in a way that allows them to grow as a multilayered epithelium resembling sulcular epithelium *in vivo,* fragments of *T. denticola* cells containing its chymotrypsin, but not the bacteria themselves, penetrate the cell layers and are taken up into epithelial vacuoles. Fewer desmosomes are detected at cell junctions, and intercellular spaces become more permeable. For periodontal pathogens that are in intimate contact with the epithelium lining the gingival sulcus, invasion subsequent to adhesion may not be critical. The initiation of cytoskeletal changes, whether by transmembrane signaling through receptor-mediated events or by the degradation of endogenous matrix components, may upset physiological pathways by which the integrity of the epithelial barrier is maintained. This is probably one of the damaging effects that is initiated by adhesin–receptor contact on oral surfaces.

8.11. CONCLUSION: KNOWLEDGE OF ADHESION LEADS TO SELECTIVE PREVENTION OF ADHESION

The ultimate goal of research into the molecular mechanisms of bacterial adhesion is to gain enough knowledge and scientific insight to prevent the adhesion, subsequent colonization, and host tissue damage by pathogens. In this context, oral microbiologists have originated many conceptual advances, both in our understanding of the host's natural protection of mucosal surfaces and teeth from infection and in developing novel strategies for boosting local immunity. Indeed, the first demonstrations that the principal protective function of secretory IgA and mucinous glycoprotein agglutinins was their inhibition of bacterial adhesion were derived from investigations of oral streptococci, saliva, and buccal epithelial cells (Williams and Gibbons, 1972, 1975). Research on caries vaccines has concentrated on *S. mutans* GTFs and on surface proteins that adhere to the salivary proteins in pellicles on teeth, which are central to *S. mutans* colonization. Research on vaccines that may eventually yield partial protection against infection with a key periodontal pathogen, *P. gingivalis,* has focused on a fimbrial protein that is also recognized as an adhesin and inflammatory cell agonist. Several investigators have also studied ways to maximize oral secretory IgA responses by generating knowledge about vaccine adjuvants, recombinant vaccine delivery systems, and stimulation of specific immunity via lymphatic tissue elsewhere in the gastrointestinal tract (McGhee and Mestecky, 1990; Michalek and Childers, 1990). At the interface between efficacy and safety, there are studies underway to define adhesion domains so that synthetic peptides that retain immunogenicity and antigenic specificity but obviate the potentially dangerous cross-reactivity of the native adhesins could be used in caries vaccines (Lehner et al., 1994; Smith et al., 1994b).

Investigations into preventing bacterial adhesion to oral surfaces have also followed nonimmunological lines. In the past there were attempts to reduce selectively the adhesion and colonization of mutans streptococci by the topical use of glucanases and even some thought to altering diets to interfere specifically with lectin-mediated adherence of some oral bacteria. An approach addressing the nonspecific nature of some attractive adhesive forces has been the recent development of a two-phase mouth rinse that desorbs bacteria from oral surfaces by taking advantage of their hydrophobicity (Rosenberg et al., 1992). The oral cavity, with its desquamating mucosal surfaces and its static saliva-bathed tooth surfaces on which plaque biofilms form, offers a challenging yet accessible model system for studying global concepts of bacterial adhesion and novel strategies for intervention. Indeed, many of the principles guiding research on infections of other mucosal surfaces and on biofilms infecting implanted medical devices originated during the past three decades of intensive investigation into the adherence of oral bacteria.

ACKNOWLEDGMENTS

Original research by the authors is supported by grant MT-5619 from the Medical Research Council of Canada (R.P.E.) and grant DE09878 from the National Institute of Dental Research (R.A.B.)

REFERENCES

Amano A, Sojar HT, Lee J-Y, Sharma A, Levine MJ, Genco RJ (1994): Salivary receptors for recombinant fimbrillin of *Porphyromonas gingivalis.* Infect Immun 62:3372–3380.

Appelbaum B, Golub S, Holt SC, Rosan B (1979): *In vitro* studies of dental plaque formation: Adsorption of oral streptococci to hydroxyapatite. Infect Immun 25:717–728.

Babu JP, Dubbous MK (1986): Interaction of salivary fibronectin with oral streptococci. J Dent Res 65:1094–1100.

Baehni PC, Song M, McCulloch CAG, Ellen RP (1992): *Treponema denticola* induces actin rearrangement and detachment of human gingival fibroblasts. Infect Immun 60:3360–3368.

Banas JA, Russell RRB, Ferretti JJ (1990): Sequence analysis of the gene for the glucan-binding protein of *Streptococcus mutans* Ingbritt. Infect Immun 58:667–673.

Barrett JF, Barrett TA, Curtiss RI (1987): Purification and partial characterization of the multicomponent dextranase complex of *Streptococcus sobrinus* and cloning of the dextranase gene. Infect Immun 55:792–802.

Beighton D (1984): The influence of saliva on the hydrophobic surface properties of bacteria isolated from oral sites of macaque monkeys. FEMS Microbiol Lett 21:239–242.

Bennick A, Cannon M, Madapallimattam G (1979): The nature of the hydroxyapatite-binding site in salivary acidic proline-rich proteins. Biochem J 183:115–126.

Birkhed D, Rosell K, Granath K (1979): Structure of extracellular water-soluble polysaccharides synthesized from sucrose by oral strains of *Streptococcus mutans, Streptococcus salivarius, Streptococcus sanguis,* and *Actinomyces viscosus.* Arch Oral Biol 24:53–61.

Bourgeau G, Mayrand D (1990): Aggregation of *Actinomyces* strains by extracellular vesicles produced by *Bacteriodes gingivalis.* Can J Microbiol 36:362–365.

Bowen WH, Schilling KM, Giertsen E, Pearson S, Lee SF, Bleiweis AS, Beeman D (1991): Role of a cell surface-associated protein in adherence and dental caries. Infect Immun 59:4606–4609.

Brennan MJ, Cisar JO, Vatter AE, Sandberg AL (1984): Lectin-dependent attachment of *Actinomyces naeslundii* to receptors on epithelial cells. Infect Immun 46:459–464.

Brennan MJ, Joralman RA, Cisar JO, Sandberg AL (1987): Binding of *Actinomyces naeslundii* to glycosphingolipids. Infect Immun 55:487–489.

Burne RA (1991): Oral ecological disasters: The role of short-term extracellular storage polysaccharides. In Bowen WH, Tabak LA (eds): Cariology for the Nineties. Rochester, NY: University of Rochester Press, pp 351–364.

Burne RA, Schilling K, Bowen WH, Yasbin RE (1987): Expression, purification and characterization of an exo-β-D-fructosidase of *Streptococcus mutans.* J Bacteriol 169:4507–4517.

Burne RA, Chen Y-YM, Penders JEC (1995): Molecular analysis of sucrose metabolism by *Streptococcus mutans* in biofilms. Abstract 1501. J Dent Res 74:199.

Busscher HJ, Cowan MM, Van der Mei HC (1992): On the relative importance of specific and non-specific approaches to oral microbial adhesion. FEMS Microbiol Rev 88:199–210.

Busscher HJ, Uyen MHWJC, Van Pelt AWJ, Weerkamp AH, Arends J (1986): Kinetics of adhesion of the oral bacterium *Streptococcus sanguis* CH3 to polymers with different surface free energies. Appl Environ Microbiol 51:910–914.

Chia J-S, Hsu T-Y, Teng L-J, Chen J-Y, Hahn L-J, Yang C-Z (1991): Glucosyltransferase gene polymorphism among *Streptococcus mutans* strains. Infect Immun 59:1656–1660.

Childs WC, Gibbons RJ (1990): Selective modulation of bacterial attachment to oral epithelial cells by enzyme activities associated with poor oral hygiene. J Periodont Res 25:172–178.

Christersson LA, Wikesjo UME, Albini B, Zambon JJ, Genco RJ (1987): Tissue localization of *Actinobacillus actinomycetemcomitans* in human periodontitis. II. Correlation between immunofluorescence and culture techniques. J Periodontol 58:540–545.

Cisar JO, Barsumian EL, Siraganian RP, Clark WB, Yeung MK, Hsu SD, Curl SH, Vatter AE, Sandberg AL (1991): Immunochemical and functional studies of *Actinomyces viscous* T14V type 1 fimbriae with monoclonal and polyclonal antibodies directed against the fimbrial subunit. J Gen Microbiol 137:1971–1979.

Cisar JO, David VA, Curl SH, Vatter AE (1984): Exclusive presence of lactose-sensitive fimbriae on a typical strain (WVU45) of *Actinomyces naeslundii.* Infect Immun 46:453–458.

Cisar JO, Sandberg AL, Clark WB (1989): Molecular aspects of adherence of *Actinomyces viscosus* and *Actinomyces naeslundii* to oral surfaces. J Dent Res 68:1558–1559.

Cisar JO, Vatter AE (1979): Surface fibrils (fimbriae) of *Actinomyces viscosus* T14V. Infect Immun 24:523–531.

Cisar JO, Vatter AE, Clark WB, Curl SH, Hurt-Calderone S, Sandberg AL (1988): Mutants of *Actinomyces viscosus* T14V lacking type 1, type 2, or both types of fimbriae. Infect Immun 56:2984–2989.

Clark WB, Beem JE, Nesbitt WE, Cisar JO, Tseng CC, Levine MJ (1989): Pellicle receptors of *Actinomyces viscosus* type 1 fimbriae in vitro. Infect Immun 57:3003–3008.

Clark WB, Lane MD, Beem JE, Bragg SL, Wheeler TT (1985): Relative hydrophobicities of *Actinomyces viscosus* and *Actinomyces naeslundii* strains and their adsorption to saliva-treated hydroxyapatite. Infect Immun 47:730–736.

Clark WB, Webb EL, Wheeler TT, Fischlschweiger W, Birdsell DC, Manshein BJ (1981): Role of surface fimbriae (fibrils) in the adsorption of *Actinomyces* species to saliva-treated hydroxyapatite surfaces. Infect Immun 33:908–917.

Clark WB, Wheeler TT, Cisar JO (1984): Specific inhibition of adsorption of *Actinomyces viscosus* T14V to saliva-treated hydroxyapatite by antibody against type 1 fimbriae. Infect Immun 43:497–501.

Cope PA, Mooser G (1993): Antibodies against active-site peptides common to gluco-syltransferases of mutans streptococci. Infect Immun 61:4814–4817.

Costello AH, Cisar JO, Kolenbrander PE, Gabriel O (1979): Neuraminidase-dependent hemagglutination of human erythrocytes by human strains of *Actinomyces viscosus* and *Actinomyces naeslundii*. Infect Immun 2:563–572.

Costerton JW, Cheng K, Geesey GG, Ladd TI, Nickel JC, Dasgupta M, et al (1987): Bacterial biofilms in nature and disease. Annu Rev Microbiol 41:435–464.

Cowan MM, Taylor KG, Doyle RJ (1986): Kinetic analysis of *Streptococcus sanguis* adhesion to artificial pellicle. J Dent Res 65:1278–1283.

Cowan MM, Taylor KG, Doyle RJ (1987a): Energetics of the initial phase of adhesion of *Streptococcus sanguis* to hydroxylapatite. J Bacteriol 169:2995–3000.

Cowan MM, Taylor KG, Doyle RJ (1987b): Role of sialic acid in the kinetics of *Streptococcus sanguis* adhesion to artificial pellicle. Infect Immun 55:1552–1557.

DaCosta T, Gibbons RJ (1968): Hydrolysis of levan by human plaque streptococci. Arch Oral Biol 13:609–617.

Davies DG, Chakrabarty AM, Geesey GG (1993): Exopolysaccharide production in biofilms: Substratum activation of alginate gene expression by *Pseudomonas aeruginosa*. Appl Environ Microbiol 59:1181–1186.

Dawson JR, Ellen RP (1990): Tip-oriented adherence of *Treponema denticola* to fibronectin. Infect Immun 58:3924–3928.

Dawson JR, Ellen RP (1994): Clustering of fibronectin adhesins toward *Treponema denticola* tips upon contact with immobilized fibronectin. Infect Immun 62:2214–2221.

DeFilippo AB, Ellen RP, McCulloch CAG (1995): Induction of cytoskeletal rearrange-ments and loss of volume regulation in epithelial cells by *Treponema denticola*. Arch Oral Biol 40:199–207.

Dickinson DP, Kubiniec MA, Yoshimura F, Genco RJ (1988): Molecular cloning and sequencing of the gene encoding the fimbrial subunit protein of *Bacteroides gingivalis*. J Bacteriol 170:1658–1665.

Douglas CWI (1990): Characterization of the alpha-amylase receptor of *Streptococcus gordonii* NCTC 7868. J Dent Res 69:1746–1752.

Doyle RJ, Rosenberg M, Drake D (1990): Hydrophobicity of oral bacteria. In Doyle RJ, Rosenberg M (eds): Microbioal Cell Surface Hydrophobicity. Washington, DC: American Society for Microbiology, pp 387–419.

Drake D, Taylor KG, Bleiweis AS, Doyle RJ (1988): Specificity of the glucan-binding lectin of *Streptococcus cricetus*. Infect Immun 56:1864–1872.

Duan Y, Fisher E, Malamud D, Golub E, Demuth DR (1994): Calcium-binding proper-ties of SSP-5, the *Streptococcus gordonii* M5 receptor for salivary agglutinin. Infect Immun 62:5220–5226.

Duncan MJ, Nakao S, Skobe Z, Xie H (1993): Interactions of *Porphyromonas gingivalis* with epithelial cells. Infect Immun 61:2260–2265.

Egli C, Leung WK, Müller KH, Hancock REW, McBride BC (1993): Pore-forming properties of the major 53-kilodalton surface antigen from the outer sheath of *Treponema denticola*. Infect Immun 61:1694–1699.

Ellen RP (1985): Specificity of attachment as a tissue-tropic influence on oral bacteria. In Mergenhagen SE, Rosan B (eds): Molecular Basis for Oral Microbial Adhesion. Washington, DC: American Society for Microbiology, pp 33–39.

Ellen RP, Bratthall D, Borgström M, Howley TP (1983): *Actinomyces viscosus* and *Actinomyces naeslundii* agglutinins in saliva. Scand J Dent Res 91:263–273.

Ellen RP, Buivids IA, Simardone JR (1989): *Actinomyces viscosus* fibril antigens detected by immunogold electron microscopy. Infect Immun 57:1327–1331.

Ellen RP, Dawson JR, Yang PF (1994a): *Treponema denticola* as a model for polar adhesion and cytopathogenicity of spirochetes. Trends Microbiol 2:114–119.

Ellen RP, Fillery ED, Chan KH, Grove DA (1980): Sialidase-enhanced lectin-like mechanism of *Actinomyces viscosus* and *Actinomyces naeslundii* hemagglutination. Infect Immun 27:335–343.

Ellen RP, Grove DA (1989): *Bacteroides gingivalis* vesicles bind to and aggregate *Actinomyces viscosus*. Infect Immun 57:1618–1620.

Ellen RP, Song M, Buivids IA (1992): Inhibition of *Actinomyces viscosus–Porphyromonas gingivalis* coadhesion by trypsin and other proteins. Oral Microbiol Immunol 7:198–203.

Ellen RP, Song M, McCulloch CAG (1994b): Degradation of endogenous plasma membrane fibronectin concomitant with *Treponema denticola* 35405 adhesion to gingival fibroblasts. Infect Immun 62:3033–3037.

Ellen RP, Veisman H, Buivids IA, Rosenberg M (1994c): Kinetics of lactose-reversible coadhesion of *Actinomyces naeslundii* WVU 398A and *Streptococcus oralis* 34 on the surface of hexadecane droplets. Oral Microbiol Immun 9:364–371.

Ellen RP, Walker DL, Chan KH (1978): Association of long surface appendages with adherence related functions of the Gram-positive species *Actinomyces naeslundii*. J Bacteriol 134:1171–1175.

Ericson D (1984): Agglutination of *Streptococcus mutans* by low-molecular-weight salivary components: Effect of β_2-microglobulin. Infect Immun 46:526–530.

Ericson D, Ellen RP, Buivids I (1987): Labeling of binding sites for β_2-microglobulin (β_2m) on nonfibrillar surface structures of mutans streptococci by immunogold and β_2m gold electron microscopy. J Bacteriol 169:2507–2515.

Ericson D, Tynelius-Bratthall G (1986): Absorption of fibronectin from human saliva by strains of oral streptococci. Scand J Dent Res 95:377–379.

Erickson PR, Herzberg MC (1993): The *Streptococcus sanguis* platelet aggregation-associated protein. Identification and characterization of the minimal platelet-interactive domain. J Biol Chem 268:1646–1649.

Evans RT, Klausen B, Sojar HT, Bedi GS, Sfintescu C, Ramamurthy NS, Golub LM, Genco RJ (1992): Immunization with *Porphyromonas* (*Bacteroides*) *gingivalis* fimbriae protects against periodontal destruction. Infect Immun 60:2926–2935.

Fenno JC, LeBlanc DJ, Fives-Taylor P (1989): Nucleotide sequence analysis of type 1 fimbrial gene from *Streptococcus sanguis* FW213. Infect Immun 57:3527–3533.

Fillery ED, Bowden GH, Hardie JM (1978): A comparison of bacteria designated *Actinomyces viscosus* and *Actinomyces naeslundii*. Caries Res 12:299–312.

Firtel M, Fillery ED (1988): Distribution of antigenic determinants between *Actinomyces viscosus* and *Actinomyces naeslundii*. J Dent Res 67:15–20.

Fives-Taylor P, Fenno JC, Holden E, Linehan L, Oligino L, Volansky M (1991): Molecular structure of fimbriae-associated genes of *Streptococcus sanguis*. In Dunny GM, Cleary PP, McKay LL (eds): Genetics and Molecular Biology of Streptococci. Washington, DC: American Society for Microbiology, pp 240–243.

Fives-Taylor PM, Thompson DW (1985): Surface properties of *Streptococcus sanguis* FW213 mutants nonadherent to saliva-coated hydroxyapatite. Infect Immun 47:752–759.

Fukushima K, Ikeda T, Kuramitsu HK (1992): Expression of *Streptococcus mutans gtf* genes in *Streptococcus milleri*. Infect Immun 60:2815–2822.

Ganeshkumar N, Arora N, Kolenbrander PE (1993): Saliva-binding protein (SsaB) from *Streptococcus sanguis* 12 is a lipoprotein. J Bacteriol 175:572–574.

Ganeshkumar N, Hannam PM, Kolenbrander PE, McBride BC (1991): Nucleotide sequence of a gene coding for a saliva-binding protein (SsaB) from *Streptococcus sanguis* 12 and possible role of the protein in coaggregation with actinomyces. Infect Immun 59:1093–1099.

Genco RJ, Sojar H, Lee J-Y, Sharma A, Bedi G, Cho M-I, Dyer DW (1994): *Porphyromonas gingivalis* fimbriae: Structure, function, and insertional inactivation mutants. In RJ Genco, Hamada S, Lehner T, McGhee J, Mergenhagen S (eds): Molecular Pathogenesis of Periodontal Disease. Washington, DC: American Society for Microbiology, pp 13–23.

Gibbons RJ, Cohen L, Hay DI (1986): Strains of *Streptococcus mutans* and *Streptococcus sobrinus* attach to different pellicle receptors. Infect Immun 52:555–561.

Gibbons RJ, Etherden I (1983): Comparative hydrophobicities of oral bacteria and their adherence to salivary pellicles. Infect Immun 41:1190–1196.

Gibbons RJ, Etherden I, Moreno EC (1983): Association of neuraminidase-sensitive receptors and putative hydrophobic interactions with high-affinity binding sites for *Streptococcus sanguis* C5 in salivary pellicles. Infect Immun 42:1006–1012.

Gibbons RJ, Etherden I, Moreno EC (1985): Contribution of stereochemical interactions in the adhesion of *Streptococcus sanguis* C5 to experimental pellicles. J Dent Res 64:96–101.

Gibbons RJ, Etherden I, Skobe Z (1983): Association of fimbriae with the hydrophobicity of *Streptococcus sanguis* FC-1 and adherence to salivary pellicles. Infect Immun 41:414–417.

Gibbons RJ, Hay DI (1988): Human salivary acidic proline-rich proteins and statherin promote the attachment of *Actinomyces viscosus* LY7 to apatitic surfaces. Infect Immun 56:439–445.

Gibbons RJ, Hay DI, Childs III WC, Davis G (1990): Role of cryptic receptors (cryptitopes) in bacterial adhesion to oral surfaces. Arch Oral Biol 35(Suppl): 107S–114S.

Gibbons RJ, Hay DI, Cisar JO, Clark WB (1988): Adsorbed salivary proline-rich protein 1 and statherin: Receptors for type 1 fimbriae of *Actinomyces viscosus* T14V-J1 on apatitic surfaces. Infect Immun 56:2990–2993.

Gibbons RJ, Hay DI, Schlesinger DH (1991): Delineation of a segment of adsorbed salivary acidic proline-rich proteins which promotes adhesion of *Streptococcus gordonii* to apatitic surfaces. Infect Immun 59:2948–2954.

Gibbons RJ, Moreno EC, Spinell DM (1976): Model delineating the effects of a salivary pellicle on the adsorption of *Streptococcus miteor* onto hydroxyapatite. Infect Immun 14:1109–1112.

Gibbons RJ, van Houte J (1971): Selective bacterial adherence to oral epithelial surfaces and its role as an ecological determinant. Infect Immun 3:567–573.

Gibbons RJ, van Houte J (1975): Bacterial adherence in oral microbial ecology. Annu Rev Microbiol 29:19–44.

Giffard PM, Allen DM, Milward CP, Simpson CL, Jacques NA (1993): Sequence of the *gtfK* gene of *Streptococcus salivarius* ATCC25975 and evolution of the *gtf* genes of oral streptococci. J Gen Microbiol 139:1511–1522.

Giffard PM, Jacques NA (1994): Definition of a fundamental repeating unit in streptococcal glucosyltransferase glucan-binding regions and related sequences. J Dent Res 73:1133–1141.

Girard AE, Jacius BH (1974): Ultrastructure of *Actinomyces viscosus* and *Actinomyces naeslundii*. Arch Oral Biol 19:71–79.

Goulbourne PA, Ellen RP (1991): Evidence that *Porphyromonas* (*Bacteroides*) *gingivalis* fimbriae function in adhesion to *Actinomyces viscosus*. J Bacteriol 173:5266–5274.

Grenier D (1992): Further evidence for a possible role of trypsin-like activity in the adherence of *Porphyromonas gingivalis*. Can J Microbiol 38:1189–1192.

Grenier D, Uitto V-J, McBride BC (1990): Cellular location of a *Treponema denticola* chymotrypsin like protease and importance of the protease in migration through the basement membrane. Infect Immun 58:347–351.

Haapasalo M, Müller K-H, Uitto V-J, Leung WK, McBride BC (1992): Characterization, cloning, and binding properties of the major 53-kilodalton *Treponema denticola* surface antigen. Infect Immun 60:2058–2065.

Haapasalo M, Singh U, McBride BC, Uitto V-J (1991): Sulfhydryl-dependent attachment of *Treponema denticola* to laminin and other proteins. Infect Immun 59:4230–4237.

Hamada N, Watanabe K, Sasakawa C, Yoshikawa M, Yoshimura F, Umemoto T (1994): Construction and characterization of a *fimA* mutant of *Porphyromonas gingivalis*. Infect Immun 62:1696–1704.

Hamada S, Slade HD (1980): Biology, immunology, and cariogenicity of *Streptococcus mutans*. Microbiol Rev 44:331–384.

Hananda N, Kuramitsu HK (1988): Isolation and characterization of the *Streptococcus mutans gtfC* gene, coding for synthesis of both soluble and insoluble glucans. Infect Immun 56:1999–2005.

Hananda N, Kuramitsu HK (1989): Isolation and characterization of the *Streptococcus mutans gtfD* gene, coding for primer-dependent glucan synthesis. Infect Immun 57:2079–2085.

Hanazawa S, Hirose K, Ohmori Y, Amano S, Kitano S (1988): *Bacteroides gingivalis* fimbriae stimulate production of thymocyte-activating factor by human gingival fibroblasts. Infect Immun 56:272–274.

Hanazawa S, Murakami Y, Hirose K, Amano S, Ohmori Y, Higuchi H, Kitano S (1991): *Bacteroides* (*Porphyromonas*) *gingivalis* fimbriae activate mouse peritoneal macrophages and induce gene expression and production of interleukin-1. Infect Immun 59:1972–1977.

Handley PS, Tipler LS (1986): An electron microscope survey of the surface structure and hydrophobicity of oral and non-oral species of the bacterial genus *Bacteroides*. Arch Oral Biol 31:325–335.

Hardy L, Jacques NA, Forester H, Campbell LK, Knox KW, Wicken AJ (1981): Effect of fructose and other carbohydrates on the surface properties, lipoteichoic acid production, and extracellular proteins of *Streptococcus mutans* Ingbritt grown in continuous culture. Infect Immun 31:78–87.

Hay DI, Bennick A, Schlesinger DH, Minaguchi K, Madapallimattam G, Schluckebier SK (1988): The primary structures of six human salivary acidic proline-rich proteins (PRP-1, RPP-2, PRP-3, PRP-4, PIF-s and PIF-f). Biochem J 255:15–21.

Herzberg MC, Gong K, MacFarlane GD, Erickson PR, Soberay AH, Krebsbach PH, Manjula G, Schilling K, Bowen WH (1990): Phenotypic characterization of *Streptococcus sanguis* virulence factors associated with bacterial endocarditis. Infect Immun 58:515–522.

Herzberg MC, MacFarlane GD, Gong K, Armstrong NN, Witt AR, Erickson PR, Meyer MW (1992): The platelet interactivity phenotype of *Streptococcus sanguis* influences the course of experimental endocarditis. Infect Immun 60:4809–4818.

Hogg SD, Manning JE (1987): The hydrophobicity of "viridans" streptococci isolated from the human mouth. J Appl Bacteriol 63:311–318.

Honda O, Cato C, Kuramitsu HK (1990): Nucleotide sequence of the *Streptococcus mutans gtfD* gene encoding the glucosyltransferase-S enzyme. J Gen Microbiol 136:2099–3105.

Hoover CI, Ng CY, Felton JR (1992): Correlation of hemagglutination activity with trypsin-like protease activity of *Porphyromonas gingivalis*. Arch Oral Biol 37:515–520.

Hoover CI, Yoshimura F (1994): Transposon-induced pigment-deficient mutants of *Porphyromonas gingivalis*. FEMS Microbiol Lett 124:43–48.

Hudson MC, Curtiss R III (1990): Regulation of expression of *Streptococcus mutans* genes important to virulence. Infect Immun 58:464–470.

Inoshita E, Amano A, Hanioka T, Tamagawa H, Shitzukuishi S (1986): Isolation and some properties of exohemagglutinin from culture medium of *Bacteroides gingivalis* 381. Infect Immun 52:421–427.

Irvin RT (1990): Hydrophobicity of proteins and bacterial fimbriae. In Doyle RJ, Rosenberg M (eds): Microbial Cell Surface Hydrophobicity. Washington, DC: American Society for Microbiology, pp 137–177.

Isogai H, Isogai E, Yoshimura F, Suzuki T, Kogota W, Takano K (1988): Specific inhibition of adherence of an oral strain of *Bacteroides gingivalis* 381 to epithelial cells by monoclonal antibodies against the bacterial fimbriae. Arch Oral Biol 33:479–485.

Jenkinson HF (1991): Phenotypic effects of inactivating the gene encoding a cell surface binding protein in *Streptococcus gordonii* challis. In Dunny GM, Cleary PP, McKay LL (eds): Genetics and Molecular Biology of Streptococci. Washington, DC: American Society for Microbiology, pp 284–288.

Kato C, Kuramitsu HK (1989): Carboxyl-terminal deletion analysis of the *Streptcoccus mutans* glucosyltransferase-I enzyme. FEMS Microbiol Lett 60:299–302.

Kato C, Kuramitsu HK (1991): Molecular basis for the association of glucosyltransferases with the cell surface of oral streptococci. FEMS Microbiol Lett 63:153–157.

Kato C, Nakano Y, Lis M, Kuramitsu HK (1992): Molecular genetic analysis of the catalytic site of *Streptococcus mutans* glucosyltransferases. Biochem Biophys Res Commun 189:1184–1188.

Kawata Y, Hanazawa S, Amano S, Murakami Y, Matsumoto T, Nishida K, Kitano S (1994): *Porphyromonas gingivalis* fimbriae stimulate bone resorption *in vitro*. Infect Immun 62:3012–3016.

Keevil CW, West A, Marsh PD, Ellwood DC (1983): Batch versus continuous culture studies of glucosyltransferase synthesis in oral streptococci. In Doyle RJ, Ciardi JE (eds): Proceedings: Glucosyltransferases, Glucans, Sucrose and Dental Caries. Washington, DC: IRL, pp 189–200.

Kiska DL, Macrina FL (1994): Genetic regulation of fructosyltransferase in *Streptococcus mutans*. Infect Immun 62:1241–1251.

Knox KW, Hardy LN, Markevics LJ, Evans JD, Wicken AJ (1985): Comparative studies on the effect of growth conditions on adhesion, hydrophobicity, and extracellular protein profile of *Streptococcus sanguis* G9B. Infect Immun 50:545–554.

Koldenbrander PE, Andersen RN, Ganeshkumar N (1994): Nucleotide sequence of the *Streptococcus gordonii* PK488 coaggregation adhesin gene, *scaA,* and ATP-binding cassette. Infect Immun 62:4469–4480.

Kuramitsu HK (1993): Virulence factors of mutans streptococci: Role of molecular genetics. Crit Rev Oral Biol Med 4:159–176.

Kuramitsu HK, Ingersoll L (1978): Interaction of glucosyltransferases with the cell surface of *Streptococcus mutans*. Infect Immun 20:652–659.

Kurashima C, Sandberg AL, Cisar JO, Mudrick LL (1991): Cooperative complement- and bacterial lectin-initiated bactericidal activity of polymorphonuclear leukocytes. Infect Immun 59:216–221.

Lamont RJ, Bevan CA, Gil S, Persson RE, Rosan B (1993): Involvement of *Porphyromonas gingivalis* fimbriae in adherence to *Streptococcus gordonii*. Oral Microbiol Immunol 8:272–276.

Lantz MS, Allen RD, Duck WL, Blume JL, Switalski LM, et al. (1991a): Identification of *Porphyromonas gingivalis* components that mediate its interactions with fibronectin. J Bacteriol 173:4263–4270.

Lantz MS, Allen RD, Vail TA, Switalski LM, Höök M (1991b): Specific cell components of *Bacteroides gingivalis* mediate binding and degradation of human fibrinogen. J Bacteriol 173:495–504.

Lawman P, Bleiweis AW (1991): Molecular cloning of the extracellular endodextranase of *Streptococcus mutans* Ingbritt. J Bacteriol 173:7423–7428.

Lee J-Y, Sojar HJ, Bedi GS, Genco RJ (1992): Synthetic peptides analogous to the fimbrillin sequence inhibit adherence of *Porphyromonas gingivalis*. Infect Immun 60:1662–1670.

Lee SF, Progulske-Fox A, Erdos GW, Piacentini DA, Ayakawa GY, Crowley PJ, Bleiweis AS (1989): Construction and characterization of isogenic mutants of *Streptococcus mutans* deficient in major surface protein antigen P1 (I/II). Infect Immun 57:3306–3313.

Lehner T, Ma JK-C, Munro G, Walker P, Childerstone A, Todryk S, Kendal H, Kelly CG (1994): T-cell and B-cell epitope mapping and construction of peptide vaccines. In Genco R, Hamada S, Lehner T, McGhee J, Mergenhagen S (eds): Molecular Pathogenesis of Periodontal Disease. Washington, DC: ASM Press, pp 279–292.

Leung K-P, Nesbitt WE, Fischlschweiger W, Hay DI, Clark WB (1990): Binding of colloidal gold-labeled salivary proline-rich proteins to *Actinomyces viscosus* type 1 fimbriae. Infect Immun 58:1986–1991.

Levine MJ, Herzberg MC, Levine MS, Ellison SA, Stinson MW, Li HC, Van Dyke T (1978): Specificity of salivary-bacterial interactions: Role of terminal sialic acid residues in the interaction of salivary glycoproteins with *Streptococcus sanguis* and *Streptococcus mutans*. Infect Immun 19:107–115.

Lépine G, Progulske-Fox A (1993): Molecular biology. In Shah HN, Mayrand D, Genco RJ (eds): Biology of the Species *Porphyromonas gingivalis*. Boca Raton: CRC Press, pp 293–319.

Li J, Ellen RP, Hoover CI, Felton JR (1991): Association of proteases of *Porphyromonas (Bacteroides) gingivalis* with its adhesion to *Actinomyces viscosus*. J Dent Res 70:82–86.

Liljemark WF, Bloomquist CG (1981): Isolation of a protein-containing cell surface component from *Streptococcus sanguis* which affects its adherence to saliva-coated hydroxyapatite. Infect Immun 34:428–434.

Liljemark WF, Bloomquist CG, Fenner LJ, Antonelli PJ, Coulter MC (1989): Effect of neuraminidase on the adherence to saliary pellicle of *Streptococcus sanguis* and *Streptococcus mitis*. Caries Res 23:141–145.

Lowrance JH, Baddour LM, Simpson WA (1990): The role of fibronectin binding in the rat model experimental endocarditis caused by *Streptococcus sanguis*. J Clin Invest 86:7–13.

Lowrance JH, Hasty DL, Simpson WA (1988): Adherence of *Streptococcus sanguis* to conformationally specific determinants in fibronectin. Infect Immun 56:2279–2285.

Lü-Lü, Singh JS, Galperin MY, Drake D, Taylor KG, Doyle RJ (1992): Chelating agents inhibit activity and prevent expression of streptococcal glucan-binding lectins. Infect Immun 60:3807–3813.

Masuda N, Ellen RP, Fillery ED, Grove DA (1983): Chemical and immunological comparison of surface fibrils of strains representing six taxonomic groups of *Actinomyces viscosus* and *Actinomyces naeslundii*. Infect Immun 39:1325–1333.

Masuda N, Ellen RP, Grove DA (1981): Purification and characterization of surface fibrils from taxonomically typical *Actinomyces viscosus* WVU627. J Bacteriol 147:1095–1104.

McBride BC, Gisslow MT (1977): Role of sialic acid in saliva-induced aggregation of *Streptococcus sanguis*. Infect Immun 18:35–40.

McBride BC, Morris EJ, Ganeshkumar N (1985): Relationship of streptococcal cell surface proteins to hydrophobicity and adherence. In Mergenhagen SE, Rosan B (eds): Molecular Basis of Oral Microbial Adhesion. Washington, DC: American Society for Microbiology, pp 85–93.

McGhee JR, Mestecky J (1990): In defense of mucosal surfaces: Development of novel vaccines for IgA responses protective at the portals of entry of microbial pathogens. Infect Dis Clin North Am 4:315–341.

McIntire FC, Crosby LK, Barlow JJ, Matta KL (1983): Structural preferences of β-galactoside–reactive lectins on *Actinomyces viscosus* T14V and *Actinomyces naeslundii* WVU 45. Infect Immun 41:848–850.

McIntire FC, Crosby LK, Vatter AE (1982): Inhibitors of coaggregation between *Actinomyces viscosus* T14 and *Streptococcus sanguis* 34: β-Galactosides, related sugars, and anionic amphipathic compounds. Infect Immun 36:371–378.

McIntire FC, Vatter AE, Boros J, Arnold J (1978): Mechanism of coaggregation between *Actinomyces viscosus* T14V and *Streptococcus sanguis* 34. Infect Immun 21:978–988.

Meyer DH, Wei J, Fives-Taylor PM (1995): Cloning of a DNA fragment associated with *Actinobacillus actinomycetemcomitans* invasion. Abstract 1508. J Dent Res 74:200.

Meyer DH, Fives-Taylor PM (1993): Evidence that extracellular components function in adherence of *Actinobacillus actinomycetemcomitans* to epithelial cells. Infect Immun 61:4933–4936.

Meyer DH, Fives-Taylor PM (1994): Characteristics of adherence of *Actinobacillus actinomycetemcomitans* to epithelial cells. Infect Immun 62:928–935.

Meyer DH, Sreenivasan PK, Fives-Taylor PM (1991): Evidence for invasion of a human oral cell line by *Actinobacillus actinomycetemcomitans*. Infect Immun 59:2719–2726.

Michalek S, Childers NK (1990): Development and outlook for a caries vaccine. Crit Rev Oral Biol Med 1:37–45.

Mintz KP, Fives-Taylor PM (1994): Adhesion of *Actinobacillus actinomycetemcomitans* to a human oral cell line. Infect Immun 62:3672–3678.

Mooser G, Hefta S, Paxton RJ, Shively JE, Lee TD (1989): Isolation and sequence of an active-site peptide containing a catalytic aspartic acid from two *Streptococcus sobrinus* alpha-glucosyltransferases. J Biol Chem 266:8916–8922.

Mooser G, Wong C (1988): Isolation of a glucan-binding domain of glucosyltransferases (1,6-α-glucan synthase) from *Streptococcus sobrinus*. Infect Immun 56:880–884.

Moreno EC, Kresak M, Hay DI (1982): Adsorption thermodynamics of acidic proline-rich salivary proteins onto calcium apatites. J Biol Chem 257:2981–2989.

Morris EJ, McBride BC (1984): Adherence of *Streptococcus sanguis* to saliva-coated hydroxyapatite: Evidence for two binding sites. Infect Immun 43:656–663.

Mouton C, Ni Eidhin D, Deslauriers M, Lamy L (1991): The hemagglutinating adhesin HA-Ag2 of *Bacterioides gingivalis* is distinct from fimbrilin. Oral Microbiol Immunol 6:6–11.

Munro C, Michalek SM, Macrina FL (1991): Cariogenicity of *Streptococcus mutans* V403 glucosyltransferase and fructosyltransferase mutants constructed by allelic exchange. Infect Immun 59:2316–2323.

Murray PA, Levine MJ, Reddy MS, Tabak LA, Bergey EJ (1986): Preparation of a sialic acid-binding protein from *Streptococcus mitis* KS32AR. Infect Immun 53:359–365.

Murray PA, Levine MJ, Tabak LA, Reddy, MS (1982): Specificity of salivary-bacterial interactions. II. Evidence for a lectin on *Streptococcus sanguis* with specificity for a NeuAcα2,3Galβ1–3GalNAc sequence. Biochem Biophys Res Commun 106:390–396.

Murray PA, Prakobphol A, Lee T, Hoover C, Fisher SJ (1992): Adherence of oral streptococci to salivary glycoproteins. Infect Immun 60:31–38.

Naito Y, Gibbons RJ (1988): Attachment of *Bacteroides gingivalis* to collagenous substrata. J Dent Res 67:1075–1089.

Nakano YJ, Kuramitsu HK (1992): Mechanism of *Streptococcus mutans* glucosyltransferase: Hybrid–enzyme analysis. J Bacteriol 174:5639–5646.

Nesbitt WE, Doyle RJ, Taylor KG (1982a): Hydrophobic interactions and the adherence of *Streptococcus sanguis* to hydroxylapatite. Infect Immun 38:637–644.

Nesbitt WE, Doyle RJ, Taylor KG, Staat RH, Arnold RR (1982b): Positive cooperativity in the binding of *Streptococcus sanguis* to hydroxylapatite. Infect Immun 35:157–165.

Nishikata M, Yoshimura F (1991): Characterization of *Porphyromonas (Bacteroides) gingivalis* hemagglutinin as protease. Biochem Biophys Res Commun 178:336–342.

Ogawa T, Hamada S (1994): Hemagglutination and chemotactic properties of synthetic peptide segments of fimbrial proteins from *Porphyromonas gingivalis*. Infect Immun 62:3305–3310.

Ogawa T, Ogo H, Uchida H, Hamada S (1994): Humoral and cellular immune responses of the fimbriae of *Porphyromonas gingivalis* and their synthetic peptides. J Med Microbiol 40:397–402.

Ogawa T, Uchida H, Hamada S (1994): *Porphyromonas gingivalis* fimbriae and their synthetic peptides induce proinflammatory cytokines in human peripheral blood monocyte cultures. FEMS Microbiol Lett 116:237–242.

Okuda K, Yamamoto A, Naito Y, Takazoe I, Slots J, Genco RJ (1986): Purification and properties of hemagglutinin from culture supernatant of *Bacteroides gingivalis*. Infect Immun 54:659–665.

Patti JM, Allen BL, McGavin MJ, Höök M (1994): MSCRAMM-mediated adherence of microorganisms to host tissues. Annu Rev Microbiol 48:585–617.

Pavloff N, Potempa J, Pike RN, Prochazka V, Kiefer MC, Travis J (1995): Molecular cloning and structural characterization of the Arg-gingipain proteinase of *Porphyromonas gingivalis*. J Biol Chem 270:1007–1010.

Perry D, Kuramitsu HK, (1990): Linkage of sucrose-metabolizing genes in *Streptococcus mutans* Infect Immun 58:3462–3464.

Pike R, McGraw W, Potempa J, Travis J (1994): Lysine- and arginine-specific proteinases from *Porphyromonas gingivalis*. Isolation, characterization, and evidence for the existence of complexes with hemagglutinins. J Biol Chem 269:406–411.

Potempa J, Pike R, Travis J (1995): The multiple forms of trypsin-like activity present in various strains of *Porphyromonas gingivalis* are due to the presence of either Arg-gingipain or Lys-gingipain. Infect Immun 63:1176–1182.

Reddy GP, Abeygunawardana C, Bush CA, Cisar JO (1994): The cell wall polysaccharide of *Streptococcus gordonii* 38: Structure and immunological comparison with the receptor polysaccharides of *Streptococcus oralis* 34 and *Streptococcus mitis* J22. Glycobiology 4:183–192.

Revis GJ, Vatter AE, Crowle AJ, Cisar JO (1982): Antibodies against the Ag2 fimbriae of *Actinomyces viscosus* T14V inhibit lactose-sensitive bacterial adherence. Infect Immun 36:1217–1222.

Rogers AH, Pilowsky K, Zilm PS (1984): The effect of growth rate on the adhesion of the oral bacteria *Streptococcus mutans* and *Streptococcus milleri*. Arch Oral Biol 29:147–150.

Rosan B, Appelbaum B, Campbell LK, Knox KW, Wicken AJ (1982): Chemostat studies of the effect of environmental control on *Streptococcus sanguis* adherence to hydroxylapatite. Infect Immun 35:64–70.

Rosan B, Baker CT, Nelson GM, Berman R, Lamont RJ, Demuth DR (1989): Cloning and expression of an adhesion antigen of *Streptococcus sanguis* G9B in *Escherichia coli*. J Gen Microbiol 135:531–538.

Rosenberg M, Buivids IA, Ellen RP (1991): Adhesion of *Actinomyces viscosus* to *Porphyromonas (Bacteroides) gingivalis*–coated hexadecane droplets. J Bacteriol 173:2581–2589.

Rosenberg M, Gelernter I, Barki M, Bar-Ness R (1992): Daylong reduction of oral malodor by a two-phase oil:water mouthrinse, as compared to chlorhexidine and placebo rinses. J Periodontol 63:39–43.

Rosenberg M, Judes H, Weiss E (1983): Cell surface hydrophobicity of dental plaque microorganisms *in situ*. Infect Immun 42:831–834.

Russell RRB (1994): The application of molecular genetics to the microbiology of dental caries. Caries Res 28:69–82.

Russell RRB, Aduse-Opoku J, Tao L, Ferretti JJ (1991): Binding protein-dependent transport system in *Streptococcus mutans*. In Dunny GM, Cleary PP, McKay LL (eds): Genetics and Molecular Biology of Streptococci. Washington, DC: American Society for Microbiology, pp 244–247.

Russell RRB, Shiroza T, Ferretti JJ, Kuramitsu HK (1988): Homology of glucosyltransferase gene and protein sequence from *Streptococcus mutans* and *Streptococcus sobrinus*. J Dent Res 67:543–547.

Sandberg AL, Ruhl S, Joralmon RA, Brennan MJ, Sutphin MJ, Cisar JO (1995): Putative glycoprotein and glycolipid polymorphonuclear leukocyte receptors for the *Actinomyces naeslundii* WVU45 fimbrial lectin. Infect Immun 63:2625–2631.

Sandros J, Papapanou P, Dahlén G (1993): *Porphyromonas gingivalis* invades oral epithelial cells *in vitro*. J Periodontal Res 28:219–226.

Scannapieco FA, Bergey EJ, Reddy MS, Levine MJ (1989): Characterization of salivary α-amylase binding to *Streptococcus sanguis*. Infect Immun 57:2853–2863.

Scannapieco FA, Bhandary K, Ramasubbu N, Levine MJ (1990): Structural relationship between the enzymatic and streptococcal binding sites of human salivary α-amylase. Bichem Biophys Res Commun 173:1109–1115.

Schilling KM, Blitzer MH, Bowen WH (1989): Adherence of *Streptococcus mutans* to glucans formed *in situ* in salivary pellicle. J Dent Res 68:1678–1680.

Schilling KM, Bowen WH (1992): Glucans synthesized *in situ* in experimental salivary pellicle function as specific binding sites for *Streptococcus mutans*. Infect Immun 60:284–295.

Shimamura A, Nakano YJ, Mukasa H, Kuramitsu HK (1994): Identification of amino acid residues in *Streptococcus mutans* glucosyltransferases influencing the structure of the glucan product. J Bacteriol 176:4845–4850.

Singh JS, Taylor KG, Doyle RJ (1993): Essential amino acids involved in glucan-dependent aggregation of *Streptococcus sobrinus*. Carb Res 244:137–147.

Slots J, Gibbons RJ (1978): Attachment of *Bacteroides melaninogenicus* subsp. *asaccharolyticus* to oral surfaces and its possible role in colonization of the mouth and of periodontal pockets. Infect Immun 19:254–264.

Smith DJ, Akita H, King WF, Taubman MA (1994a): Purification and antigenicity of a novel glucan-binding protein of *Streptococcus mutans*. Infect Immun 62:2545–2552.

Smith DJ, Taubman MA, King WF, Eida S, Powell JR, Eastcott J (1994b): Immunological characteristics of a synthetic peptide associated with a catalytic domain of mutans streptococcal glucosyltransferase. Infect Immun 62:5470–5476.

Sreenivasan PK, Meyer DH, Fives-Taylor PM (1993): Requirements for invasion of epithelial cells by *Actinobacillus actinomycetemcomitans*. Infect Immun 61:1239–1245.

Strömberg N, Borén T (1992): *Actinomyces* tissue specificity may depend on differences in receptor specificity for GalNAcβ-containing glycoconjugates. Infect Immun 60:3268–3277.

Strömberg N, Borén T, Carlen A, Olsson J (1992): Salivary receptors for GalNAcβ-sensitive adherence of *Actinomyces* spp.: Evidence for heterogeneous GalNAcβ and proline-rich protein receptor properties. Infect Immun 60:3278–3286.

Strömberg N, Karlsson K-A (1990): Characterization of the binding of *Actinomyces naeslundii* (ATCC 12104) and *Actinomyces viscosus* (ATCC 19246) to glycosphingolipids using a solid-phase overlay approach. J Biol Chem 265:11251–11258.

Sun JW, Wanda SY, Camilli A, Curtiss R (1994): Cloning and DNA sequencing of the dextranase inhibitor gene (*dei*) from *Streptococcus sobrinus*. J Bacteriol 176:7213–7222.

Svanberg M, Westergren G, Olsson J (1984): Oral implantation in humans of *Streptococcus mutans* strains with different degrees of hydrophobicity. Infect Immun 43:817–821.

Uitto V-J, Pan Y-M, Leung WK, Larjava H, Ellen RP, Finlay BB, McBride BC (1995): Cytopathic effects of *Treponema denticola* chymotrypsin-like proteinase on migrating and stratified epithelial cells. Infect Immun 63:3401–3410.

Umemoto T, Nakatani Y, Nakamura Y, Namikawa I (1993): Fibronectin-binding proteins of a human oral spirochete, *Treponema denticola*. Microbiol Immunol 37:75–78.

Vacca-Smith AM, Jones CA, Levine MJ, Stinson MW (1994): Glucosyltransferase mediates adhesion of *Streptococcus gordonii* to human endothelial cells in vitro. Infect Immun 62:2187–2194.

Van der Mei HC, Cox SD, Geertsema-Doornsusch GI, Doyle RJ, Busscher HJ (1993): A critical appraisal of positive cooperativity in oral streptococcal adhesion: Scatchard analyses of adhesion data versus analyses of the spatial arrangement of adhering bacteria. J Gen Microbiol 139:937–948.

von Eichel-Streiber C, Sauerborn M, Kuramitsu HK (1992): Evidence for a modular structure of the homologous repetitive C-terminal carbohydrate-binding sites of *Clostridium difficile* toxins and *Streptococcus mutans* glucosyltransferases. J Bacteriol 174:6707–6710.

Walker GJ, Brown RA, Taylor C (1984): Activity of *Streptococcus mutans* α-D-glucosyltransferases released under various growth conditions. J Dent Res 63:397–400.

Walker GJ, Hare MD, Morrey-Jones JG (1982): Effect of variation in the growth conditions on endo-dextranase production by *Streptococcus mutans*. Carb Res 107:111–122.

Weerkamp AH, van der Mei, Slot JW (1987): Relationship of cell surface morphology and composition of *Streptoccocus salivarius* K$^+$ to adherence and hydrophobicity. Infect Immun 55:438–445.

Weinberg A, Holt SC (1990): Interaction of *Treponema denticola* TD-4, GM-1, and MS25 with human gingival fibroblasts. Infect Immun 58:1720–1729.

Weinberg A, Holt SC (1991): Chemical and biological activities of a 64-kilodalton outer sheath protein from *Treponema denticola* strains. J Bacteriol 173:6935–6947.

Weiss E, Rosenberg M, Judes H, Rosenberg E (1982): Cell-surface hydrophobicity of adherent oral bacteria. Curr Microbiol 7:125–128.

Wexler DL, Hudson MC, Burne RA (1993): *Streptococcus mutans* fructosyltransferase (*ftf*) and glucosyltransferase (*gtfBC*) operon fusion strains in continuous culture. Infect Immun 61:1259–1267.

Williams RC, Gibbons RJ (1972): Inhibition of bacterial adherence by secretory immunoglobulin A: A mechanism of antigen disposal. Science 177:697–699.

Williams RC, Gibbons RJ (1975): Inhibition of streptococcal attachment to receptors on human buccal epithelial cells by antigenically similar salivary glycoproteins. Infect Immun 11:711–718.

Winkler JR, Matarese V, Hoover CI, Kramer RH, Murray PA (1988): An *in vitro* model to study bacterial invasion of periodontal tissues. J Periodontol 59:40–45.

Wren BW (1991): A family of clostridial and streptococcal ligand-binding proteins with conserved C-terminal repeat sequences. Mol Microbiol 5:797–803.

Wu-Yuan CD, Gill RE (1992): An 87-kilodalton glucan-binding protein of *Streptococcus sobrinus* B13. Infect Immun 60:5291–5295.

Yang P, Ellen RP (1995): *Treponema denticola*-induced actin rearrangement in the ventral cortex of human gingival fibroblasts. Abstract 940. J Dent Res 74:129.

Yeung MK (1992): Conservation of an *Actinomyces viscosus* T14V type 1 fimbrial subunit homolog among divergent groups of *Actinomyces* spp. Infect Immun 60:1047–1054.

Yeung MK, Cisar JO (1990): Sequence homology between the subunits of two immunologically and functionally distinct types of fimbriae of *Actinomyces* spp. J Bacteriol 172:2462–2468.

Yeung MK, Kozelsky CS (1994): Transformation of *Actinomyces* spp by a Gram-negative broad-host-range plasmid. J Bacteriol 176:4173–4176.

Yoshimura F, Takahashi D, Nodosaka Y, Suzuki T (1984): Purification and characterization of a novel type of fimbriae from the oral anaerobe *Bacteroides gingivalis*. J Bacteriol 160:949–957.

Yoshimura F, Takasawa T, Yoneyama M, Yamaguchi T, Shiokawa H, Suzuki T (1985): Fimbriae from the oral anaerobe *Bacteroides gingivalis:* Physical, chemical, and immunological properties. J Bacteriol 163:730–734.

9

COAGGREGATION: ENHANCING COLONIZATION IN A FLUCTUATING ENVIRONMENT

JACK LONDON
PAUL E. KOLENBRANDER

Laboratory of Microbial Ecology, National Institute of Dental Research, National Institutes of Health, Bethesda, Maryland 20892-4350

9.1. INTRODUCTION

The oral cavity of humans is a dynamic environment in which the surfaces of all tissues are continually exposed to endogenously and exogenously supplied nutrients. The former is characterized by saliva and crevicular fluid, and the latter includes the intermittent uptake of beverages and food. The host's normal respiration and secretory activity maintain a relatively stable gas content and pH, respectively. These chemostat-like conditions are conducive for microbial development, and it is not surprising that the human mouth is host to over 20 genera of bacteria and over 500 definable taxa (Moore et al., 1988); any individual may harbor as many as 200 bacterial species (Moore et al., 1988). Sites at which bacterial populations develop are subject to local pertubations in gas content, pH, and chemical composition. In general the sites become more anaerobic and, depending on the types of microbes at these sites, the pH will rise or decline. In the absence of proper prophylaxis, uncontrolled growth at these sites can result in the onset of several pathological states; caries and the various forms of periodontal disease are the most frequently observed

Bacterial Adhesion: Molecular and Ecological Diversity, pages 249–279
© *1996 Wiley-Liss, Inc.*

pathologies. Although some members of the oral flora have been shown to cause infections in other loci of the human body, most do not (Moore and Moore, 1994).

The failure to isolate significant numbers of oral bacteria in the intestinal tract and, reciprocally, to find no intestinal bacteria in the oral cavity attests to the specificity of the respective colonization processes (Moore and Moore, 1994). Colonization of soft and hard oral tissues by oral bacteria is a highly evolved, highly specific process. The process itself follows a relatively consistent time line, with oral bacteria being categorized as early or late colonizers. Initial attachment may be mediated by nonspecific physical forces (Weiss et al., 1982; Rosenberg et al., 1983; Busscher et al., 1992, 1993), but permanence is achieved by interactions between adhesive proteins on the surfaces of colonizing microorganisms and receptor molecules on the colonized surface. Membrane-associated receptors on epithelial cells or salivary proteins deposited on a cleansed tooth provide the substrata for the early colonizing bacteria. A number of these molecules have been described already. In saliva, they include mucins, glycoproteins, proline-rich proteins, histidine-rich proteins, enzymes like α-amylase, and phosphate-containing proteins like statherin (Kolenbrander and London, 1992). Glyco-proteins and glycolipids constitute the more common receptor molecules found on resident host cells (Jones and Isaacson, 1983; Sharon and Lis, 1989; see also Chapter 8, this volume.)

A simple method to investigate cell–cell interactions among oral bacteria is to mix two cell types and observe the cell suspension for visible coaggregates. Microscopic examination reveals that the coaggregates are a network composed of both cell types. Each cell type coaggregates with a specific set of partner cell types, and almost always the partners belong to a different genus. In some well-studied coaggregations, it is apparent that functionally identical adhesins are expressed on cells of different genera.

As an ecologically delineated group, the oral bacteria are unique, having evolved adherence mechanisms that permit intricate and highly specific cell–cell interactions. Many strains bear more than one type of adhesin, and it is not unusual to find both lectin-like (e.g., for protein–sugar interactions) and nonlectin adhesins (e.g., for protein–protein interactions) simultaneously ex-pressed on an individual oral bacterium (London, 1991; Hasty et al., 1992). Presently, the scientific literature is replete with prefatory characterizations of these cell–cell interactions, and summaries can be found in recent reviews (London, 1991; Kolenbrander and London, 1992, 1993; Kolenbrander, 1993; Kolenbrander et al., 1993; Ellen et al., 1994; Jenkinson, 1994a,b; Sutcliffe and Russell, 1995). The evolutionary relationships involved in the colonization process within the plaque community and the evolutionary forces that may play a role in acquisition of adhesive proteins and their specificities are discussed in this chapter.

9.2. COLONIZATION

9.2.1. Diversity

Over the past 25 years two areas of study have conspicuously changed our perception of dental plaque and the oral cavity as a microbial habitat. First, application of transmission and scanning electron microscopy to plaque deposits increased our understanding of microbial deposition. It was apparent that these deposits were not random arrangements of bacteria; rather, they exhibited a substantial degree of organization (Jones, 1972; Listgarten et al., 1973; Listgarten, 1976; Keyes and Rams, 1983). In addition, a large variety of morphotypes were readily discernible in the plaque, suggesting that these communities were composed of large numbers of different types of bacteria. Moreover, many of the bacteria in the plaque deposits were intimately associated with one another, a clear distinction from simply being entrapped within some glycocalyx-like matrix.

Second, newly developed anaerobic technologies and a more sophisticated understanding of the growth requirements of anaerobic oral bacteria have vastly improved recovery of the total microflora present at various sites in the mouth and especially in supra- and subgingival plaque deposits. Identification of the resident oral morphotypes by molecular taxonomic techniques established the diversity of this econiche and has greatly extended the list of its known inhabitants (Table 9.1).

The first bacteria to adhere and colonize a freshly cleaned tooth surface are primarily *Streptococcus* and *Actinomyces* species (Nyvad and Kilian, 1987). Other bacteria, including haemophili, veillonellae, and propionibacteria, comprise a minor portion of the primary colonizers in the first 4 h. After 12 h, the population becomes quite diverse and includes most of the 20 predominant genera, which continue to increase in numbers as dental plaque matures over a period of a few days.

Among the most complete studies of bacterial populations occupying subgingival sites of human teeth are those of W.E.C. Moore and L.V.H. Moore (1994) and colleagues at Virginia Polytechnic Institute and of A. Tanner and coworkers (1994) and S.S. Socransky and A.D. Haffajee (1994) and colleagues at Forsyth Dental Center. The top 25 most prevalent species of bacteria isolated from subgingival plaque from healthy sites are listed in Table 9.1 in descending order from the most to least prevalent and are based on recently summarized data (Moore and Moore, 1994). For each, the corresponding percentage of total isolates is given in parentheses adjacent to the species ranking for each species in the "Health" column only (Table 9.1, column on left). Each of the species' respective rankings in five other periodontal health conditions are given in the columns to the right. Since only about 40 species were given in the abbreviated list for each periodontal health condition (Moore and Moore, 1994), the ranks with a plus sign indicate a ranking lower than that rank for that respective clinical condition. More than 1,000 isolates in each condition

TABLE 9.1. The 25 Most Prevalent Species Determined by Percentage of Isolates Taken at Random From Subgingival Crevices of Healthy Sites of Subjects[a]

Species	Rank in different periodontal health conditions					
	Health	Gingivitis	Moderate	Adult	Juvenile	Severe
A. naeslundii[b]	1 (18)[c]	2	1	1	2	4
F. nucleatum	2 (11)	1	2	2	3	1
S. sanguis	3 (9)	14	32	3	23	41+
S. oralis	4 (7)	24	17	8	8	41
S. intermedius	5 (5)	19	14	11	36	9
P. micros	6 (2)	7	3	4	10	6
A. meyeri	**7 (2)**	**43+**	**30**	**31**	**20**	**31**
S. gordonii	**8 (2)**	**40**	**35**	**47+**	**42+**	**41+**
G. morbillorum	**9 (2)**	**31**	**44+**	**27**	**37**	**41+**
A. odontolyticus	**10 (2)**	**18**	**22**	**12**	**18**	**41+**
L. rimae	11 (2)	10	29	13	13	12
C. gingivalis	12 (2)	43+	44+	36	42+	41+
P. acnes	13 (2)	43+	40	47+	33	18
V. atypica	14 (1)	37	44+	21	42+	41+
C. ochracea	15 (1)	28	19	18	27	22
A. georgiae	16 (1)	27	44+	44	42+	41+
B. gracilis	17 (1)	6	21	5	25	25
Actinomyces sv. WVA963	18 (1)	43+	37	47+	14	41+
V. parvula	19 (1)	30	8	15	6	16
H. segnis	20 (1)	43+	44+	47+	42+	41+
A. israelii	21 (1)	9	38	22	34	21
P. nigrescens	22 (1)	16	18	16	30	27
V. dispar	23 (1)	43+	44+	47+	42+	41+
S. mutans	24 (1)	43+	44+	32	42+	41+
Streptococcus SM	25 (1)	43+	24	47+	42	41+

[a] The rank of each species among the isolates from other periodontal health conditions is also given. Species rank is based on data obtained from Moore and Moore (1994).

[b] *Actinomyces georgiae, Actinomyces israelii, Actinomyces meyeri, Actinomyces naeslundii, Actinomyces odontolyticus, Bacteroides gracilis, Bifidobacterium dentium, Campylobacter concisus, Campylobacter rectus, Capnocytophaga gingivalis, Capnocytophaga ochracea, Eubacterium alactolyticum, Eubacterium nodatum, Eubacterium saphenum, Eubacterium timidum, Fusobacterium alocis, Fusobacterium nucleatum, Gemella morbillorum, Haemophilus segnis, Lactobacillus rimae, Lactobacillus uli, Peptostreptococcus anaerobius, Peptostreptococcus micros, Porphyromonas gingivalis, Prevotella nigrescens, Propionibacterium acnes, Selenomonas infelix, Selenomonas sputigena, Streptococcus gordonii, Streptococcus intermedius, Streptococcus mutans, Streptococcus oralis, Streptococcus sanguis, Veillonella atypica, Veillonella dispar, Veillonella parvula.*

[c] Numbers in parentheses are the percentages of isolates taken at random from subgingival samples.

were identified (Moore and Moore, 1994). At a glance it can be seen that the four species in bold type that rank number 7, 8, 9, and 10 in healthy sites do not appear in the top 10 again in other periodontal health conditions, and this is an example of the large flux in numbers of a given species in different oral environments. Further examination of Table 9.1 reveals that many other bacteria in the bottom 15 rankings in "Health" also drop to even lower rankings in other periodontal health conditions, indicating that other species are taking their place and, thus, illustrating the broad diversity of species in the oral microflora.

In health, *Actinomyces naeslundii* is the most numerous species in the top 10 species rank, comprising 18% of the isolates (Moore and Moore, 1994), and it remains among the most frequently isolated species in samples taken from different periodontal health conditions (top row of Table 9.1). Two other actinomyces, *Actinomyces meyeri* and *Actinomyces odontolyticus,* equal 2% each of the flora. The streptococci, *Streptococcus sanguis, Streptococcus oralis, Streptococcus intermedius,* and *Streptococcus gordonii,* in the top 10 are collectively 23% of the isolates found in healthy sites. In fact, 9 of the top 10 species found in plaque from healthy sites are gram positive. The exception is *Fusobacterium nucleatum,* which is present as 11% of the population. All the other species of the top 25 are significant in numbers, and together the top 25 species constitute about 80% of the isolates identified. Although there are six genera each of gram-positive and gram-negative bacteria, the gram-positive bacteria are about 75% of the population.

This population changes dramatically in gingivitis (Moore and Moore, 1994), where four of the top six species isolated are gram negative and they comprise 62% (21/34) of that group (Table 9.2). The column for "Health" has been moved to the right in Table 9.2. By emphasizing their ranking in bold type, it is easy to see that 14 of the most numerous 25 species in gingivitis (Table 9.2, left column) were ranked 30th or much lower in health (Table 9.2, right column). The emergence of 56% (14/25) of new species into the top 25 species documents the size of the population flux with the change from health to gingivitis.

Considering that the total number of bacteria recovered in a sample from a gingivitis condition may be 10,000 times more than the 10^3 or 10^4 cultural count recovered from a healthy site (Moore and Moore, 1994), the overwhelming occupation of the econiche by gram-negative cell bodies is remarkable. The fusobacteria in particular increase to the highest numbers. Their production of butyric acid, sulfides, and mercaptans as metabolic end-products is destructive to gingival tissues. This event may initiate sequelae that include release of mitogenic factors, which encourage more bacterial proliferation and diversity. The genera often considered relevant to the onset and continued advancement of periodontal disease are *Bacteroides, Campylobacter, Eubacterium, Fusobacterium, Peptostreptococcus, Porphyromonas, Prevotella,* and *Selenomonas* (Table 9.2). Although not shown in Table 9.2, *Actinobacillus actinomycetemcomitans, Eikenella corrodens,* and *Treponema* spp. are thought to contribute to

TABLE 9.2. **The 25 Most Prevalent Species Determined by Percentage of Isolates Taken at Random From Subgingival Crevices of Gingivitis-Affected Sites of Subjects**[a]

Species	Rank in different periodontal health conditions					
	Gingivitis	Moderate	Adult	Juvenile	Severe	Health
F. nucleatum[b]	1 (12)[c]	2	2	3	1	2
A. naeslundii	2 (9)	1	1	2	4	1
L. uli	**3 (4)**	**9**	**9**	**7**	**5**	**38+**
C. concisus	**4 (3)**	**44+**	**6**	**42+**	**41+**	**30**
S. sputigena	**5 (3)**	**15**	**10**	**39**	**23**	**36**
B. gracilis	6 (3)	21	5	25	25	17
P. micros	7 (3)	3	4	10	6	6
P. anaerobius ID	**8 (2)**	**20**	**23**	**16**	**29**	**38+**
A. israelii	9 (2)	38	22	34	21	21
L. rimae	10 (2)	29	13	13	12	11
P. gingivalis	**11 (2)**	**6**	**20**	**28**	**41+**	**38+**
E. timidum	**12 (2)**	**10**	**7**	**9**	**3**	**33**
E. nodatum	**13 (2)**	**11**	**14**	**4**	**2**	**38+**
S. sanguis	14 (2)	32	3	23	41+	3
E. saphenum	**15 (2)**	**5**	**25**	**17**	**20**	**38+**
P. nigrescens	16 (2)	18	16	30	27	22
F. alocis	**17 (2)**	**16**	**17**	**12**	**14**	**37**
A. odontolyticus	18 (2)	22	12	18	41+	10
S. intermedius	19 (2)	14	11	36	9	5
E. alactolyticum	**20 (1)**	**44+**	**29**	**40**	**24**	**38+**
C. rectus	**21 (1)**	**4**	**19**	**5**	**26**	**38+**
S. infelix	**22 (1)**	**26**	**28**	**42+**	**41+**	**38+**
B. dentium	**23 (1)**	**34**	**35**	**42+**	**41+**	**38+**
S. oralis	24 (1)	17	8	8	41	4
P. anaerobius II	**25 (1)**	**33**	**26**	**32**	**28**	**38+**

[a] The rank of each species among the isolates from other periodontal health conditions is also given. Species rank based on data obtained from Moore and Moore (1994).

[b] *Actinomyces georgiae, Actinomyces israelii, Actinomyces meyeri, Actinomyces naeslundii, Actinomyces odontolyticus, Bacteroides gracilis, Bifidobacterium dentium, Campylobacter concisus, Campylobacter rectus, Capnocytophaga gingivalis, Capnocytophaga ochracea, Eubacterium alactolyticum, Eubacterium nodatum, Eubacterium saphenum, Eubacterium timidum, Fusobacterium alocis, Fusobacterium nucleatum, Gemella morbillorum, Haemophilus segnis, Lactobacillus rimae, Lactobacillus uli, Peptostreptococcus anaerobius, Peptostreptococcus micros, Porphyromonas gingivalis, Prevotella nigrescens, Propionibacterium acnes, Selenomonas infelix, Selenomonas sputigena, Streptococcus gordonii, Streptococcus intermedius, Streptococcus mutans, Streptococcus oralis, Streptococcus sanguis, Veillonella atypica, Veillonella dispar, Veillonella parvula.*

[c] Numbers in parentheses are the percentages of isolates taken at random from subgingival samples.

periodontal disease. Clearly, more than one organism participates, and this suggests that periodontal disease accompanies a change in the entire microbial community as an econiche and is not caused by a single species. It is worth noting that all but *Eubacterium* and *Peptostreptococcus* of this latter list of 11 genera are gram negative.

Seventeen genera and 36 species are listed in Tables 9.1 and 9.2. Closer examination of the rankings in the two tables will show that all but three of the top 10 species in all of the other four periodontal health conditions combined are already identified in samples from healthy (Table 9.1) and gingivitis-affected (Table 9.2) sites. The species not listed are *Prevotella intermedia* (number 7 in moderate; number 8 in severe), *Eubacterium* D06 (number 1 in juvenile; number 7 in severe), and *Streptococcus* D39 (number 10 in severe) (Moore and Moore, 1994). Only eight additional species are found in the lower 15 species of the top 25 species in the combined other four periodontal health conditions. Thus, the greatest change in the bacterial population occurs between health and gingivitis, since most of the prevalent species in subgingival plaque of moderate, juvenile, adult, and severe periodontal disease are already present in either gingivitis or health.

At least 24 of the 36 species (Tables 9.1 and 9.2) have been tested, and all coaggregate with one or more partners. Besides these species, numerous others, including *Rothia dentocariosa, Corynebacterium matruchotii, Eikenella corrodens, Actinobacillus actinomycetemcomitans, Prevotella intermedia, Haemophilus parainfluenzae, Capnocytophaga sputigena, Treponema denticola, Treponema pectinovorum, Treponema socranskii* subsps. and *Treponema vincentii*, have been tested, and they coaggregate with their own specific partner strains. Collectively, most of the commonly found oral bacteria have been identified, and most, if not all, have coaggregation partners.

An important caveat must be discussed here. Dental microbiologists still remain divided on the causative agents of most forms of periodontal disease; one school of thought favors specific bacterial pathogens, and the other views the diseases as a community process. Similarly, it may be argued that numbers of specific bacterial types are not accurate indicators of potential pathogens. Some of the species found in low numbers may be the true pathogens inflicting damage despite their lesser numbers. Resolution of these conundrums is still not at hand.

9.2.2. Nutritional Interrelationships

Despite its compactness, an active periodontal lesion is a microcosm of other significantly larger anaerobic environments like marine sediments, rumen, and anaerobic digestors. All contain morphologically and biochemically heterogeneous communities (Savage, 1983, 1986; Mohn and Tiedje, 1992; Lowe et al., 1993). The periodontal pocket hosts members of both bacterial kingdoms, the *Proteobacteria* and *Archaebacteria*. Methane bacteria, e.g., *Methanobrevibacter* species, presumably grow at the expense of hydrogen generated by formate-producing gram-negative bacteria (Kemp et al., 1983) fixing carbon dioxide, another metabolic end-product. As the pocket develops and becomes increasingly anaerobic, multiple food chains, or perhaps more appropriately food webs, are established (Theilade, 1990) that give rise to lactate and acetate utilizers; menadione-, fumarate-, and succinate-requiring bacteria; and, at the bottom of the chain, hydrogen- and carbon dioxide-utilizing archebacteria. Causal relationships between the establishment of food chains and the appearance of certain oral bacteria in plaque have not been firmly established.

From the preceding discussion, it may be argued that in most periodontal lesions competition for access to the community is balanced against a broad degree of commensalism within the community. Several true symbiotic relationships between plaque microorganisms have been reported. Coaggregation can foster such intimate alliances. For example, bacteria like *Veillonella atypica* and *Veillonella dispar* coaggregate with several species of oral streptococci that produce lactate; the end-product can be captured easily and used as an energy source by the veillonellae. The veillonellae, in turn, produce menadione (vitamin K) (Kolenbrander and Moore, 1991), a growth factor required by certain gram-negative oral bacteria. In mixed culture experiments, growth of *Wolinella* and *Porphyromonas* were shown to be dependent on the sharing of the metabolites formate and succinate (Grenier and Mayrand, 1986). Growth of oral treponemes is stimulated by porphyromonads (Grenier, 1992) and eubacteria (ter Steeg and van der Hoeven, 1990). The nutritional interrelationships of complex, multispecies, oral bacterial communities have been studied *in vitro* and shown to be particularly sensitive to changes in the pH of the community (Bradshaw et al., 1989). Whether this sharing actually occurs in plaque as an exclusionary or localized event has yet to be established. Nevertheless, it is clear that the intimacy provided by the adherence of these bacteria to one another would certainly enhance the chances for the conserving and sharing of nutrients.

A corollary to the existence of a commensal state is that the plaque residents tolerate growing in close proximity to one another. The numbers of any given cell type are subject to fluctuation (Moore and Moore, 1994; Socransky and Haffajee, 1994; Tanner et al., 1994), but, variation in numbers does not appear to result from the production of bacteriostatic or bacteriocidal products (e.g., bacteriocins or antibiotics); despite the fact that there are a number of reports of certain plaque bacteria producing these compounds. Presumably, the antimicrobials are not produced in sufficient quantity or with a sufficiently broad spectrum to permit one cell type to become dominant. Selection of a specific biotype during the development of a plaque community is more likely a reflection of the alterations of local factors, including the degree of anaerobiosis and change in the pH and the nutritional content at the site, e.g., where serum-rich crevicular fluid replaces saliva as a major source of nutrients.

9.3. ADHESIVE PROTEINS

The process of colonization is directly dependent on the ability of a resident bacterium to adhere to an inanimate or animate substratum. To this end, essentially all bacteria have acquired adhesive molecules on their surfaces that facilitate attachment to and ultimately control of a localized environment. Many bacteria have evolved protein adhesins whose specificity limits their choice of econiches, but simultaneously offers a selective advantage within predetermined sites. A striking example of such specificity is found among strains of *Escherichia coli;* some strains preferentially colonize the human urinary tract, while others are found in the intestinal tract. The genes encoding the lectin-like adhesins are part of large, complex transcriptional units named for the fimbriae that bear them: Pap and type 1 operons, respectively.

Adhesins are displayed on the surface of bacterial cells in two major motifs. Many are intercalated into the bacterium's cell wall or outer membrane (Kagermeier and London, 1986; Weiss et al., 1987b; Tempro et al., 1989; London, 1991), and others are displayed on fimbriae (pili) (Weiss et al., 1988b; London, 1991). The latter category may be divided into two subgroups. The most common mode reported to date has adhesins occurring as distinct entities strategically placed at or near the distal portion of the fimbriae (London, 1991). The second, and significantly smaller, group consists of those adhesins that are integral parts of the fimbrial subunit (Irvin et al., 1989; Farinha et al., 1994; Lee et al., 1994). Having the adhesin displayed on fimbriae is purported to be advantagous by virtue of reducing the charge repulsion effects created by the natural negative charge on cell surfaces (Jones and Isaacson, 1983). However, as examples discussed later will show, there appears to be little difference in efficacy of adhesion between the two modes of adhesin.

Among oral bacteria, members of virtually every genus found in plaque deposits have been shown to possess one or more adherence mechanisms that facilitates attachment to soft tissues, hard tissues, or resident bacteria already bound to these surfaces.

9.3.1. Lectin-Like Adhesins

As mentioned earlier, sugar receptors are abundant in the oral environment. Among these, galactose and related sugars, including lactose (McIntire et al., 1978), galactosamine, *N*-acetylgalactosamine, and *N*-acetylneuraminic acid, are the most common adhesin cognates. All of these sugars inhibit cell–cell interactions mediated by both established and putative lectin-like adhesins reported for oral bacteria. Thus, "lactose sensitivity" has become the universally accepted test for galactoside-specific interactions (McIntire et al., 1978; Kolenbrander and London, 1992).

9.3.1.1. Prevotella loescheii *Adhesins*

Prevotella loescheii, a gram-negative bacteroid, possesses multiple adhesins that implement cell–cell attachment with *Streptococcus oralis* and two biotypes of *Actinomyces israelii* (Kolenbrander and Andersen, 1984; Kolenbrander et al., 1985). A galactoside-specific adhesin mediates attachment of *Prevotella loescheii* to *Streptococcus oralis* and neuraminidase-treated erythrocytes (Weiss et al., 1987a, 1989). Immunological studies established that two different nonlectin-type adhesins are responsible for the interactions with the two biotypes of *Actinomyces israelii* (Weiss et al., 1987a).

In a series of studies, the galactoside-specific adhesin was localized on the surface of the cell, purified to homogeneity, characterized biochemically, and its gene cloned and sequenced. A family of highly specific and exquisitely inhibitory monoclonal antibodies were prepared against the adhesin. These identified, under denaturing conditions, a 75 kDa polypeptide as the adhesin (Weiss et al., 1988a). Indirect and direct immunoelectron microscopy carried out with 10 or 20 nm gold particles coated with antimouse IgG or antiadhesin monoclonal antibody revealed that the adhesin was associated with the distal portion of the bacterium's fimbriae. The IgG-coated gold beads were observed singly, in pairs, and in clusters of three or more beads (Fig. 9.1), suggesting that distribution of the adhesive protein occurred in a random fashion during synthesis or export of the fimbrial monomers (Weiss et al., 1988b). The monoclonal antibodies were used to demonstrate that two distinct adhesins can be

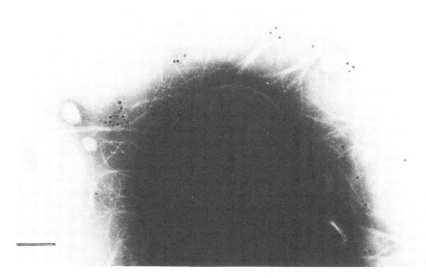

Fig. 9.1. Arrangement of galactoside-specific adhesin molecules on the surface of *Prevotella loescheii* PK1295. *Prevotella loescheii* cells were treated with 15 nm gold particles coated with monoclonal antibodies (clone 5BB1-1A2) specific for the lectin-like adhesin. Stained with phosphotungstic acid. Bar = 0.5 μm.

present simultaneously on the surface of the same cell (London et al., 1989). Quantitative antiadhesin monoclonal antibody–binding experiments were used to estimate the number of adhesins per cell; these studies yielded a *theoretical* maximum value of between 300 and 500 molecules per bacterium (Weiss et al., 1988b). This range of values agreed roughly with estimates made from the electron microscopy data and is approximately half the value estimated for the pap adhesin of *Escherichia coli* (Lindberg et al., 1989; Kuehn et al., 1992).

The monoclonal antibodies were covalently coupled to Sepharose 4B particles and used to isolate small quantities of the pure adhesin (200 to 300 mg protein per 10 g cells, wet weight) (London and Allen, 1990). The purified polypeptide migrated with a molecular weight of 75 kDa under denaturing conditions and manifested a basic pI of between 8.3 to 8.5. A portion of the purified material was used to prepare rabbit polyclonal antiadhesin serum, which, as expected, inhibited coaggregation between *Prevotella loescheii* and *Streptococcus oralis* (London and Allen, 1990). The homogeneous preparation, like many commercially available plant lectins, retained its ability to interact with galactoside-specific receptor molecules. Under slightly acidic conditions (pH 6.6 to 6.8) the adhesin bound to partner streptococcal cells and prevented coaggregation with *Prevotella loescheii.* Basic conditions (pH 7.6) caused the adhesin to aggregate sufficiently to agglutinate both streptococcal cells and neuraminidase-treated erythrocytes. As the pH approached or exceeded the protein's pI of 8.5, autoaggregation increased to the point where the adhesin became insoluble and was rendered inactive (London and Allen, 1990).

Some of the purified adhesin was used to determine the N-terminal amino acid sequence of the protein. After identifying the first 28 residues, degenerate probes were synthesized by reverse translation of the amino acid sequence employing codon usage of the *Dichelobacter nodosus* and *Porphyromonas gingivalis* fimbrial genes (Manch-Citron et al., 1992). The probes were used to screen a *Prevotella loescheii* GEM-11 bacteriophage gene library, and the adhesin gene, designated *plaA,* or segments thereof, was cloned into pGEM7Zf(+) and transformed into an appropriate *Escherichia coli* strain (Manch-Citron et al., 1992). DNA sequencing quickly revealed a putative 22 amino acid signal sequence; however, further sequencing of the gene became problematic. The codon representing the 28th amino acid residue of the mature protein was followed by an ochre stop signal (UAA). A second stop codon was found 21 nucleotides downstream of the first, and it was followed by a large open reading region in frame two.

To establish that the second large open reading frame (ORF) was, in fact, the remainder of the adhesin gene, the purified adhesin was subjected to endopeptidase treatment, and six peptide fragments were purified and sequenced. All six amino acid sequences were found in the deduced amino acid sequence of the second ORF, and these spanned most of the region in question (Manch-Citron et al., 1992). The correlation between the two sets of data could be reconciled only by postulating that a frameshift was essential for the complete translation of the *plaA* mRNA transcript.

A structural analysis of the region around the stop codons and frameshift revealed that virtually all of the components required for a ribosomal hop were also present. Naturally occurring ribosomal frameshifting hops, a process whereby a section of mRNA is neither read nor translated and a frameshift takes place, are very uncommon events. Gene 60, which encodes the topoisomerase of bacteriophage T4, is the prototype for a frameshifting hop (Weiss et al., 1990c). The hop bypasses 50 nucleotides (nt 519 to 569) and has a stop codon in all three frames; it requires identical takeoff and landing sites, a slippery sequence between the takeoff and landing site, and a stem loop structure as well as a pseudoknot downstream of the landing site (Weiss et al., 1990c). The *Prevotella loescheii* adhesin gene, *plaA,* possesses the same elements around its frameshift/hop site. The identical ochre stop codons can serve as takeoff/landing sites, the region contains a UAA in frame 2 prior to the shift, the appropriate slip regions are located between the ochre codons, and there is a downstream stem loop and pseudoknot (Manch-Citron and London, 1994). The presumed action of the combined elements is to slow or destabilize the ribosome sufficiently to bypass the nucleotides situated between the ochre terminators and advance a single base (+1 frameshift) entering the second reading frame. A number of functional genetic constructs of other genes, which perform forward or reverse frameshifts with varying degrees of efficacy while bypassing one or several nucleotides, have been fabricated (Weiss et al., 1990c). It is easy to visualize a mechanism for ribosomal slippage across a few nucleotides; however, it is not clear how the two long distance hops described above occur. It may be that secondary structure of the mRNA (e.g., loops or bubbles) juxaposes the takeoff/landing sites, physically shortening the distances the ribosome must actually travel.

The purpose of frameshifting in general appears to be conservation of genetic material, e.g., encoding more than one protein on a region of DNA (Weiss et al., 1990b) or fusing proteins for export to the cell envelope. Fusions are common occurrences among viruses where the fused protein complex can act as multifunctional catalytic unit on the cell surface or where it may be separated into its individual components. The role of the frameshifting hop in the *plaA* gene is not yet clear—no second protein has been found encoded by the gene, and it is unlikely that the adhesin polypeptide fuses to another functional protein. It is possible that the frameshifting hop is an evolutionary triviality representing the cell's successful attempt at coupling an intracellular protein with adhesive potential to a leader sequence that would facilitate its export. Presumably the coding information for the first 50 amino acid residues (leader plus the first 28 amino acids of the mature protein) would be acquired from some other region of the chromosome. However, the expectation is that the cumbersome hop region would be quickly disposed of since there is no obvious selective advantage in maintaining it after perfecting an export system.

When comparing the deduced amino acid sequence of the mature adhesin with other protein sequences in various databases, several significant correlations were found. A region spanning residues 319 to 532 of the adhesin exhibited 44% similarity with *Escherichia coli* β-galactosidase. Smaller runs, e.g., 120 residues or less, showed homology to a glucosyltransferase and glucose-6-phosphate dehydrogenase (J. Manch-Citron and J. London, unpublished data). Whether these regions actually represent a portion of the lectin-binding domain has not yet been determined.

9.3.1.2. Lectin-Like Adhesins of Capnocytophaga Species

Members of the genus *Capnocytophaga,* a group of gram-negative gliding bacteria routinely found in developing plaque deposits, bear adhesins that are intercalated into their outer membranes (Tempro et al., 1989). The most diverse arrays are found on *Capnocytophaga ochracea* and recognize receptors on three different bacteria, *Streptococcus oralis* ATCC55229 (formerly *S. sanguis* H1), *Actinomyces naeslundii* PK984, and *Actinomyces israelii* PK16 (Kolenbrander and Andersen, 1984). The receptor recognition sites are represented by a completely different set of sugars than those described for the *Prevotella loescheii* adhesin. For example, the interaction between the *Capnocytophaga ochracea* adhesin and *Streptococcus oralis* or *Actinomyces naeslundii* receptor is inhibited by L-rhamnose or D-fucose, with the latter requiring a twofold higher concentration of the methyl sugar to achieve a 50% inhibition. Effective inhibition of coaggregation between *Capnocytophaga ochracea* and *Actinomyces israelii* is achieved only by adding a combination of sugars, L-rhamnose and N-acetylneuraminic acid. *Capnocytophaga sputigena* coaggregates only with the two actinomyces, and the sugar-inhibition patterns are the same as those described with *Capnocytophaga ochracea* (E.I. Weiss and P.E. Kolenbrander, unpublished data). *Capnocytophaga gingivalis* coaggregates only with *Actinomyces israelii* PK16 via an aminogalactoside- or neuraminlactose-sensitive molecule; neuraminlactose was the most effective inhibitor (Kagermeier et al., 1984). The similarities in sugar-inhibition patterns, sequential loss of adhesin activity in coaggregation-negative strains of *Capnocytophaga ochracea,* and synergies between sugars required to inhibit some interactions suggested that the receptors may be clustered into a polysaccharide complex (Weiss et al., 1987b). As is discussed below, the rhamnose receptor is a single entity with no associated aminogalactosides.

Immunological procedures were used to identify the outer membrane-associated adhesin. The source of the adhesin was purified, washed membrane preparations of the *Capnocytophaga gingivalis* that retained the ability to interact with and aggregate partner cells. Polyclonal antiserum prepared against intact cells of the wild-type strain was repeatedly adsorbed with a coaggregation-negative mutant, and the resultant IgG abrogated coaggregation when preincubated with wild-type *Capnocytophaga gingivalis* cells. After the denatured adhesin polypeptide had been transferred to cellulose filters, screening was performed by using antiadhesin IgG preparations conjugated to radiolabeled iodine or by using the IgG as primary antibody in Western blot analyses. Both types of antibody identified a 150 kDa polypeptide as the putative adhesin; the protein's pI was estimated to be 8.6 (Kagermeier and London, 1986). Monoclonal antibody studies confirmed the size of the polypeptide (Tempro et al., 1989). In a separate study, monoclonal antibodies that inhibited coaggregation between *Capnocytophaga ochracea* and *Streptococcus oralis* reacted with a large membrane-associated polypeptide (Weiss et al., 1990a). From these preliminary studies, the putative capnocytophagae adhesins appear to be relatively large proteins. Isolation of active, membrane-free, capnocytophagae adhesins has yet to be reported.

Quantitative studies with antiadhesin monoclonal antibodies, similar to those performed with *Prevotella loescheii,* show that the number of adhesin molecules per cell is roughly the same. These experiments yielded values of between 200 and 300 adhesin molecules per cell (Tempro et al., 1989). Unless the affinity of the capnocytophagae adhesin for its receptor is unusually greater or the net charge on the cell's surface is more positive than prevotellae, the similarity in numbers of adhesins found on *Capnocytophaga gingivalis* and *Prevotella loescheii* suggests that membrane associated adhesins are as efficacious as fimbrial associated adhesins.

9.3.1.3. Other Lectin-Like Adhesins

Table 9.3 lists the reports describing lectin and nonlectin adhesins involved in the attachment of oral bacteria to one another or to other surfaces. In most instances, the criterion used to implicate lectin activity was inhibition or reversal of the specific interactions by one or more sugars. Since much of the data are descriptive in nature, e.g., establishing which partner possesses adhesin and which possesses receptor, few substantive comparisons or comments can be made. Attempts to isolate and characterize many of the lectin adhesins listed in Table 9.3 have met with limited or no success. The *Actinomyces naeslundii* (formerly *Actinomyces viscosus*) T14V gal-galNAc–specific adhesin is one of the more avidly sought after proteins (Cisar et al., 1983; Donkersloot et al., 1985; Yeung and Cisar, 1990). Paradoxically, the early studies describing the specificities of this adhesin–receptor pair provided much of the impetus for the plethora of current studies. It was initially proposed that the lectin activity resided in the subunit of the type 2 fimbriae (Cisar et al., 1980). While certain polyclonal antibodies prepared against purified intact fimbriae inhibited coaggregation (Revis et al., 1982), none of the monoclonal preparations had any effect on the interaction (Cisar, 1986). It would appear that, like most other fimbrial associated adhesins, the lectin-like *Actinomyces naeslundii* protein is genetically distinct from its support system and constitutes a minor constituent of the total surface protein. Studies with the *Actinomyces naeslundii* T14V type 1 fimbrial nonlectin adhesin seem to support this conclusion (see below).

9.3.2. Nonlectin Adhesins

Protein–protein interactions comprise another major group of adhesin–receptor complexes. A large number of pathogenic bacteria have acquired adhesins that recognize receptors like integrins (RDG family) or selectins, making them tissue specific; these have been reviewed in detail elsewhere (Falkow et al., 1992; Isberg and Nhieu, 1994). While many oral bacteria live in close association with epithelial surfaces, a search of the literature failed to unearth reports of RGD-specific adhesins. This finding is surprising in light of the fact that some gram-negative oral bacteria attach to and degrade proteins known to contain integrin receptors (Ciborowski et al., 1994; Pike et al., 1994).

9.3.2.1. The Proline-Rich Protein Adhesin of A. naeslundii (Formerly A. viscosus)

The importance of type 1 fimbriae in the attachment of the *A. naeslundii* to the tooth surface model, saliva-coated spheroidal hydroxyapatite (sSHA), was described a decade ago (Clark et al., 1984). Subsequently, acidic proline-rich protein (PRP) was shown to be the specific salivary component that binds to hydroxyapatite and serves as the receptor for the type 1 fimbrial adhesin (Gibbons et al., 1988). In binding to the inert substratum, the PRP undergoes a conformational shift that makes the receptor site available to the bacterial adhesin. The availability of pure PRP made it possible to localize the adhesive protein on the surface of *Actinomyces naeslundii*. Actinomyces cells were mixed with PRP-coated gold particles and examined by transmission electron microscopy. Electron micrographs showed the sparsely scattered gold particles to be associated with the peripheral regions or tips of the fimbriae (Leung et al., 1990). The isolation of *Actinomyces naeslundii* mutants that possess type 1 fimbriae but exhibit a reduced sSHA binding capacity support the concept that the adhesins are associated with, but genetically distinct from, the fimbrial subunit (Nesbitt et al., 1992). The idea is confirmed by the evidence that monoclonal antibodies against the cloned 54 kDa type 1 fimbrial subunit fail to block adherence of *A. naeslundii* T14V to sSHA (Cisar et al., 1991).

9.3.2.2. Nonlectin Adhesins of Oral Streptococci

The search for adhesins that participate in the attachment of viridans streptococci to hydroxyapatite surfaces has been carried out in several laboratories. Early on, fimbriae were associated with the adherence of *Streptococcus sanguis* FC1 (Gibbons et al., 1983), *Streptococcus parasanguis* (formerly *Streptococcus sanguis*) FW213 (Fives-Taylor, 1982; Fachon-Kalweit et al., 1985; Fives-Taylor and Thompson, 1985), and *Streptococcus sanguis* 12 (Morris et al., 1985) to sSHA particles. Much of the supporting evidence was based on inhibition of cellular attachment using polyclonal antisera made against fimbriae preparations. Antibody prepared against a purified 36 kDa surface protein from *Streptococcus sanguis* 12 and designated SsaB effectively, but not completely,

TABLE 9.3. Adhesins Mediating Human Oral Bacterial Coaggregation

Bacterial strain	Subunit size in kDa (name)	Cell surface location	Receptor-bearing cells	Sugar Inhibitor[a]	Reference
Gram negative					
Capnocytophaga gingivalis DR2001	150	Outer membrane	*Actinomyces israelii* PK16	NANA	Kagermeier and London (1986); Tempro et al. (1989)
Capnocytophaga ochracea ATCC33596	155	Outer membrane	*Streptococcus oralis* H1; *Actinomyces naeslundii* PK984; *Actinomyces israelii* PK16	Rha; Rha; NANA	Weiss et al. (1990a); Weiss et al. (1990a); Weiss et al. (1990a)
Fusobacterium nucleatum ATCC10953	39.5	Cell envelope	*Streptococcus sanguis* CC5A		Kaufman and DiRienzo (1989)
Fusobacterium nucleatum T18	42	Outer membrane	*Porphyromonas gingivalis* T22	Lac	Kinder and Holt (1993)
Haemophilus parainfluenzae HP-28	34	Outer membrane	*Streptococcus sanguis* SA-1	GalNAc	Lai et al. (1990)
Porphyromonas gingivalis ATCC3277 (*Bacteroides gingivalis* ATCC3277)[b]	40	Fimbriae	*Actinomyces naeslundii* WVU627 (*Actinomyces viscosus* WVU627)[b]		Goulbourne and Ellen (1991)
Prevotella loescheii PK1295 (*Bacteroides loescheii* PK1295)[b]	75 (PlaA)	Fimbriae	*Streptococcus oralis* 34	Lac	Weiss et al. (1988a); London and Allen (1990)
Veillonella atypica PK1910	43; 45	Fimbriae	*Actinomyces israelii* PK14; *Streptococcus oralis* 34	Lac	Weiss et al. (1988a); Hughes et al. (1992)

Gram positive					
Actinomyces naeslundii T14V (*Actinomyces viscosus* T14V)[b]		Fimbriae	*Streptococcus oralis* 34	Lac	Cisar et al. (1983)
Actinomyces naeslundii ATCC12104		Fimbriae	*Streptococcus oralis* 34	Lac	Kolenbrander and Andersen (1985) Cisar (1986)
Streptococcus gordonii PK488	35 (ScaA)	Cell wall	*Actinomyces naeslundii* PK606		Kolenbrander and Andersen (1990) Kolenbrander et al. (1994)
Streptococcus gordonii DL1	76 (SarA)	Cell wall	*Actinomyces naeslundii* WVU627 (*Actinomyces viscosus* WVU627)[b]		Jenkinson and Easingwood (1990) Jenkinson (1992)
	205 (SspA)	Cell wall	*Actinomyces naeslundii*		Jenkinson et al. (1993)
	260 (CshA)	Cell wall	*Actinomyces naeslundii*		McNab and Jenkinson (1992)
	245 (CshB)	Cell wall	*Actinomyces naeslundii*		McNab et al. (1994)
Streptococcus gordonii G9B	45, 62/60	Cell wall	*Porphyromonas gingivalis* 33277		Lamont et al. (1994)
	170	Cell wall	*Porphyromonas gingivalis* 33277		Lamont et al. (1994)
Streptococcus salivarius HB	380	Fibrils (91 nm)	*Veillonella parvula* V1 (*Veillonella alcalescens* V1)[b]		Weerkamp et al. (1986a) Weerkamp et al. (1986b)

[a] NANA, *N*-acetylneuraminic acid; Rha, L-rhamnose; Lac, lactose; GalNAc, *N*-acetyl-β-D-galactosamine.

[b] Previously used name is given in parentheses.

prevented attachment of the streptococci to sSHA. Immunogold staining suggested that the surface protein was not a fimbrial subunit. Its gene was cloned, and its amino acid sequence was deduced from the DNA nucleotide sequence by reverse translation (Morris et al., 1987; Ganeshkumar et al., 1988, 1991). The mature protein was preceded by a putative 19 amino acid signal sequence. The cloned *Streptococcus sanguis* 12 surface protein gene shared 73 percent identity with the *Streptococcus parasanguis* FW213 type 1 fimbrial gene, which had also been cloned and sequenced in the interim (Fenno et al., 1989). Meanwhile, the antibodies recognizing the 36 kDa surface protein of *Streptococcus sanguis* 12 (Ganeshkumar et al., 1991) cross-reacted strongly with a putative 36 kDa adhesin of *Streptococcus gordonii* PK488, which mediates coaggregation with *Actinomyces naeslundii* PK606 (Kolenbrander and Andersen, 1990). The gene encoding the cross-reacting *Streptococcus gordonii* surface component (designated *ScaA*) was cloned, sequenced, and its amino acid composition deduced (Kolenbrander et al., 1994).

Surprisingly, a comparison of the amino acid sequences of the sSHA adhesin (SsaB), the type 1 fimbrial subunit (fimA), and the coaggregation-relevant surface protein (ScaA) revealed that all three were very similar proteins (Kolenbrander et al., 1994). The ScaA–SsaB pair were 91% identical, whereas the ScaA–FimA pair were only 80% identical. SsaB was shown to be a lipoprotein (Ganeshkumar et al., 1993) incorporating palmitate at the cys-20 acylation site and, like SsaB, ScaA, and fimA, also contain the lipoprotein consensus sequence, LxxC. All three proteins have been linked to adherence, and it is not unusual to find significant homology between isofunctional proteins of related bacteria. The first two are presumed to recognize some factor in saliva that binds to SHA; the third member of this group, ScaA, is believed to function as a nonlectin mediator of coaggregation. Differences in the respective DNA sequence may reflect alterations in the function of the gene products.

Sequencing of the region upstream of *scaA* established that the gene was part of an operon containing an ATP-binding protein and potential membrane spanning protein, not unlike the ATP-binding cassettes responsible for oligo-peptide import/export (Alloing et al., 1989; Alloing et al., 1994). In a very recent report, the genetic organization of the *fimA* gene was described (Fenno et al., 1995). Like the *scaA, fimA* is part of a polycistronic unit that includes an ATP-binding protein and a membrane translocating protein. At this time it is not clear what purpose these conserved genetic systems actually serve. Nor is it clear that any of them function primarily as transport proteins, with adherence being a secondary function. However, the covalent palmitoyl residues permit this trio of related proteins to be anchored to the cell's membrane or lipoteichic acid, protrude through the peptidoglycan layer, and be strategically positioned to perform any of their proposed functions. A recent review of lipoproteins of gram-positive bacteria accentuates this rapidly advancing arena of surface proteins (Sutcliffe and Russell, 1995).

9.3.2.3. *Proteases as Adhesins*

Many of the gram-negative oral bacteria, including treponemes, synthesize a number of serine-, cysteine-, and metalloproteases; some of these are also hemolysins capable of providing a source of iron. These enzymes are generally considered to be virulence agents that participate in the destruction of host tissues. Cysteine proteases from several strains of *Porphyromonas gingivalis* may serve as a paradigm for a correlative function, adherence. When treated with surfactants, *P. gingivalis* W12 cells release two cysteine protease activities associated with 150 and 120 kDa polypeptides (called *porphypains*), respectively. Both exhibit activity toward arginine and lysine residues, and both appear to be derived from a 180 kDa precursor (Ciborowski et al., 1994). Specific binding sites were distinguished by temperature differential experiments; at 0°C, the proteases bound to the matrix protein but did not hydrolyze it (Ciborowski et al., 1994).

Several small, distinct cysteine proteases isolated from the culture medium of *Porphyromonas gingivalis* H66 exhibited either specifically for arginine (a 50 kDa polypeptide, called *gingipain*) or lysine (a 60 kDa polypeptide, called *lys-gingipain*) (Pike et al., 1994). Both are thought to be noncovalently associated to 44 kDa (or smaller) hemagglutinins to form large-molecular-weight complexes that facilitate attachment to and lysis of erythrocytes. Recent work has resolved the bothersome size differences, and it appears that proteases of *Porphyromonas gingivalis* H66 are synthesized as larger entities, which include the hemagglutinin domain (Potempa et al., 1995). Similarly, the original descriptions of the *Porphyromonas gingivalis* ATCC53977 53 kDa trypsin-like protease and the 1.5 kb gene encoding it were modified upon further study (Otogoto and Kuramitsu, 1993). The cysteine protease was actually a 96 to 98 kDa enzyme encoded within a 2.7 kb gene (Madden et al., 1995). More significant, a hemagglutinin gene, initially believed to be downstream of the protease gene, is encompassed within the 2.7 kb region of DNA. The genetic arrangement of the hemagglutinin and cysteine protease activity appears to be identical for all of the extensively studied strains. The separation of the two activities first reported for the H66 strain can be attributed to be a post-translational event, probably an autolytic cleavage of the larger protein.

These reports may provide some useful insights into alternate paths for the acquisition of adhesins that mediate cell–cell interactions. The fact that protease inhibitors interfere with attachment of the protease suggests that the two sites are structurally similar (Nishikata and Yoshimura, 1991; Shah et al., 1992). The adhesive domains of two forms of porphypain and the ATCC53977 cysteine protease appear to be transcribed as a single gene product. If the two sites are indeed similar, the simplest scenario to explain the acquisition of an adhesive site is (1) the gene originally contained multiple active sites or (2) some transpositional event duplicated the active site within the gene. In either event, one of the sites was modified to provide a binding domain. Other explanations are possible, but they involve more complex evolutionary gymnastics. Once the two domains are separated, it is highly unlikely that reassociation occurs, especially in a natural environment.

The intriguing aspect of this type of attachment process is its intrinsically temporal nature. Since proteases are ultimately released from the cell and the adhesive domains released with them, anchorage of the microbes to host cells may occur only for finite periods of time, especially during periods of starvation or nutrient depletion. Survival may depend on the ability of the bacterium to exit a nutrient-depleted environment by detachment or destruction of adhesins. The protease domain may also degrade the adhesive domain to provide amino acids during periods of stress. A secondary role for a *Prevotella loescheii* metalloprotease, similar to that just described, has been suggested. The protease activity was partially purified from clarified culture fluids and was shown to degrade the bacterium's galactoside-specific, lectin-like adhesin (Cavedon and London, 1993). It was postulated that the enzyme might serve as a release mechanism to free the bacterium from its coaggregation partner or host tissues.

9.4. RECEPTORS

Salivary proteins and certain polysaccharides on viridans streptococci are the two groups of receptors receiving the greatest attention in recent years. The former group has been mentioned briefly in preceeding sections of this chapter and discussed in several current reviews (Genco et al., 1988; Hirsch and Clarke, 1989); therefore, they will not be discussed here. As alluded to previously, the early colonizing viridans streptococci synthesize cell-associated, extracellular polysaccharides that serve as receptors for a significant number of oral bacteria. Five polysaccharides from the following strains, *Streptococcus gordonii* strain 38, *Streptococcus oralis* strains H1, J22, and 34, and *Streptococcus sanguis* K103 were analyzed and shown to be composed of repeating linear hexasaccharide backbones. The hepta- and octasaccharides in the group are the result of branch-chain sugars covalently linked to the six-sugar backbone (Cassels and London, 1989; Abeygunawardana et al., 1990, 1991a,b; Cassels et al., 1990; Glushka et al., 1992; Reddy et al., 1993, 1994). The sugars comprising the units include glucose, galactose, rhamnose, *N*-acetylgalactosamine, or *N*-acetylglucosamine.

The isolated *Streptococcus oralis* H1 and 34 polysaccharides and their corresponding six sugar units were definitively shown to be receptors for adhesins on the surfaces of *Capnocytophaga ochracea* and the group consisting of *Actinomyces naeslundii, Prevotella loescheii, Veillonella atypica,* and *Streptococcus gordonii* DL1, respectively. Inhibition studies with the H1 polysaccharide, hexasaccharide, rhamnose, and fucose identified the rhamnose α-1,2-rhamnose disaccharide portion of the backbone unit as the most likely receptor site. Inhibition studies supported by immunological analyses characterized both the receptor region and antigenic determinant of the strain 34 linear unit. The *N*-acetylgalactosamine on the nonreducing end of the unit was immunodominant while the *N*-acetylgalactosamine β-1,3-galactose disaccharide at the reducing end of the chain appears to be the lectin receptor region (Cisar, 1986; McIntire et al., 1988).

Strains J22 and 38 are heptasaccharides that differ only in the arrangement of the reducing sugars at the reducing end of the chain unit; the disaccharide of the former is galactose β-1,3-*N*-acetylgalactosamine, while the latter resembles strain 34 with *N*-acetylgalactosamine β-1,3-galactose. The lectins of the *Actinomyces naeslundii* recognize the receptors on strains 34, 38, and J22 regardless of the arrangement of the disaccharide. However, only the *N*-acetylgalactosamine β-1,3-galactose configuration found on strains 34 and 38 are recognized by the intrageneric coaggregation partners *Streptotoccus gordonii* DL1 and *Streptococcus gordonii* PK488 (Kolenbrander et al., 1990; Hsu et al., 1994; Clemans and Kolenbrander, 1995).

The role of extracellular polysaccharides on bacterial surfaces is generally thought to be either protective or adhesive in nature. The latter seems to be the case for the water-insoluble glucans produced by the "mutans" group of streptococci. In other instances, a glycocalyx layer provides an effective means of damping or avoiding host defensive responses. The capacity actually to alter a capsular or extracellular polysaccharide in response to a host defense mechanism amplifies the level of protection markedly. Also, polysaccharides are often less antigenic than proteins. It has been reported that the polysaccharides of the viridans streptococci are major surface antigens (Cisar, 1986; McIntire et al., 1988; Abeygunawardana et al., 1991b; Reddy et al., 1993, 1994; Hsu et al., 1994). It has also been reported that the antigenic make-up of the viridans streptococci is subject to rapid change *in vivo* (Bratthall and Gibbons, 1975; Gibbons and Howell, 1978; Howell et al., 1979). However, it has not been established that this is due to rapidly occurring mutations in the polysaccharide-synthesizing enzymes. These immunologically distinct polysaccharides also may contribute to the antigenic diversity so characteristic of plaque deposits housing a large variety of bacterial types that reduce or overwhelm localized immune responses.

9.5. CONCLUSIONS

The hallmark of dental plaque, especially plaque associated with periodontal disease, is its biological diversity. Representatives of the three phylogenetic kingdoms are found in periodontal pockets. Unicellular eukaryotic predators that graze on the bacteria flora are readily observed in wet mounts of plaque, but rarely discussed in recent review articles. The successful elective cultivation of methane-producing Archaea raises the possibility that other anaerobic members of this kingdom, e.g., sulfate-reducing bacteria, may be present in these deposits. Many of the bacterial types found in these surroundings bear multiple adhesins that potentially permit them to interact with neighboring bacteria to form a closely knit community. Microbicides do not appear to be significant determinants in the population dynamic unless introduced externally.

An important unresolved question is whether coaggregation patterns reflect random, benign couplings or whether they are predictors of commensal or symbiotic relationships. If the latter is the case, it is essential to know if predisposing factors, e.g., nutritional requirements, have selected for certain cell-to-cell interactions. Did the evolution of specific adhesins coincide with the evolution of commensal relationships? For example, the veillonellae have evolved such that they utilize a restricted number of energy sources, namely, three carbon compounds like lactate. This was apparently accomplished rather simply by the loss of the glycolytic enzyme hexokinase (Rogosa et al., 1965). Presumably the mutation involved the loss of the gene encoding this key enzyme. The bacterium's sugar transport system, which would have been rendered redundant by the loss of hexokinase, also appears to be missing (Winter and Delwiche, 1975). Over 98% of the oral veillonellae isolates possess adhesins specific for *Streptococcus salivarius* or the viridans group (Hughes et al., 1988). It is fascinating to speculate whether the loss of the hexokinase occurred prior to or as a result of the oral veillonellae's intimate association with lactic acid–producing bacteria. Such an association would have allowed a random mutation in a component of an essential energy-yielding system to be sustained.

The potential for metabolic exchange to occur in plaque deposits and its role in creating commensal states was discussed above (section 9.2.2). The sharing of nutrients represents cell-to-cell communication in its simplest form. More sophisticated forms of communication have been reported within homogeneous populations of prokaryotes. Here, signal transduction via systems comprising two or more components facilitate a coordinated synthesis of proteins that enhance or are required for the survival of the community as a whole (Surette and Stock, 1994). The next level of complexity, if it exists, would find transcriptional regulation of this sort occurring between genetically distinct bacteria in a heterogeneous population like that found in plaque. Information exchange at the genetic level has probably been demonstrated already. Penicillin-binding proteins with almost identical amino acid sequences have been found in strains of *Streptococcus pneumoniae* and *Streptococcus gordonii*. Both species are transformable, and the genes encoding these proteins are believed to have been passed from one to the other by this process (Smith et al., 1991; Spratt, 1994). The sequence identity between a *Porphyromonas gingivalis* cysteine protease and *Streptococcus pyogenes* exotoxin B was rationalized by invoking genetic exchange via conjugal plasmid transfer (Madden et al., 1995). Any environment that places heterogeneous bacteria in contact with one another for protracted periods of time has the potential to accelerate horizontal evolution or encourage the unexpected development of signal transduction pathways. Such contacts may even play a role in encouraging vertical evolutionary processes.

A discussion of possible predecessor molecules for the present day adhesins that mediate protein–protein interactions was presented above (section 9.3). The paradigm drew, as analogues, protease catalytic sites that were subsequently modified to perform adherence functions; recognition and hydrolysis were replaced with recognition and binding. Adaptive models have also been suggested for the lectin-like adhesins (Hultgren et al., 1988; Hultgren et al., 1989) and the ATP-binding cassette lipoproteins (Kolenbrander et al., 1994). Candidates for ancestral lectin-like molecules have been drawn from sugar transport systems and glycolytic enzymes; however, the data supporting these models are very tenuous. None of the deduced amino acid sequences of enteric bacterial adhesins shows similarities to known sugar transporting or metabolizing proteins. Using hemagglutination as a test system, the gal-α-1,4-Gal–specific adhesin gene (pap-3 G gene product) was subjected to oligonucleotide-directed mutagenesis to determine the recognition sites. A number of mutations introduced throughout the gene (from positions 88 to 316) disrupted hemagglutination, but no specific binding region was identified (Klann et al., 1994). The *Prevotella loescheii* adhesin has two large, separate regions that show homology to *Escherichia coli* β-galactosidase and glucosyltransferase, respectively (Manch-Citron and London, 1994). The former region is the best prospect for future insertional inactivation studies to establish a link between lectin-like adhesins and sugar-metabolizing enzymes. By utilizing molecular approaches to problems in ecology, microbiologists are finally poised to define the issues that contribute to the evolution of complex communities of bacteria.

ACKNOWLEDGMENTS

We thank C. Gentry-Weeks and J. Thompson for helpful comments during the preparation of this manuscript.

REFERENCES

Abeygunawardana C, Bush CA, Cisar JO (1990): Complete structure of the polysaccharide from *Streptococcus sanguis* J22. Biochemistry 29:234–248.

Abeygunawardana C, Bush CA, Cisar JO (1991a): Complete structure of the cell surface polysaccharide of *Streptococcus oralis* ATCC 10557: A receptor for lectin-mediated interbacterial adherence. Biochemistry 30:6528–6540.

Abeygunawardana C, Bush CA, Cisar JO (1991b): Complete structure of the cell surface polysaccharide of *Streptococcus oralis* C104: A 600-MHz NMR study. Biochemistry 30:8568–8577.

Alloing G, dePhilip P, Claverys J-P (1994): Three highly homologous membrane-bound lipoproteins participate in oligopeptide transport by the Ami system of the Gram-positive *Streptococcus pneumoniae*. J Mol Biol 241:44–58.

Alloing G, Trombe M-C, Claverys J-P (1989): Cloning of the *amiA* locus of *Streptococcus pneumoniae* and identification of its functional limits. Gene 76:363–368.

Bradshaw DJ, McKee AS, Marsh PD (1989): Effects of carbohydrate pulses and pH on population shifts within oral microbial communities *in vitro.* J Dent Res 68:1298–1302.

Bratthall D, Gibbons RJ (1975): Changing agglutination activities of salivary immunoglobulin A preparations against oral streptococci. Infect Immun 11:603–606.

Busscher HJ, Cowan MM, van der Mei HC (1992): On the relative importance of specific and non-specific approaches to oral microbial adhesion. FEMS Microbiol Rev 8:199–209.

Busscher HJ, Geertsema-Doornbusch GI, van der Mei HC (1993): On mechanisms of oral microbial adhesion. J Appl Bacteriol 74:136S–142S.

Cassels FJ, Fales HM, London J, Carlson RW, van Halbeek H (1990): Structure of a streptococcal adhesin carbohydrate receptor. J Biol Chem 265:14127–14135.

Cassels FJ, London J (1989): Isolation of a coaggregation-inhibiting cell wall polysaccharide from *Streptococcus sanguis* H1. J Bacteriol 171:4019–4025.

Cavedon K, London J (1993): Adhesin degradation: A possible function for a *Prevotella loescheii* protease? Oral Microbiol Immunol 8:283–287.

Ciborowski P, Nishikata M, Allen RD, Lantz MS (1994): Purification and characterization of two forms of a high-molecular-weight cysteine proteinase (porphypain) from *Porphyromonas gingivalis.* J Bacteriol 176:4549–4557.

Cisar JO (1986): Fimbrial lectins of the oral *Actinomyces.* In Mirelman D (ed): Microbial Lectins and Agglutinins: Properties and Biological Activity. New York: John Wiley & Sons, pp 183–196.

Cisar JO, Barsumian EL, Curl SH, Vatter AE, Sandberg AL, Siraganian RP (1980): The use of monoclonal antibodies in the study of lactose-sensitive adherence of *Actinomyces viscosus* T14V. J Reticuloendothel Soc 28:73s–79s.

Cisar JO, Barsumian EL, Siraganian RP, Clark WB, Yeung MK, Hsu SD, Curl SH, Vatter AE, Sandberg AL (1991): Immunochemical and functional studies of *Actinomyces viscosus* T14V type 1 fimbriae with monoclonal and polyclonal antibodies directed against the fimbrial subunit. J Gen Microbiol 137:1971–1979.

Cisar JO, Curl SH, Kolenbrander PE, Vatter AE (1983): Specific absence of type 2 fimbriae on a coaggregation-defective mutant of *Actinomyces viscosus* T14V. Infect Immun 40:759–765.

Clark WB, Wheeler TT, Cisar JO (1984): Specific inhibition of adsorption of *Actinomyces viscosus* T14V to saliva-treated hydroxyapatite by antibody against type 1 fimbriae. Infect Immun 43:497–501.

Clemans DL, Kolenbrander PE (1995): Isolation and characterization of coaggregation-defective (Cog−) mutants of *Streptococcus gordonii* DL1 (Challis). J Ind Microbiol 15:193–197.

Donkersloot JA, Cisar JO, Wax ME, Harr RJ, Chassy BM (1985): Expression of *Actinomyces viscosus* antigens in *Escherichia coli:* Cloning of a structural gene (*fimA*) for type 2 fimbriae. J Bacteriol 162:1075–1078.

Ellen RP, Dawson JR, Yang PF (1994): *Treponema denticola* as a model for polar adhesion and cytopathogenicity of spirochetes. Trends Microbiol 2:114–119.

Fachon-Kalweit S, Elder B, Fives-Taylor P (1985): Antibodies that bind to fimbriae block the adhesion of *Streptococcus sanguis* to saliva-coated hydroxyapatite. Infect Immun 48:617–624.

Falkow S, Isberg RR, Portnoy DA (1992): The interaction of bacteria with mammalian cells. Annu Rev Cell Biol 8:333–363.

Farinha MA, Conway BD, Glasier LMG, Ellert NW, Irvin RT, Sherburne R, Paranchych W (1994): Alteration of the pilin adhesin of *Pseudomonas aeruginosa* PAO results in normal pilus biogenesis but a loss of adherence to human pneumocyte cells and decreased virulence in mice. Infect Immun 62:4118–4122.

Fenno JC, LeBlanc DJ, Fives-Taylor P (1989): Nucleotide sequence analysis of a type 1 fimbrial gene of *Streptococcus sanguis* FW213. Infect Immun 57:3527–3533.

Fenno JC, Shaikh A, Spatafora G, Fives-Taylor P (1995): The *fimA* locus of *Streptococcus parasanguis* encodes an ATP-binding membrane transport system. Mol Microbiol 15:849–863.

Fives-Taylor P (1982): Isolation and characterization of a *Streptococcus sanguis* FW213 mutant nonadherent to saliva-coated hydroxyapatite beads. In Schlessinger D (ed): Microbiology—1982. Washington, DC: American Society for Microbiology, pp 206–209.

Fives-Taylor PM, Thompson DW (1985): Surface properties of *Streptococcus sanguis* FW213 mutants non-adherent to saliva coated hydroxyapatite. Infect Immun 47:752–759.

Ganeshkumar N, Arora N, Kolenbrander PE (1993): Saliva-binding protein (SsaB) from *Streptococcus sanguis* 12 is a lipoprotein. J Bacteriol 175:572–574.

Ganeshkumar N, Hannam PM, Kolenbrander PE, McBride BC (1991): Nucleotide sequence of a gene coding for a saliva-binding protein (SsaB) from *Streptococcus sanguis* 12 and possible role of the protein in coaggregation with *Actinomyces*. Infect Immun 59:1093–1099.

Ganeshkumar N, Song M, McBride BC (1988): Cloning of a *Streptococcus sanguis* adhesin which mediates binding to saliva-coated hydroxyapatite. Infect Immun 56:1150–1157.

Genco RJ, Zambon JJ, Christersson LA (1988): The origin of periodontal infections. Adv Dent Res 2:245–259.

Gibbons RJ, Etherden I, Skobe Z (1983): Association of fimbriae with the hydrophobicity of *Streptococcus sanguis* FC-1 and adherence to salivary pellicles. Infect Immun 41:414–417.

Gibbons RJ, Hay DI, Cisar JO, Clark WB (1988): Adsorbed salivary proline-rich protein 1 and statherin: Receptors for type 1 fimbriae of *Actinomyces viscosus* T14V-J1 on apatitic surfaces. Infect Immun 56:2990–2993.

Gibbons RJ, Howell TH (1978): Antigenic variation in population of oral streptococci. Adv Exp Med Biol 107:829–838.

Glushka J, Cassels FJ, Carlson RW, van Halbeek H (1992): Complete structure of the adhesin receptor polysaccharide of *Streptococcus oralis* ATCC 55229 (*Streptococcus sanguis* H1). Biochemistry 31:10741–10746.

Goulbourne PA, Ellen RP (1991): Evidence that *Porphyromonas* (*Bacteroides*) gingivalis fimbriae function in adhesion to *Actinomyces viscosus*. J Bacteriol 173:5266–5274.

Grenier D (1992): Nutritional interactions between two suspected periodontopathogens, *Treponema denticola* and *Porphyromonas gingivalis*. Infect Immun 60:5298–5301.

Grenier D, Mayrand D (1986): Nutritional relationships between oral bacteria. Infect Immun 53:616–620.

Hasty DL, Ofek I, Courtney HS, Doyle RJ (1992): Multiple adhesins of streptococci. Infect Immun 60:2147–2152.

Hirsch RS, Clarke NG (1989): Infection and periodontal diseases. Rev Infect Dis 11:707–715.

Howell TH, Spinell DM, Gibbons RJ (1979): Antigenic variation in populations of *Streptococcus salivarius* isolated from the human mouth. Arch Oral Biol 24:389–397.

Hsu SD, Cisar JO, Sandberg AL, Kilian M (1994). Adhesive properties of viridans streptococcal species. Microb Ecol Health Dis 7:125–137.

Hughes CV, Andersen RN, Kolenbrander PE (1992): Characterization of *Veillonella atypica* PK1910 adhesin-mediated coaggregation with oral *Streptococcus* spp. Infect Immun 60:1178–1186.

Hughes CV, Kolenbrander PE, Andersen RN, Moore LVH (1988): Coaggregation properties of human oral *Veillonella* spp.: Relationship to colonization site and oral ecology. Appl Environ Microbiol 54:1957–1963.

Hultgren SJ, Lindberg F, Magnusson G, Kihlberg J, Tennent JM, Normark S (1989): The PapG adhesin of uropathogenic *Escherichia coli* contains separate regions for receptor binding and for the incorporation into the pilus. Proc Natl Acad Sci USA 86:4357–4361.

Hultgren SJ, Lindberg F, Magnusson G, Tennent JM, Normark S (1988): Isolation of the pre-assembled gal-alpha (1–4) gal-specific pilus-associated adhesin from the periplasm in uropathogenic *Escherichia coli*. In Switalski L, Hook M, Beachey E (eds): Molecular Mechanisms of Microbial Adhesion. New York: Springer-Verlag, pp 36–43.

Irvin RT, Doig P, Lee KK, Sastry PA, Paranchych W, Todd T, Hodges RS (1989): Characterization of the *Pseudomonas aeruginosa* pilus adhesin: Confirmation that the pilin structural protein subunit contains a human epithelial cell-binding domain. Infect Immun 57:3720–3726.

Isberg RR, Nhieu GTV (1994): Binding and internalization of microorganisms by integrin receptors. Trends Microbiol 2:10–14.

Jenkinson HF (1992): Adherence, coaggregation, and hydrophobicity of *Streptococcus gordonii* associated with expression of cell surface lipoproteins. Infect Immun 60:1225–1228.

Jenkinson HF (1994a): Adherence and accumulation of oral streptococci. Trends Microbiol 2:209–212.

Jenkinson HF (1994b): Cell surface protein receptors in oral streptococci. FEMS Microbiol Lett 121:133–140.

Jenkinson HF, Easingwood RA (1990): Insertional inactivation of the gene encoding a 76-kilodalton cell surface polypeptide in *Streptococcus grodonii* Challis has a pleiotropic effect on cell surface composition and properties. Infect Immun 58:3689–3697.

Jenkinson HF, Terry SD, McNab R, Tannock GW (1993): Inactivation of the gene encoding surface protein SspA in *Streptococcus gordonii* DL1 affects cell interactions with human salivary agglutinin and oral *Actinomyces*. Infect Immun 61:3199–3208.

Jones GW, Isaacson RE (1983): Proteinaceous bacterial adhesins and their receptors. Crit Rev Microbiol 10:229–260.

Jones SJ (1972): A special relationship between spherical and filamentous microorganisms in mature human dental plaque. Arch Oral Biol 17:613–616.

Kagermeier A, London J (1986): Identification and preliminary characterization of a lectinlike protein from *Capnocytophaga gingivalis* (cmended). Infect Immun 51:490–494.

Kagermeier AS, London J, Kolenbrander PE (1984): Evidence for the participation of N-acetylated amino sugars in the coaggregation between *Cytophaga* species strain DR2001 and *Actinomyces israelii* PK16. Infect Immun 44:299–305.

Kaufman J, DiRienzo JM (1989): Isolation of a corncob (coaggregation) receptor polypeptide from *Fusobacterium nucleatum.* Infect Immun 57:331–337.

Kemp CW, Curtis MA, Robrish SA, Bowen WH (1983): Biogenesis of methane in primate dental plaque. FEBS Lett 155:61–64.

Keyes PH, Rams TE (1983): A rationale for management of periodontal diseases: Rapid identification of microbial "therapeutic agents" with phase-contrast microscopy. J Am Dent Assoc 106:803–812.

Kinder SA, Holt SC (1993): Localization of the *Fusobacterium nucleatum* T18 adhesin activity mediating coaggregation with *Porphyromonas gingivalis* T22. J Bacteriol 175:840–850.

Klann AG, Hull RA, Palzkill T, Hull SI (1994): Alanine-scanning mutagenesis reveals residues involved in binding of pap-2–encoded adhesins. J Bacteriol 176:2312–2317.

Kolenbrander PE (1993): Coaggregation of human oral bacteria: Potential role in the accretion of dental plaque. J Appl Bacteriol Symp Suppl 74:79S–86S.

Kolenbrander PE, Andersen RN (1984): Cell-to-cell interactions of *Capnocytophaga* and *Bacteroides* species with other oral bacteria and their potential role in development of plaque. J Periodontal Res 19:564–569.

Kolenbrander PE, Andersen RN (1985): Use of coaggregation-defective mutants to study the relationships of cell to-cell interactions and oral microbial ecology. In Mergenhagen SE, Rosan B (eds): Molecular Basis of Oral Microbial Adhesion. Washington, DC: American Society for Microbiology, pp 164–171.

Kolenbrander PE, Andersen RN (1990): Characterization of *Streptococcus gordonii* (*S. sanguis*) PK488 adhesin-mediated coaggregation with *Actinomyces naeslundii* PK606. Infect Immun 58:3064–3072.

Kolenbrander PE, Andersen RN, Ganeshkumar N (1994): Nucleotide sequence of the *Streptococcus gordonii* PK488 coaggregation adhesin gene, scaA, and ATP-binding cassette. Infect Immun 62:4469–4480.

Kolenbrander PE, Andersen RN, Holdeman LV (1985): Coaggregation of oral *Bacteroides* species with other bacteria: Central role in coaggregation bridges and competitions. Infect Immun 48:741–746.

Kolenbrander PE, Andersen RN, Moore LVH (1990): Intrageneric coaggregation among strains of human oral bacteria: Potential role in primary colonization of the tooth surface. Appl Environ Microbiol 56:3890–3894.

Kolenbrander PE, Ganeshkumar N, Cassels FJ, Hughes CV (1993): Coaggregation: Specific adherence among human oral plaque bacteria. FASEB J 7:406–413.

Kolenbrander PE, London J (1992): Ecological significance of coaggregation among oral bacteria. Adv Microb Ecol 12:183–217.

Kolenbrander PE, London J (1993): Adhere today, here tomorrow: Oral bacterial adherence. J Bacteriol 175:3247–3252.

Kolenbrander PE, Moore LVH (1991): The genus *Veillonella*. In Balows HGTA, Dworkin M, Harder W, Schleifer KH (eds): The Prokaryotes. A Handbook on the Biology of Bacteria: Ecophysiology, Isolation, Identification, Applications. New York: Springer-Verlag, pp 2034–2047.

Kuehn MJ, Heuser J, Normark S, Hultgren SJ (1992): P pili in uropathogenic *Escherichia coli* are composite fibres with distinct fibrillar adhesive tips. Nature 356:252–255.

Lai C-H, Bloomquist C, Liljemark WF (1990): Purification and characterization of an outer membrane protein adhesin from *Haemophilus parainfluenzae* HP-28. Infect Immun 58:3833–3839.

Lamont RJ, Gil S, Demuth DR, Malamud D, Rosan B (1994): Molecules of *Streptococcus gordonii* that bind to *Porphyromonas gingivalis*. Microbiology 140:867–872.

Lee KK, Sheth HB, Wong WY, Sherburne R, Paranchych W, Hodges RS, Lingwood CA, Krivan H, Irvin RT (1994): The binding of *Pseudomonas aeruginosa* pili to glycosphingolipids is a tip-associated event involving the C-terminal region of the structural pilin subunit. Mol Microbiol 11:705–713.

Leung K-P, Nesbitt WE, Fischlschweiger W, Hay DI, Clark WB (1990): Binding of colloidal gold-labeled salivary proline-rich proteins to *Actinomyces viscosus* type 1 fimbriae. Infect Immun 58:1986–1991.

Lindberg F, Tennent JM, Hultgren SJ, Lund B, Normark S (1989): PapD, a periplasmic transport protein in P-pilus biogenesis. J Bacteriol 171:6052–6058.

Listgarten MA (1976): Structure of the microbial flora associated with periodontal health and disease in man. J Periodontol 47:1–18.

Listgarten MA, Mayo H, Amsterdam M (1973): Ultrastructure of the attachment device between coccal and filamentous microorganisms in "corn cob" formations of dental plaque. Arch Oral Biol 18:651–656.

London J (1991): Bacterial adhesins. Annu Rep Med Chem 26:239–247.

London J, Allen J (1990): Purification and characterization of a *Bacteroides loeschei* adhesin that interacts with procaryotic and eucaryotic cells. J Bacteriol 172:2527–2534.

London J, Hand AR, Weiss EI, Allen J (1989): *Bacteroides loeschei* PK1295 cells express two distinct adhesins simultaneously. Infect Immun 57:3940–3944.

Lowe SE, Jain MK, Zeikus JG (1993): Biology, ecology, and biotechnological applications of anaerobic bacteria adapted to environmental stresses in temperature, pH, salinity, or substrates. Microbiol Rev 57:451–509.

Madden TE, Clark VL, Kuramitsu HK (1995): Revised sequence of the *Porphyromonas gingivalis* PrtT cysteine protease/hemagglutinin gene: Homology with streptococcal pyrogenic exotoxin B/streptococcal proteinase. Infect Immun 63:238–247.

Manch-Citron JN, Allen J, Moos JM, London J (1992): The gene encoding a *Prevotella loescheii* lectin-like adhesin contains an interrupted sequence which causes a frameshift. J Bacteriol 174:7328–7336.

Manch-Citron JN, London J (1994): Expression of the *Prevotella* adhesin gene (*plaA*) is mediated by a programmed frameshifting hop. J Bacteriol 176:1944–1948.

McIntire FC, Crosby LK, Vatter AE, Cisar JO, McNeil MR, Bush CA, Tjoa SS, Fennessey PV (1988): A polysaccharide from *Streptococcus sanguis* 34 that inhibits coaggregation of *S. sanguis* 34 with *Actinomyces viscosus* T14V. J Bacteriol 170:2229–2235.

McIntire FC, Vatter AE, Baros J, Arnold J (1978): Mechanism of coaggregation between *Actinomyces viscosus* T14V and *Streptococcus sanguis* 34. Infect Immun 21:978–988.

McNab R, Jenkinson HF (1992): Gene disruption identifies a 290 kDa cell-surface polypeptide conferring hydrophobicity and coaggregation properties in *Streptococcus gordonii*. Mol Microbiol 6:2939–2949.

McNab R, Jenkinson HF, Loach DM, Tannock GW (1994): Cell-surface–associated polypeptides CshA and CshB of high molecular mass are colonization determinants in the oral bacterium *Streptococcus gordonii*. Mol Microbiol 14:743–754.

Mohn WW, Tiedje JM (1992): Microbial reductive dehalogenation. Microbiol Rev 56:482–507.

Moore WEC, Moore LVH (1994): The bacteria of periodontal diseases. Periodontol 2000 5:66–77.

Moore WEC, Moore LVH, Cato EP (1988): You and your flora. US Fed Culture Collections Newslett 18:7–22.

Morris EJ, Ganeshkumar N, McBride BC (1985): Cell surface components of *Streptococcus sanguis:* Relationship to aggregation, adherence, and hydrophobicity. J Bacteriol 164:255–262.

Morris EJ, Ganeshkumar N, Song M, McBride BC (1987): Identification and preliminary characterization of a *Streptococcus sanguis* fibrillar glycoprotein. J Bacteriol 169:164–171.

Nesbitt WE, Beem JE, Leung K-P, Clark WB (1992): Isolation and characterization of *Actinomyces viscosus* mutants defective in binding salivary proline-rich proteins. Infect Immun 60:1095–1100.

Nishikata M, Yoshimura F (1991): Characterization of *Porphyromonas* (*Bacteroides*) *gingivalis* hemagglutinin as a protease. Biochem Biophys Res Commun 178:336–342.

Nyvad B, Kilian M (1987): Microbiology of the early colonization of human enamel and root surfaces *in vivo*. Scand J Dent Res 95:369–380.

Otogoto J, Kuramitsu H (1993): Isolation and characterization of the *Porphyromonas gingivalis prtT* gene, coding for protease activity. Infect Immun 61:117–123.

Pike R, McGraw W, Potempa J, Travis J (1994): Lysine- and arginine-specific proteinases from *Porphyromonas gingivalis*. J Biol Chem 269:406–411.

Potempa J, Pike R, Travis J (1995): The multiple forms of trypsin-like activity present in various strains of *Porphyromonas gingivalis* are due to the presence of either arg-gingipain or lys-gingipain. Infect Immun 63:1176–1182.

Reddy GP, Abeygunawardana C, Bush CA, Cisar JO (1994): The cell wall polysaccharide of *Streptococcus gordonii* 38: Structure and immunochemical comparison with the receptor polysaccharides of *Streptococcus oralis* 34 and *Streptococcus mitis* J22. Glycobiology 4:183–192.

Reddy GP, Chang C-C, Bush CA (1993): Determination by heteronuclear NMR spectroscopy of the complete structure of the cell wall polysaccharide of *Streptococcus sanguis* strain K103. Anal Chem 65:913–921.

Revis GJ, Vatter AE, Crowle AJ, Cisar JO (1982): Antibodies against the Ag2 fimbriae of *Actinomyces viscosus* T14V inhibit lactose-sensitive bacterial adherence. Infect Immun 36:1217–1222.

Rogosa M, Krichevsky MI, Bishop FS (1965): Truncated glycolytic system in *Veillonella.* J Bacteriol 90:164–171.

Rosenberg M, Rosenberg E, Judes H, Weiss E (1983): Bacterial adherence to hydrocarbons and to surfaces in the oral cavity. FEMS Microbiol Lett 20:1–5.

Savage DC (1983): Morphological diversity among members of the gastrointestinal microflora. Int Rev Cytol 82:305–334.

Savage DC (1986): Gastrointestinal microflora in mammalian nutrition. Annu Rev Nutr 6:155–178.

Shah NH, Ghargia SE, Progulske-Fox A, Brocklehurst K (1992): Evidence for independent molecular identity and functional interaction of the hemagglutinin and cysteine proteinase (gingivain) of *Porphyromonas gingivalis.* J Med Microbiol 36:239–244.

Sharon N, Lis H (1989): Lectins as cell recognition molecules. Science 246:227–246.

Smith JM, Dowson CG, Spratt BG (1991): Localized sex in bacteria. Nature 349:29–31.

Socransky SS, Haffajee AD (1994): Microbiology and immunology of periodontal diseases. Periodontol 2000 5:7–168.

Spratt BG (1994): Resistance to antibiotics mediated by target alterations. Science 264:388–393.

ter Steeg PF, van der Hoeven JS (1990): Growth stimulation of *Treponema denticola* by periodontal microorganisms. Autonie Leeuwenhoek 57:63–70.

Surette MG, Stock JB (1994): Transmembrane signal transducing proteins. In Ghuysen J-M, Hakenbeck R (eds): Bacterial Cell Wall. New York: Elsevier Science BV, pp 465–483.

Sutcliffe IC, Russell RRB (1995): Lipoproteins of Gram-positive bacteria. J Bacteriol 177:1123–1128.

Tanner A, Maiden MFJ, Paster BJ, Dewhirst FE (1994): The impact of 16S ribosomal RNA-based phylogeny on the taxonomy of oral bacteria. Periodontol 2000. 5:26–51.

Tempro P, Cassels F, Siraganian R, Hand AR, London J (1989): Use of adhesin-specific monoclonal antibodies to identify and localize an adhesin on the surface of *Capnocytophaga gingivalis* DR2001. Infect Immun 57:3418–3424.

Theilade E (1990): Factors controlling the microflora of the healthy mouth. In Hill MJ, Marsh PD (eds): Human Microbial Ecology. Boca Raton, FL: CRC Press, pp 1–56.

Weerkamp AH, Handley PS, Baars A, Slot JW (1986a): Negative staining and immunoelectron microscopy of adhesion-deficient mutants of *Streptococcus salivarius* reveal that the adhesive protein antigens are separate classes of cell surface fibril. J Bacteriol 165:746–755.

Weerkamp AH, van der Mei HC, Liem RSB (1986b): Structural properties of fibrillar proteins isolated from the cell surface and cytoplasm of *Streptococcus salivarius* (K+) cells and nonadhesive mutants. J Bacteriol 165:756–762.

Weiss E, Rosenberg M, Judes H, Rosenberg E (1982): Cell-surface hydrophobicity of adherent oral bacteria. Curr Microbiol 7:125–128.

Weiss EI, Eli I, Shenitzki B, Smorodinsky N (1990a): Identification of the rhamnose-sensitive adhesin of *Capnocytophaga ochracea* ATCC 33596. Arch Oral Biol 35:127s–130s.

Weiss EI, Kolenbrander PE, London J, Hand AR, Andersen RN (1987a): Fimbria-associated proteins of *Bacteroides loescheii* PK1295 mediate intergeneric coaggregations. J Bacteriol 169:4215–4222.

Weiss EI, London J, Kolenbrander PE, Andersen RN (1989): Fimbria-associated adhesin of *Bacteroides loeschei* that recognizes receptors on procaryotic and eucaryotic cells. Infect Immun 57:2912–2913.

Weiss EI, London J, Kolenbrander PE, Andersen RN, Fischler C, Siraganian RP (1988a): Characterization of monoclonal antibodies to fimbria-associated adhesins of *Bacteroides loescheii* PK1295. Infect Immun 56:219–224.

Weiss EI, London J, Kolenbrander PE, Hand AR, Siraganian R (1988b): Localization and enumeration of fimbria-associated adhesins of *Bacteroides loescheii*. J Bacteriol 170:1123–1128.

Weiss EI, London J, Kolenbrander PE, Kagermeier AS, Andersen RN (1987b): Characterization of lectinlike surface components on *Capnocytophaga ochracea* ATCC33596 that mediate coaggregation with Gram-positive oral bacteria. Infect Immun 55:1198–1202.

Weiss RB, Dunn DM, Atkins JF, Gesteland RF (1990b): Ribosomal frameshifting from 2 to +50 nucleotides. Prog Nucleic Acids Res Mol Biol 39:159–183.

Weiss RB, Huang WM, Dunn DM (1990c): A nascent peptide is required for ribosomal bypass of the coding gap in bacteriophage T4 gene 60. Cell 62:117–126.

Winter PF, Delwiche EA (1975): Cell wall composition and incorporation of radio-labelled compounds by *Veillonella alcalescens*. Can J Microbiol 21:2039–2047.

Yeung MK, Cisar JO (1990): Sequence homology between the subunits of two immunologically and functionally distinct types of fimbriae of *Actinomyces* spp. J Bacteriol 172:2462–2468.

10

SENSING, RESPONSE, AND ADAPTATION TO SURFACES: SWARMER CELL DIFFERENTIATION AND BEHAVIOR

ROBERT BELAS

Center of Marine Biotechnology, University of Maryland Biotechnology Institute, Baltimore, Maryland 21202

10.1. INTRODUCTION

Attachment to surfaces is of paramount importance to many species of bacteria. In nutrient-limited habitats, for example, in open ocean waters, surfaces are often places where nutrients accumulate, brought to those surfaces by a combination of electrostatic and hydrophobic interactions (see Chapter 3, this volume). Bacteria may be drawn by chemotactic sensing to these nutrient-rich surfaces, which provide an ideal niche for bacterial colonization and growth. These surfaces provide the cells with sufficient metabolic resources to fuel the metabolic requirements of the organisms. This in turn allows for bacterial division and the subsequent production of microcolonies on these surfaces.

One of the fundamental aspects of bacterial attachment is how a surface is recognized by the bacterial cell and, once recognized, how the bacterium responds and adapts its metabolic processes to life in this new niche. Some bacteria come predisposed for the purpose of attachment: they are essentially

Bacterial Adhesion: Molecular and Ecological Diversity, pages 281–331
© *1996 Wiley-Liss, Inc.*

"glueballs" covered with a glycocalyx composed of carbohydrates and proteins (see Chapter 1, this volume). For these bacteria, little or no response to the surface is necessary. Their interaction is always assured and beyond their control. Such assuredness comes at the metabolic cost that the cell bears to synthesize continually the necessary components of the glycocalyx required for attachment. This strategy for survival may not be appropriate for all bacteria or under all environmental circumstances.

There is, however, an alternative strategy to ensure survival of bacterial cells on a submerged surface. This strategy requires both the sensing and response to appropriate surface-derived signals, which then induce the expression of a very specific set of genes whose products are required for the adaptation and survival of the cell. The advantage of this approach is that the cell does not needlessly expend carbon, nitrogen, and energy reserves to synthesize a glycocalyx constitutively, which may not be needed if a surface is not present. Instead, only when a surface is sensed are those reserves used to prepare the cell for attachment and adaptation, thus saving valuable resources for use at the most appropriate time.

This chapter is focused on the "responsive strategy" for bacterial attachment to surfaces. As an example, a very specific mechanism of bacterial sensing, response, and, ultimately, adaptation to life on a submerged surface— which is known as swarmer cell differentiation, motility, and behavior—are discussed.

Unlike swimming behavior, which occurs in liquid environments, swarming is the organized movement of bacterial cells over a surface that is dependent on a specialized cell possessing unique characteristics that provide an adaptive advantage for living on such surfaces (Henrichsen, 1972). Swarming motility has been demonstrated in many different bacterial genera, both gram negative and gram positive. It is the objective of this chapter to describe the current state of knowledge regarding swarming behavior, as a form of bacterial differentiation and behavior, and, specifically, as a prokaryotic sensory phenomenon responsive to surfaces. To do so, *Proteus mirabilis* will be used as the model swarming cell, as it exemplifies many of the features common to the various swarming bacteria.

10.2. A GENERAL DESCRIPTION OF *P. MIRABILIS* SWARMING

P. mirabilis is a motile gram-negative bacterium, similar in many aspects of its physiology to other members of the family Enterobacteriaceae, such as *Escherichia coli* and *Salmonella typhimurium*. It was originally described and named by Hauser in 1885 for the character in Homer's *Odyssey* who "has the power of assuming different shapes in order to escape being questioned" (Hoeniger, 1964). *P. mirabilis* is considered to be an opportunistic pathogen and one of the principal causes of urinary infections in hospital patients with urinary catheters (Bidnenko et al., 1985a,b; Dlugovitzky et al., 1988; Ebringer et al., 1989; Frolov et al., 1986; McLean et al., 1985; Mobley and Warren, 1987; Senior, 1983; Story, 1954; Toshkov et al., 1977). It is believed that the ability of *P. mirabilis* to colonize the surfaces of catheters and the urinary tract may be aided by the characteristic first described over a century ago and currently referred to as *swarmer cell differentiation.*

When grown in suitable liquid media *P. mirabilis* exists as a 1.5–2.0 μm motile cell with 6 to 10 peritrichous flagella. These bacteria, called *swimmer cells,* display characteristic swimming and chemotactic behavior, moving toward nutrients and away from repellents (Adler, 1983). However, a dramatic change in cell morphology takes place when cells grown in liquid are transferred to a nutrient medium solidified with agar. Shortly after encountering an agar surface, the cells begin to elongate (Hoeniger, 19864, 1965, 1966; Hoeniger and Cinits, 1969; Hughes, 1956). This is the first step in the production of a morphologically and biochemically differentiated cell, referred to as the *swarmer cell* (Fig. 10.1A). The process of elongation takes place with only a slight increase in cell width and is due to an inhibition in the normal septation mechanism (Belas et al., 1995). Elongation of the swarmer cell can give rise to cells 60–80 μm in length. During this process, DNA replication proceeds without significant change in rate compared with that in the swimmer cell (Gmeiner et al., 1985). Not surprisingly, the rate of synthesis of certain proteins, e.g., flagellin, is altered markedly in the swarmer cell (Armitage, 1981; Armitage and Smith, 1978; Armitage et al., 1979; Falkinham and Hoffman, 1984; Hoffman and Falkinham, 1981; Jin and Murray, 1988). The result of this process is a very long, nonseptate, polyploid cell, with 10^3–10^4 flagella per cell (Fig. 10.1A). The number of chromosomes in the swarmer cell is roughly proportional to the increase in length, such that a 40 μm swarmer cell has about 20 chromosomes (Belas, 1992). Eventually septation and division do take place at the ends of the long swarmer cells, producing a microcolony of differentiated cells. Table 10.1 compares the physical features of swimmer and swarmer cells.

Concurrent with cellular elongation, changes take place in the rate of synthesis of flagella on the swarmer cell (Fig. 10.2). While swimmer cells have only a few flagella, the elongated swarmer cells are profusely covered by hundreds to thousands of new flagella synthesized specifically as a consequence of growth on the surface (Armitage and Smith, 1978; Hoeniger, 1965, 1966; Houwink and van Iterson, 1950; Leifson et al., 1955). The term *flagellin factories* was first used by Hoeniger (1965) to describe the tremendous synthesis of new flagella (compound of the protein flagellin) in swarmer cell differentiation. The newly synthesized surface-induced flagella are composed of the same flagellin subunit as the swimmer cell flagella, indicating that the same flagellar species is overproduced upon surface induction. The result of the surface-induced differentiation process is a swarmer cell, which differs from the swimmer cells by having the unique ability to move over solid media in a translocation process referred to as *swarming* (Henrichsen, 1972). However, individual

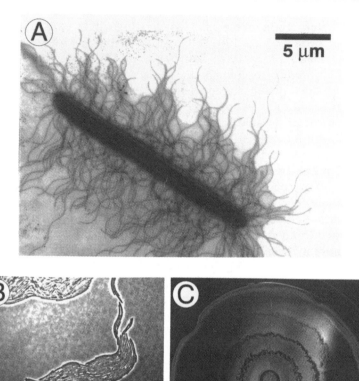

Fig. 10.1. *P. mirabilis* swarming colony formation and swarmer cell morphology. **A:** A full differentiated swarmer cell as shown through electron microscopy of a negative stained specimen taken from the swarming periphery. The cell shown is approximately 25 μm in length. **B:** Movement of a mass of cells from the swarming periphery onto uncolonized agar. The swarming periphery is shown in the upper left. **C:** The characteristic "bulls-eye" colony of latticed zones of bacteria produced by the cyclic events of differentiation, movement, cessation of movement, and dedifferentiation.

swarmer cells by themselves do not have the ability to swarm (Bisset, 1973a,b; Bisset and Douglas, 1976; Brogan et al., 1971; Douglas, 1979; Douglas and Bisset, 1976). As shown in Figure 10.1B, swarming is the result of a coordinated, multicellular effort of groups of differentiated swarmer cells (Bisset, 1973a,b; Dienes, 1946). The process begins when a group of differentiated swarmer cells move outward as a mass and continues until the swarming mass

TABLE 10.1. Physical Features of Swimmer and Swarmer Cells

Characteristic	Swimmer cells	Swarmer cells	Reference
Flagella			
Number per cell	1–10	500–5,000	Hoeniger (1965)
Length (μm)	0.75	5.25	Hoeniger (1965)
Number per unit cell length	3	150	Hoeniger (1965)
Motility	Tumbling	Smooth	Dick et al. (1985; Hoeniger (1965)
Cell dimensions (μm)	0.7×1–2	0.7×10–80+	Hoeniger (1965)
Chromosome number	1–2	Multiple; correlated with cell length	Hoeniger (1966)

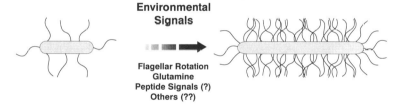

Swimmer Cell	Characteristic	Swarmer Cell
1.5 to 2.0 μm	*Length*	10 to >80 μm
4 to 10	*Flagella*	10^3 to 10^4
1 to 2	*Genomes*	Polyploid
Swimming & Chemotaxis	*Motile Behavior*	Swarming, Chemotaxis & Coordinated Cell-to-Cell Communication

Environmental Signals

Flagellar Rotation
Glutamine
Peptide Signals (?)
Others (??)

Fig. 10.2. Characteristics of *P. mirabilis* swarmer cell differentiation and swarming motility. Swarmer cell differentiation is controlled through the combined sensing of environmental conditions that reduce wild-type flagellar filament rotation and through the sensing of a specific chemical stimulus, the amino acid glutamine. Other signals may play a role in this process, e.g., iron availability, peptide signals, and perhaps other compounds. The differentiated swarmer cell is characterized by an elongated, polyploid cell that synthesizes numerous flagella in response to the aforementioned signals.

of bacteria is reduced in number due to loss of constituent cells that fall behind on the surface or when the mass reverses direction.

Swarming of *P. mirabilis* is cyclic in nature (in Fig. 10.3). Once swarmer cells have fully differentiated, the swarming colony moves outward in unison from all points along the periphery for a period of several hours and then stops (Bisset, 1973a,b; Bisset and Douglas, 1976; Douglas, 1979; Douglas and Bisset, 1976). This cessation of movement is accompanied by a dedifferentiation of the swarmer cell back to swimmer cell morphology, a process referred to as *consolidation* (Bisset, 1973a,b; Hoeniger, 1964, 1965, 1966; Hoeniger and Cinits, 1969; Williams and Schwarzhoff, 1978). The cycle of swarming and consolidation is then repeated several times until the agar surface is covered by concentric rings formed by the swarming mass of bacteria (Douglas and Bisset, 1976). This cycle of events gives rise to the characteristic "bulls-eye" appearance of *P. mirabilis* colonies (Figs. 10.1C, 10.4). The swarmer cell requires contact at all times with the surface to maintain the differentiated state. When removed from the surface of an agar plate and suspended in liquid medium, cells quickly begin to septate and divide into short cells, and the

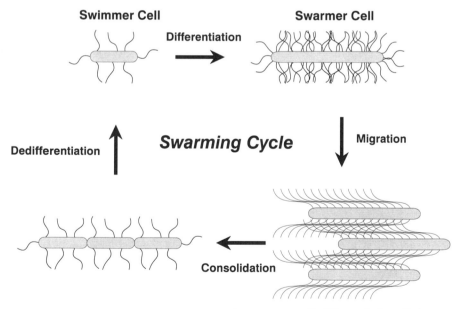

Fig. 10.3. Temperal cycling of *P. mirabilis* swarmer cell differentiation. Swimmer cells differentiate into swarmer cells upon sensing conditions that induce the expression of swarmer-cell-specific genes. Once fully differentiated, the swarmer cells then move *en masse* and migrate away from the initial point of inoculation. Such migration is punctuated by cyclic events of dedifferentiation (referred to as *consolidation*), whereupon migration stops and the cells revert to a morphology similar to that of the swimmer cell. This process of differentiation and dedifferentiation then continues until the agar plate is covered by the bacterial mass.

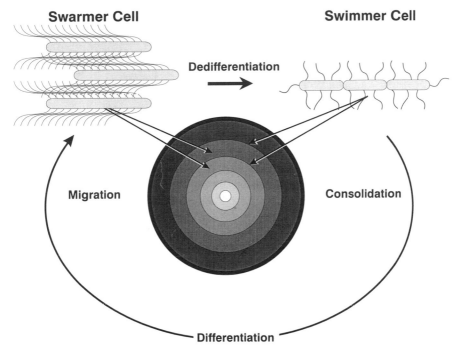

Fig. 10.4. Spatial cycling of *P. mirabilis* swarmer cell differentiation. On agar media, swarmer cell differentiation is observed as a series of ever-expanding concentric rings (see Fig. 10.1C). These rings are composed of morphologically and biochemically distinct cell types. During swarming the leading edge of the swarming zone is composed of completely differentiated swarmer cells. At consolidation, swarming motility stops, and the cells revert to a swimmer cell morphology.

synthesis of flagella returns to the level observed in swimmer cells (Jeffries and Rogers, 1968). Thus, the differentiation process is reversible, as a result of both the consolidation process and the removal of the inducing stimulus from the surface (Fig. 10.4).

10.3. OTHER SWARMING BACTERIA

Swarming has been observed in both gram-negative and gram-positive genera (Fig. 10.5), including *Vibrio* (Belas et al., 1986; McCarter et al., 1988; McCarter and Silverman, 1990; Ulitzer, 1975), *Serratia* (Alberti and Harshey, 1990), *E. coli, Sal. typhimurium, Yersinia, Aeromonas, Rhodospirillum* (Harshey, 1994; Harshey and Matsuyama, 1994), *Bacillus,* and *Clostridium* (Allison and Hughes, 1991a; Harshey, 1994; Harshey and Matsuyama, 1994; Henrichsen, 1972). These organisms can swarm on media solidified with agar, usually 1%–2% agar for *Vibrio parahaemolyticus* and *P. mirabilis* and lower concentra-

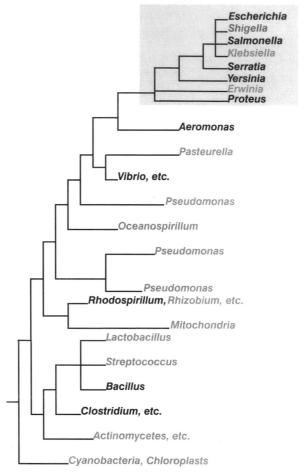

Fig. 10.5. A phylogenetic tree showing the relationships between bacterial genera in which swarming motility has been demonstrated. Those genera with at least one swarming species are indicated in black lettering and are compared with other non-swarming species (light grey lettering). A light grey box surrounds the Enterobacteria-ceae species. (Adapted from Ochman and Wilson, 1987, with permission of the publisher.)

tions for *Serratia marcescens, E. coli,* and *Sal. typhimurium* (Alberti and Har-shey, 1990; Harshey and Matsuyama, 1994). The ubiquitous occurrence of swarming among eubacteria suggests that this mode of surface translocation must play an important role in the colonization of natural environments by microorganisms.

Of the swarming genera, the swarming of *P. mirabilis* has been the most intensively investigated, with studies of *V. parahaemolyticus* and *Ser. marces-cens* swarming following closely thereafter (Alberti and Harshey, 1990; Belas et al., 1986; McCarter et al., 1988; McCarter and Silverman, 1989, 1990; McCarter and Wright, 1993; Sar et al., 1990). Sufficient information is now available to compare the swarming of *Proteus* species, *V. parahaemolyticus,* and *Ser. marcescens* 274 (Table 10.2). Like *Proteus* swarmer cells, *Vibrio* and *Serratia* swarmers are elongated, nonseptate, highly flagellated, occur peripherally but not centrally in swarming colonies, and are reversibly differentiated from the short, nonswarmer cells on transfer between liquid and solid media (Alberti and Harshey, 1990; Belas et al., 1986). In *P. mirabilis, V. parahaemolyticus,* and *Ser. marcescens* the introduction of mutations affecting chemotaxis abolishes or impairs normal swarming behavior. Thus, as described earlier, all three genera monitor flagellar filament rotation to determine external conditions and, when the filament rotation becomes impaired, induce swarmer cell differentiation. Beyond these similarities there are many differences between the three genera.

Wild-type *V. parahaemolyticus* growing in a liquid medium have one sheathed, polar flagellum composed of four flagellin subunits encoded by a set of four *fla* (flagellin) genes (McCarter, 1995; McCarter and Silverman, 1990). This flagellum is responsible for swimming in liquid media. In addition to this constitutively produced polar flagellum, *Vibrio* swarmer cells exhibit multiple, unsheathed peritrichously arranged flagella, referred to as *lateral flagella,* that are responsible for the swarming phenotype exhibited on the surface of solid media. The lateral flagella filament is coded for by the *laf* set of genes, one of which, *lafA,* codes for the single flagellin structural gene. Thus, lateral flagella have a different subunit composition from polar flagella, and both flagella have independent motor structures (Sar et al., 1990) and are energized through different mechanisms (Atsumi et al., 1992). Consequently, *V. parahaemolyticus* is said to possess "mixed flagellation." At least two conditions are required to induce transcription of the genes coding for lateral flagella and the other phenotypic changes associated with swarmer cell differentiation: the physical restriction of the rotation of the polar flagellum (by solid surfaces, liquids of high viscosity, or anti-cell surface antibodies) and the iron-limiting growth conditions (McCarter et al., 1988; McCarter and Silverman, 1989, 1990).

Lateral flagella regulation at the level of transcription of *laf* genes in *V. parahaemolyticus* has been deduced from elegant genetic work using the *lux* genes from the luminescent bacterium *Vibrio fischeri.* These genes encode the catalytic components of the bioluminescence system, and promoterless versions have been inserted into a transposon, mini-Mu *lux* (Tet^R), which in turn was transduced by bacteriophage P1 into *V. parahaemolyticus* (Belas et al., 1986). Such mini-Mu *lux* (Tet^R) insertions can produce transcriptional fusions between *laf* genes and the *lux* genes carried on the transposon, the result of which is the construction of *V. parahaemolyticus* strains where the *lux* genes are under the control of the *laf* gene promoter (Belas et al., 1986; McCarter et al., 1988). The resulting luminescence facilitates measurement

TABLE 10.2 Comparison of Swarmer Cell Differentiation and Motile Behavior of *P. mirabilis*, *V. parahaemolyticus*, and *Ser. marcescens* 274

Characteristic	P. mirabilis	V. parahaemolyticus	Ser. marcescens
Swarmer cell length (μm)	10–80+	30–40	5–30
Nonseptate swarmer cells	Yes	Yes	Yes
Flagella per swarmer cell	10^3–10^4	10^2–10^3 (including 1 polar)	10^2–10^3
Flagella per swimmer cell	4–10	1 polar	1–2
Flagellation	Uniform	Mixed	Uniform
Flagellin gene(s)	*flaA–C*, only 1 expressed at a time	*flaA-D* for polar flagellin; *lafA* for lateral flagellin	*hag*
Evident consolidation zones	Yes	No	No
Dedifferentiation during swarming?	Yes	Yes	Yes
Physical stimulus for differentiation	Inhibition of flagellar rotation	Inhibition of flagellar rotation	Inhibition of flagellar rotation
Chemical stimulus for differentiation/motility	Glutamine, iron (?), zinc (?), chemotaxis sensing machinery	Iron limitation, chemotaxis sensing machinery	Chemotaxis sensing machinery
Agar concentration required to induce swarming	1.5%	1.5%	0.35%–0.75%
Extracelluar slime/surfactant	Yes	No	Yes

and is noninvasive, thus permitting the cells to continue their processes while the level of *laf* activity is measured.

In contrast to *V. parahaemolyticus,* swarming *Proteus* strains do not possess a typical pattern of mixed flagellation, where two distinct flagellar types are produced simultaneously. In wild-type *P. mirabilis* cells, the flagella of both swimmer and swarmer cells are composed of the same flagellin protein encoded by the *flaA* gene (Belas and Flaherty, 1994). Interestingly, *P. mirabilis,* like *V. parahaemolyticus,* possesses multiple copies of flagellin-encoding genes (referred to as *flaA, flaB,* and *flaC*) (Belas and Flaherty, 1994). Normally, *flaA* is expressed and *flaB* and *flaC* are silent copies of flagellin-encoding genes. However, if *flaA* is mutated, spontaneous deletions fusing the promoter region of *flaA* to either *flaB* or *flaC* may occur. This gene conversion produces a hybrid flagellin protein where FlaA sequences comprise the N-terminal end of the protein and either FlaB or FlaC sequences comprise the C-terminal portion. So, while *P. mirabilis* produces only a single flagellin species at any time, the type of flagellin produced may undergo antigenic variation.

Other differences between *P. mirabilis* and *V. parahaemolyticus* are also evident. For example, glutamine is necessary for swarmer cell differentiation in *P. mirabilis,* while in *V. parahaemolyticus* iron limitation is required for the induction of differentiation. Also, *P. mirabilis* can differentiate and consolidate in repetitive cycles to form concentric swarming zones, while *V. parahaemolyticus* and *Ser. marcescens* only rarely show this cyclic event. Because of this, it has been suggested that the swarming of *V. parahaemolyticus* and *Proteus* species is an example of convergent rather than divergent evolution (Belas, 1992; Belas et al., 1986; Williams and Schwarzhoff, 1978).

In *Ser. marcescens* 274, *E. coli,* and *Sal. typhimurium,* swarmer cell differentiation occurs only on the surface of media solidified by 0.75%–0.85% agar (Alberti and Harshey, 1990; Harshey and Matsuyama, 1994). Although this requirement for a low agar concentration differs from that of *Proteus* and *Vibrio,* it is not unique among swarming bacteria. *Azospirillum brasilense* (which possesses mixed flagellation) swarms optimally on 0.75% agar and not at all on 1.5% agar (Hall and Krieg, 1983). *Ser. marcescens* 274 does not possess mixed flagellation. Instead, swimmer cells growing in liquid media possess one to two flagella composed of the same flagellin protein found in the flagella of swarmer cells. A single gene, *hag,* codes for this flagellin protein. Lastly, while glutamine (for *P. mirabilis*) and iron limitation (for *V. parahaemolyticus*) are required for differentiation, none of a wide range of chemicals tested has been implicated in the differentiation of *Ser. marcescens* 274 (Alberti and Harshey, 1990).

10.4. A DESCRIPTION OF *P. MIRABILIS* SWARMING

10.4.1. Swarmer Cell Differentiation

As a result of the work of Kvittingen, Hoeniger, and other investigators, the process and functional motile mechanism of swarming at the observational level became well described (Hoeniger, 1964, 1965, 1966; Hoeniger and Cinits, 1969; Kvittingen, 1949a,b, 1953). To expand upon the brief description provided in section 10.1, broth cultures of swarming *Proteus* species consist of short rods and coccobacilli (0.6–0.7 by 0.7–2.0 μm) with 1–10 peritrichous flagella (Hoeniger, 1965) and 1–2 copies of the chromosome (Hoeniger, 1966). On transfer to a solid medium, the cells widen slightly and lengthen substantially, entering into a period of active growth and division that lasts approximately 30–90 min to form a primary colony (Hoeniger, 1965). Division then ceases in some of the cells, which undergo a dramatic morphogenesis (Hoeniger, 1964, 1966; Jones and Park, 1967; Kvittingen, 1949b, 1953). These cells become elongated to lengths of 20 to greater than 80 μm (Hoeniger, 1965). Elongation is thought to occur by intercalation of new cell wall with old cell wall (Hoeniger and Cinits, 1969). Although cell septation ceases, DNA synthesis continues and the chromosomes increase proportionately to cell elongation (Hoeniger, 1966; Jones and Park, 1967), thereby producing a polyploid cell. The DNA is uniformly distributed along the length of the cell and is interspersed with evenly spaced areas of cytoplasm to resemble discreet "cellular" units with no peptidoglycan cross walls (Hoeniger, 1966; Jones and Park, 1967) or with incomplete cross walls (Fuscoe, 1973). There is also an increase in the length and number of flagella to up to several thousand per cell (Hoeniger, 1965). The flagella on swarmer cells are composed of the same flagellin subunit and have the same serological specificity as those on the short cells (Belas et al., 1991a; Dick et al., 1985). These elongated, highly flagellated, polyploid swarmer cells (Fig. 10.1A) start to move away from the parent colony in interacting, multicellular groups known as *rafts* to form, first, an open mesh and, ultimately, a random swirling mass (Belas, 1992; Bisset, 1973a; Fuscoe, 1973; Hoeniger, 1964; Kvittingen, 1949a; Morrison and Scott, 1966), the motion of which extends the boundary of the colony (fig. 10.1B). The long swarmer cells at the leading edge of the swarming colony form protrusions at right angles to the periphery of the mass. During swarmer cell advance, the swirling mass may sometimes turn back in a wave-like motion toward the parent colony (Belas, 1992; Bisset, 1973a; Fuscoe, 1973; Hoeniger, 1964; Kvittingen, 1949a; Morrison and Scott, 1966). Swarming may stop and restart at regular intervals, in which case, under favorable conditions, individual swarms may be several millimeters in width and the duration of each swarming and each intervening resting period and may be in the order of 2 hours (Hoeniger, 1964) (Figs. 10.1C, 10.2–10.4).

The morphogenesis of swimmer cells to swarmer cells is reversible (Figs. 10.3, 10.4). Reversion to short swimmer cells occurs throughout the course of swarming across an agar plate concomitant to the temporary and cyclic stops in swarming migration. This process is referred to as *consolidation*. As swarmer cells are only induced and maintained by surfaces, reversion also occurs if swarmer cells are transferred from solid to liquid media (Fig. 10.3). Thus, dedifferentiation is a result of consolidation or removal from the inducing effects of a surface (Belas, 1992; Belas et al., 1991a). As diagrammed in Figure

10.4, on an agar surface, the short cell type is the only form found in the center of colonies (Kvittingen, 1949a). In an active swarming colony, the swarmer cells may extend some distance inward from the periphery, and swarmer cells have been observed to invade from the rear and advance over and between those closer to the colonial edge (Bisset, 1973b; Hoeniger, 1964; Kvittingen, 1949a), suggesting a dynamic exchange of cells between all parts of the swarming colony.

After overnight incubation, the swarming colony can cover an entire agar plate and appear either as a continuous, even film of thin growth or, more often, as a series of regularly spaced concentric zones of thickened growth (Fig. 10.1C). The type of pattern that develops depends on a variety of factors, including the degree of softness, dryness, and freshness of the medium, the prior subculture history of the organism (Kvittingen, 1949a), atmospheric humidity (Hughes, 1957), the temperature of incubation (Hoeniger, 1964), the amount of surface moisture (Henrichsen, 1972), and the strain of the organism (Falkinham and Hoffman, 1984; Williams and Schwarzhoff, 1978). As shown in Figure 10.6 mutant strains that do not form wild-type consolidation patterns have also been documented (Allison and Hughes, 1991b; Belas, 1992; Belas et al., 1991a).

10.4.2. Differences Between Swimmer and Swarmer Cells

10.4.2.1. Cell Wall and Membrane Composition Differences

The suspicion that the cell envelope of swarmer cells may differ significantly from that of swimmer cells arose because of the highly flagellate, elongate, and polyploid nature of the swarmer cells and the known involvement of the cell envelope in flagella function and in normal division (Armitage et al., 1975, 1979; Armitage and Smith, 1978). A variety of direct and indirect evidence confirmed this suspicion (Table 10.3). In comparative studies between swimmer and swarmer cells, the following properties of swarmer cells provided indirect evidence: the increased permeability demonstrated by leakage of intracellular amino acids and pentose sugars (Armitage et al., 1975); the enhanced sensitivity to deoxycholate (Armitage et al., 1975) and rifampicin (hydrophobic molecules normally excluded by the hydrophilic outer membrane of gram-negative organisms) (Armitage et al., 1975); the enhanced sensitivity to physical disruption by sonication (Jin and Murray, 1988); the freeze-fracture cleavage and fluidity of the outer membrane (Armitage, 1982); and the reduction in the number of inner–outer membrane adhesions (Bayer patches) and phage binding (Williams and Schwarzhoff, 1978). Changes in the actual composition of the lipopolysaccharide (LPS) (Armitage et al., 1979) and outer membrane protein (Falkinham and Hoffman, 1984) and in the cytochromes of the inner membrane (Falkinham and Hoffman, 1984) provided direct evidence.

To explain these differences, and the apparent contradiction that an increase in the hydrophilic long side chain LPS in the outer membrane of

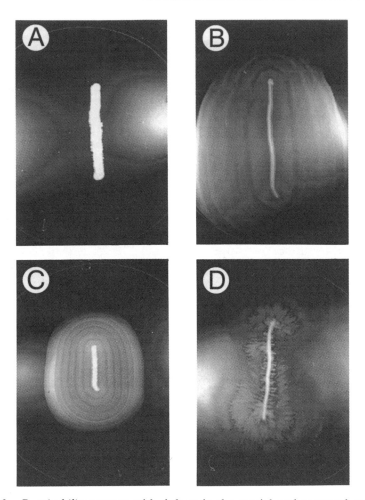

Fig. 10.6. *P. mirabilis* mutants with defects in the spatial and temperal control of swarming behavior. Each strain was inoculated as a 3 cm line at the center of fresh L agar medium and incubated for 48 h at 30°C. **A:** Swarming null mutant (Swr⁻ Swm⁻ Elo⁻). **B:** Swarming mutant with infrequent and variable consolidation (Swr^{cr} Swm⁺ Elo⁺). **C:** Swarming mutant with increased consolidation frequency (Swr^{cr} Swm⁺ Elo⁺). **D:** Swarming mutant lacking clearly defined consolidation, referred to as an *indeterminate swarming mutant* (Swr^{cr} Swm⁺ Elo⁺).

swarmer cells conferred enhanced hydrophobicity, a redistribution of outer membrane LPS associated with the 50-fold increase in flagella concentration in swarmer cells was proposed. It was suggested that the long side chain LPS could aggregate at the base of the flagella to provide outer membrane stability and prevent membrane damage due to flagellar rotation. As a result, LPS deficiency would occur in other regions of the membrane and create localized exposure of phospholipid on the surface and areas of lipid bilayer in the outer membrane. Both of these features are normally absent in gram-negative

TABLE 10.3. Cell Envelope Features of Swimmer and Swarmer Cells

Characteristics	Swimmer cells	Swarmer cells	Reference
Permeability	Normal	Increased	Armitage et al. (1975)
Sensitivity to chemical and physical disruption	Normal	Increased	Armitage et al. (1975), Jin and Murray (1988)
Sensitivity to hydrophobic agents	Normal	Increased	Armitage et al. (1975), Armitage et al. (1979)
Outer membrane			
LPS composition	Normal	Higher proportion with long oligosaccharide sidechains	Armitage et al. (1979)
Fluidity	Normal	Increased	Armitage (1982)
Protein composition	Normal	Changes in proportion	Falkinham and Hoffman (1984)
Bayer patches	Normal	Reduced	Williams and Schwarzhoff (1978)
Cytochromes	High in b, a, d	Low in b, deficient in a and d	Falkinham and Hoffman (1984)

bacteria (Armitage, 1981, 1982; Armitage et al., 1974, 1979). The exposed phospholipid would create a more hydrophobic surface, more permeable to hydrophobic antibiotics and more sensitive to surfactants (Armitage, 1982; Armitage et al., 1975, 1979). The regions of lipid bilayer would create a cleavable and more flexible outer membrane (Armitage, 1982). The LPS aggregation could possibly restrict the LPS available for Bayer patches and the LPS-dependent phage-binding sites. Furthermore, as the initial increase in flagellin synthesis has been reported to occur before the inhibition of cell division in swarmer cell development, it was proposed that it is the outer membrane reorganization associated with flagella development that impairs septation (Armitage and Smith, 1978). Some of these speculations are supported by recent data demonstrating that genetic lesions directly affecting LPS biosynthesis and, more specifically, chain length result in abnormal cellular elongation during swarming cell differentiation (Belas et al., 1995).

10.4.2.2. Metabolic Differences
Differences have also been found in some of the metabolic features of swimmer and swarmer cells. Swarmer cells exhibit decreased uptake and incorporation of macromolecular precursors (Armitage, 1981; Armitage et al., 1975) and decreased oxygen consumption (Armitage, 1981; Falkinham and Hoffman, 1984). The decreased uptake was attributed to the effect of envelope changes on active transport processes (Armitage et al., 1975). Due to the decreased oxygen consumption, swarmers were initially regarded as nongrowing bacteria of low metabolic activity, with most energy expended on movement (Armitage, 1981).

During the search for other indicators of decreased metabolic activity, the enzyme tryptophanase was found to be repressed and uninducible in swarmer cells (Hoffman and Falkinham, 1981). Other enzymes, cytochromes, and outer membrane proteins (Table 10.3) were also examined for differing patterns of expression in the two cell types (Falkinham and Hoffman, 1984). This work revealed not only differences between swarmer and swimmer cells, but a variety in the pattern of protein expression within one cell type. In swarmer cells, for example, tryptophanase is uninducible, and phenylalanine deaminase is inducible, as is urease. It was this variation in the expression of a number of determinants that resulted in swarming being regarded as an example of prokaryotic differentiation, with a common regulatory signal acting at the level of transcription (Falkinham and Hoffman, 1984). This opinion has subsequently received substantial support (Allison and Hughes, 1991a,b; Allison et al., 1992b; Belas, 1992, 1994; Belas et al., 1991a; Dick et al., 1985; Jin and Murray, 1987, 1988). In addition, the inducible nature of phenylalanine deaminase and urease indicated that swarmers possess functional transcription and translation. The higher activities of both of these enzymes in swarmer cells indicated that these cells do possess the capacity for metabolic activity. Despite this capacity, transcription and translation do not appear to be necessary for the persistence of migration, for, once swarming has been initiated, it continues even in the presence of the transcription and translation inhibitors rifampicin and chloramphenicol, respectively (Falkinham and Hoffman, 1984). Neither is an exogenous energy source necessary for migration once it has been initiated, as swarming cells will continue to swarm if transferred to non-nutrient agar (Williams et al., 1976). In light of this demonstration of functional transcription and translation and at least the capacity of, although not the necessity for, metabolic activity, the decreased respiratory activity and cyto-chrome content of swarmer cells was not interpreted, as it was earlier, as indicating low metabolic activity or a nongrowing status. Rather, it was proposed that the energy required for swarming is derived not from respiration, but from fermentation. Other evidence supports this contention, as normal swarming motility also occurs both aerobically in the presence of cyanide and anaerobically (Falkinham and Hoffman, 1984). Utilization of some kind of fermentable storage material and the catabolism of certain amino acids have been proposed to provide both ATP and the transmembrane proton potential used to drive the flagellar motors (Armitage, 1981; Falkinham and Hoffman, 1984).

The continuation of swarming when swarmer cells are transferred to non-nutrient agar with low surface tension (agar–water–detergent) (Armitage and Smith, 1978) could also possibly be related to a stored energy source (VanderMolen and Williams, 1977), and the periodicity of swarming could be associated with the depletion and reaccumulation of such a source (Williams and Schwarzhoff, 1978). Consumption of stored energy as the swarm proceeds could account for observed differences in swarmer cells at different times during a single swarming event. Swarmer cells transferred to fresh solid medium continue to swarm, but the distance traveled after transfer depends on the distance traveled before transfer, where early swarmer cells travel further after transfer than do late swarmer cells (Williams et al., 1976). Early swarmer cells are very motile but become progressively slower as a swarm spreads (Douglas, 1979). It has been proposed that the depletion of the energy source could act by affecting the production of slime (Williams and Schwarzhoff, 1978) or by causing a drop in membrane potential, thence a decrease in flagellar rotation and cessation of the movement of swarmers (Armitage, 1981). Alcian blue and sudan black B–stainable inclusions could possibly constitute a storage material (Williams and Schwarzhoff, 1978), as could glucuronic acid (Falkinham and Hoffman, 1984). However, conclusive evidence for the existence of a storage material is lacking.

10.4.3. Swarming Motility

The association between swarming and the development of filamentous cells was recognized in the earliest investigations of the phenomenon (Williams and Schwarzhoff, 1978), but the factors responsible for inducing the morpho-genesis, and the mechanisms involved, are still largely a mystery. The investigations of Kvittingen (1949a,b) were among the first that specifically addressed the aspect of swarmer cell formation, and he concluded that swarmers represented a stage in the normal life cycle of *Proteus*. His conclusion was based partially on the observation of filamentous cells in young broth cultures. From microscopic observations of living and stained cells he maintained that only a small fraction of the swarmer cell population remained viable, and he speculated that this was composed of gametes derived from short cells following nuclear reduction or meiosis (Kvittingen, 1949b). Subsequent attempts to demonstrate genetic exchange between swarming strains of *Proteus* were not conclusive (Kvittingen, 1953).

Possibly the most influential work on swarming has been that of Lominski and Lendrum (1947), which was reported shortly before Kvittingen's first paper. They proposed that swarming of *Proteus* was a negative chemotactic response to toxic metabolites that accumulated in areas of high population density on a solid medium. Their hypothesis applied specifically to the outward migration of swarmer cells from a central inoculum (see also section 10.4.4). However, they implied that the toxic metabolites that acted as repellents to the swarmers also influenced their development, and this concept has persisted in the literature for some time (Coetzee, 1972). That filament formation in bacteria could be induced by several treatments, including exposure to penicillin and toxic chemicals, was well established (Hughes, 1956, 1957), and Hughes (1956) considered the swarmers of *Proteus* as examples of such involutionary forms. He confirmed the earlier report (Kvittingen, 1949b) of a high mortality rate among swarmers and interpreted the observation as evidence that their formation was induced by toxic metabolites (Hughes, 1957). It was suggested that swarmer cell formation was caused by a volatile metabolite similar in its mechanism to penicillin, but a specific compound was not isolated or identified (Hughes, 1957). That the toxic metabolite that induced the formation of swarmer cells also served as a repellent was not suggested; in fact, it was noted that no evidence of negative chemotaxis was detected. Hydrogen peroxide added to broth cultures of *P. vulgaris* induced the formation of elongated cells (up to seven times the normal maximum length), but the effect was noted only after 17 h of incubation, and it was not detected on agar-solidified media (Hughes, 1957). By assuming a value of 2.5 μm as a normal maximum length for broth-grown cells (Hoeniger, 1966), the peroxide-induced cells would have been approximately 18 μm long, a value significantly less than the maximum cell length of 80 μm reported for swarmer cells (Hoeniger, 1965). The observation that peroxide inhibited motility of the organism (Hughes, 1957) also seems to preclude its consideration as the swarmer cell–inducing agent. That certain *Proteus* species produce characteristic volatile amines (Proom and Woiwod, 1951) and specific growth-inhibiting metabolites (Grabow, 1972) has been reported. Although certain properties of these compounds are similar to those of the swarmer-inducing metabolites proposed by Lominski and Lendrum (1947), they have not been shown to induce swarmer cell formation.

An extracellular slime postulated to facilitate migration (Fuscoe, 1973; Stahl et al., 1983) appears to be associated with swarmer cells. The existence of slime was postulated after tracks of a different refractive index were observed on the agar surface behind migrating edge cells (Fuscoe, 1973). Slime has since been visualized in fixed impression smears of the swarm edge by phase-contrast light microscopy and on agar surfaces and converging swarming cells by scanning and transmission electron microscopy. The slime appears to blanket rafts of swarm cells rather than envelop individual cells and to persist on the agar surface behind the moving rafts (Stahl et al., 1983; VanderMolen and Williams, 1977). Because of its ruthenium-red–staining properties, slime is thought to consist of an acidic polysaccharide (Stahl et al., 1983) and a large amount of water (Fuscoe, 1973; Stahl et al., 1983). Whether slime is absolutely essential for swarming has not been established, but recent evidence from mutants defective in capsular polysaccharide production suggests that it may play an important role in the migration of the swarming colony (C. Hughes, personal communication, 1994).

10.4.4. Swarming Behavior

The concentric zone pattern is formed by crowding and the piling up of growth to form the zone edges, and two mechanisms have been described (Douglas and Bisset, 1976). According to the classic description of zone formation, concentric zones occur when the colony (not the cells) grows in a stepwise manner as a result of cyclic events of differentiation and dedifferentiation (Figs. 10.3 and 10.4). After the edge of the swarm comes to rest, the swarmer cells divide into short rods delineated by the repeating cellular units along their length. They then multiply for a few generations to form a zone of thickened growth. When swarming is about to restart, once again some of the cells transform into swarmer cells and a new mass of swarmer cells moves out over the agar to form a new zone. This process may be repeated several times to cover the whole surface of an agar plate with concentric zones (Hoeniger, 1964; Kvittingen, 1949a). Variation in the sharpness of the concentric rings depends on the changes in the velocity of colonial advance, which in turn depends on the changes in velocity of swarmer cells at the colonial edge. If the velocity of the swarmer cells at the edge is generally high but periodically undergoes dramatic reductions, the associated cessations of colonial advance are prolonged and the concentric rings are sharply defined. If the velocity of edge cells is generally lower and less variable, the periodic pause in colonial advance is brief and the rings so formed are more diffuse and less well defined (Douglas, 1979).

It is not universally believed that zones are always formed in the classic manner, and, according to an alternative description, the thickening arises as a fold of growth inside the colony behind the leading edge. In this pattern, variation in the rate of advance of the swarm edge with occasional cessation of colonial growth does not equate with the formation of zones. Instead, the outermost edge of swarming colonies is constantly composed of long forms irrespective of swarm advance. When the colonial edge is rapidly advancing, a thickened zone of growth may arise behind the colony edge when more centripetal swarmers collide with more advanced cells and create congestion. In this thickened zone crowding causes constriction of the movement of the long swarmer cells, and their movement ceases before the morphological reversion to short cells occurs. This mode of zone formation gives rise to diffuse zones. The zonation pattern may vary between strains and in the same strain on different occasions (Bisset, 1973b; Bisset and Douglas, 1976). Despite the elucidation of a second pattern of zone formation the classic description persisted (Williams and Schwarzhoff, 1978).

10.5. THE MOLECULAR DETERMINANTS OF SWARMING

The advent of modern molecular biology techniques has renewed interest in understanding *P. mirabilis* swarmer cell differentiation as it affects the invasiveness and pathogenicity of this organism. In my laboratory, we have used Tn5 transposon mutagenesis to construct mutations in the swarmer cell genes (Belas et al., 1991a,b; De Lorenzo et al., 1990). A similar methodology has been employed in other laboratories (Allison and Hughes, 1991b). We have found that Tn5 transposition is random, and, once integrated into the *P. mirabilis* chromosome, the transposon is stably maintained (Belas et al., 1991b). Characterization of the subsequent swarming-defective mutants (Swr⁻) has revealed several phenotypic classes, including mutants that are defective in (1) swimming motility (Swm⁻; including defects in flagellin synthesis, Fla⁻; motor rotation, Mot⁻; and chemotaxis, Che⁻); (2) swarmer cell elongation (both null mutants [Elo⁻] and constitutively elongated cells [Eloᶜ]); and (3) a large group of mutants with an impairment of swarming (Swrᶜʳ) caused by an uknown defect (Belas et al., 1991a). While this last group does manifest nonwild-type swarming motility, many of these mutants were altered in their ability to form discrete consolidation zones. It has been speculated that these mutants have arisen from transposon insertions in genes coding for production of extracellular signal molecules or the receptors of those signals (Belas, 1992). Localization with pulse gel electrophoresis of the various Tn5 insertions on the chromosomes of swarming mutants has revealed that the genes necessary for swarmer cell differentiation and swarming behavior occur on closely linked genetic loci (Allison and Hughes, 1991b).

The genetic lesions that result in the various mutant phenotypes are now being characterized and the defective genes analyzed for regulation and expression during swarmer cell differentiation and multicellular motility and behavior. The results of these studies are presented in the next section.

10.5.1. Differentiation to Swarmer Cells

Swarmer cell differentiation and swarming motility of *P. mirabilis* are the result of at least four separate phenomena, including (1) the ability to sense cues from the environment; (2) the production of an elongated swarmer cell; (3) the synthesis of vastly increased amounts of flagellin (a hallmark of the differentiated swarmer cell); and (4) the coordinated multicellular interactions resulting in cyclic waves of cellular differentiation and dedifferentiation (Belas, 1992). A defect in any of these events results in either abnormal swarming behavior or the complete lack of differentiation and/or motility. The following section 10.6 then focuses on what is known concerning the fourth aspect of swarming.

10.5.1.1. The Surface Signal—Flagellar Rotational Tethering

Studies measuring gene expression in *P. mirabilis, Ser. marcescens,* and *V. parahaemolyticus* have convincingly demonstrated that the central environmental stimulus sensed by the cell is physical in nature (Alberti and Harshey, 1990; Allison et al., 1993; Belas et al., 1986, 1995; McCarter et al., 1988). Swarmer cell genes are induced when swimmer cells are transferred to the surface of solidified media, suspended in media of high viscosity, or agglutinated with antibody to the cell surface. Growth in media amended with polymers such as Ficoll 400 or polyvinylpyrrolidone 360 is as effective at inducing swarmer gene expression as growth on an agar surface (Alberti and Harshey, 1990; Allison et al., 1993; Belas et al., 1986, 1995; McCarter et al., 1988). These branched polymers increase the microviscosity of the medium, thus creating a matrix that interferes with the movement of swimming bacteria. An even more effective inducing condition is the addition of antibody (but not preimmune serum) to the growth medium, which results in cell tethering and agglutination (McCarter et al., 1988). McCarter and coworkers employed an antibody raised against the cell surface of *V. parahaemolyticus* that recognized primarily the polar flagellum sheath (McCarter et al., 1988). Agglutination with this antibody produced abnormal expression of swarmer cell differentiation in *V. parahaemolyticus.* What these conditions have in common is that they inhibit the normal rotation of the flagellar filament(s). This observation has led to the hypothesis that the flagella function as tactile sensors of external conditions and directly transfer that information into the cell where the signal is transduced to control transcriptionally the expression of the genes associated with swarmer cell differentiation.

Genetic techniques have strengthened the hypothesis that the rotation of the flagellar filament is the main factor controlling expression of swarmer cell differentiation. McCarter and Silverman (1990) have shown that genetic defects that impair the structure of the polar flagellum of *V. parahaemolyticus* affect regulation of swarmer cell genes. The composition of the *V. parahaemolyticus* polar filament is complex, being encoded by separate genes: *flaA, flaB, flaC,* and *flaD.* The genes are organized in two separate loci, with *flaA–flaB* and *flaC–flaD* being physically paired, respectively (McCarter, 1995). Mutants with defects in *flaA* and/or *flaB* are indistinguishable from the wild-type cell with respect to swimming motility. Thus FlaA and FlaB flagellins are not critical to polar flagellar function. In contrast, all mutants with lesions in *flaC* (i.e., *flaC*, *ΔflaAB flaC* and *ΔflaAB ΔflaCD*) have a defective polar flagellum. All strains with *flaC* mutations move poorly in semisolid motility media, and, when viewed in the light microscope, they swam aberrantly with a slow and wobbly movement. Examination in the electron microscope of strains with *flaC* defects grown in liquid medium showed "hairy" bacteria with many unsheathed flagella (McCarter and Silverman, 1990). Flagellar preparations confirmed that the unsheathed flagella were lateral flagella. The lateral filaments are polymerized from a single protein, LafA, which has a molecular size different from the polar flagellins. Thus, the FlaC flagellin is important for filament structure, and regulation of *laf* genes is released from surface dependence in mutants with *flaC* defects.

Genetic analyses of the flagellin-encoding genes of *P. mirabilis* have also demonstrated a complex genetic system, which is unlike *V. parahaemolyticus*. Swimmer cells and swarmer cells of *P. mirabilis* produce the same flagellin protein encoded by the same gene. Belas and Flaherty have shown that there are three copies of flagellin-encoding genes in *P. mirabilis* (Belas, 1994; Belas and Flaherty, 1994). Only one of these flagellin-encoding genes, *flaA*, is expressed in wild-type cells, whether in the swimmer or swarmer cell phase (Belas and Flaherty, 1994). Mutations that disrupt the function of *flaA* result in abnormal expression of swarmer cells, demonstrating the central role of the flagellar filament in signal transduction (Belas and Flaherty, 1994). Defects in other essential flagellar genes of *P. mirabilis* also result in abnormal swarmer cell differentiation (see below). Recently, Harshey and Matsuyama (1994) described a type of swarmer cell differentiation that occurs in *E. coli* and *Sal. typhimurium*. In these studies, mutations that affected flagellar filament function also produced abnormal expression of the swarmer cell phenotype, thus demonstrating the universal nature of sensing via monitoring of flagellar rotation to produce the differentiated swarmer cell.

10.5.1.2. Glutamine

Many attempts have been made to determine whether chemotactic behavior plays a role in swarmer cell differentiation and swarming behavior. Evidence both supporting and discounting chemotaxis in swarming behavior have been presented over the years, but recent evidence has demonstrated the importance of chemotaxis in swarming motility and differentiation (see also section 10.6.1.4). Most have been acquired through the careful analysis of swarming mutants that have clear defects in genes encoding protein components of the chemotaxis mechanism.

Although chemotaxis is now considered to be essential for swarmer cell differentiation and motility in all of the swarming bacteria thus far studied, only in *P. mirabilis* has a very specific interaction been described linking the amino acid glutamine and swarmer cell differentiation (Allison et al., 1993). By supplementing minimal agar medium unable to support swarming migration, Allison and coworkers identified a single amino acid, glutamine, as sufficient to signal initiation of cell differentiation and migration (Allison et al., 1993). In contrast, addition of the other 19 common amino acids (excluding glutamine) individually or in combination did not initiate differentiation, even after prolonged incubation. In liquid minimal media amended with polyvinylpyrrolidone to increase viscosity and inhibit flagellar rotation (as described above), addition of glutamine also induces swarmer cell differentiation. Furthermore, the induction can be completely inhibited by glutamine analogues, indicating the specificity of the glutamine signal in swarmer cell differentiation. Interestingly, glutamine is only chemotactic to the differentiated swarmer cell and not to the swimmer cell (Allison et al., 1993). These data suggest that glutamine functions in a dual role, both initiating differentiation and directing the migration of swarming cells.

This study indicates that *P. mirabilis* swarmer cells are chemotactic but respond to a more limited range of attractants than swimmer cells. Moreover, the discovery that amino acid chemoattractants that are mutually exclusive for either swimming (e.g., glutamate) or swarming (glutamine) cells may indicate the presence of separate sensory components coupled with the two forms of motility. These results are very tantalizing and suggest that a combination of signals, including inhibition of flagellar rotation and the presence of glutamine, are necessary for the induction of *P. mirabilis* swarmer cell differentiation.

10.5.1.3. Iron Limitation in V. parahaemolyticus

Swarmer cell differentiation in *V. parahaemolyticus* is controlled by a second signal, limitation for iron in the growth medium (McCarter and Silverman, 1989, 1990). Mutants with defects in *flaC* (one of four genes encoding polar flagellin and required for polar flagellum synthesis) produce lateral flagella when grown in 2216 marine broth; however, these mutants failed to synthesize lateral flagella when grown in heart infusion broth. One pertinent difference in the composition of these two media is the availability of iron (McCarter and Silverman, 1989, 1990). Addition of iron chelators to heart infusion broth results in the synthesis of lateral flagella by strains with *flaC* mutations. Furthermore, growth of the wild-type strain in viscous, iron-rich broth did not result in production of lateral flagella, while growth of the wild-type strain in viscous, iron-poor broth did elicit lateral flagella synthesis. Therefore, expression of *V. parahaemolyticus* swarmer cell genes requires both iron limitation and interference with polar flagellar rotation. Iron deficiency alone is not sufficient to induce swarmer cell differentiation, since lateral flagella synthesis is not induced when the wild type is grown in iron-limited broth (McCarter and Silverman, 1989, 1990). Regulation by the availability of iron affects transcription of *laf* (lateral flagellin) genes. Direct measurements of mRNA encoding lateral flagellin subunit (i.e., *lafA* mRNA) and light production of *laf*::*lux* transcriptional fusion strains revealed that transcription of *laf* genes requires perturbation of polar flagellar function and iron-limiting growth conditions (McCarter and Silverman, 1989, 1990). The effect of iron limitation on the regulation of swarmer cell differentiation has not been described in either *P.*

mirabilis or *Ser. marcescens* 274; however, it would not come as a surprise if either or both of these bacteria have specialized requirements for iron that may tie into the regulation of swarmer cell differentiation.

10.5.2. Genes and Proteins Involved in Differentiation (Table 10.4)

10.5.2.1. Flagellar Genes—The P. mirabilis flaA Locus

During differentiation from swimmer to swarmer cell, the rate of synthesis of certain proteins undergoes a dramatic increase. The most evident of these overproduced proteins is flagellin (FlaA, the subunit of the flagellar filament). Although swimmer cells have only a few flagella, the elongated swarmer cells are profusely covered by thousands of newly synthesized flagella (Hoeniger, 1965). In many ways the increase in flagellin expression is the hallmark of the differentiated swarmer cell, as has been shown not just for *P. mirabilis* but also for *Ser. marcescens* (Alberti and Harshey, 1990; O'Rear et al., 1992), and *V. parahaemolyticus* (Belas et al., 1986; McCarter et al., 1988; McCarter and Silverman, 1990).

Biochemical studies of flagella isolated from *P. mirabilis* swimmer and swarmer cells (Bahrani et al., 1991; Belas et al., 1991a), along with genetic data from Tn*5* mutageneses (Belas et al., 1991a,b), suggest that *P. mirabilis* synthesizes only a single flagellin species. Recently, Belas and Flaherty (1994) reported on the cloning of a region of *P. mirabilis* chromosomal DNA capable of complementing *E. coli* FliC⁻ mutants. Nucleotide and deduced amino acid sequence analyses revealed that this region contains three open reading frames (ORFs), which were identified based on their homology to other known flagellar genes. The region included the 5′ end of a putative homologue of *E. coli fliD* (an essential flagellar gene responsible for filament assembly) and two nearly identical copies of flagellin-encoding genes, *flaA* and *flaB*. (A third copy of a possible flagellin-encoding gene, referred to as *flaC*, was also found in this study by DNA:DNA homology.) *flaA* encodes a protein that complements *E. coli* FliC⁻ mutants, but *flaB* fails to complement any *E. coli* defects (Belas and Flaherty, 1994). Moreover, *flaB* encodes a protein that is larger than predicted from its deduced amino acid sequence when the gene was used in *E. coli* minicell protein programming experiments. These data suggest that, although *flaA* is functional in *E. coli, flaB* is not.

The discovery of multiple flagellin genes of *P. mirabilis* (Belas and Flaherty, 1994) has raised provocative questions concerning their regulation and function in association with swarmer cell differentiation. These issues gain further significance as the overproduction of flagella during swarmer cell differentiation is considered to be an important prerequisite for the colonization, invasion, and ultimate pathogenicity of this organism (Allison et al., 1992a,b). Many bacterial species have been observed to possess multiple copies of flagellin genes. For example, *Sal. typhimurium fliC* and *fljB* (Scott and Simon, 1982; Szekely and Simon, 1983; Zieg et al., 1977) and *Campylobacter coli flaA* and *flaB* (Alm et al., 1992; Guerry et al., 1988, 1991) each encode two separate

TABLE 10.4. Genetic Loci Associated With *P. mirabilis* Swarming Differentiation, Motility, and Behavior

Genetic locus or gene	Type of mutation	Mutant phenotype	Reference
flaA	Allelic exchange	Swm⁻ Swr⁻ Elo⁻	Belas (1994), Belas and Flaherty (1994)
flaB	Allelic exchange	Wild type	Belas (1994), Belas and Flaherty (1994)
flaD	Allelic exchange	Swm⁻ Swr⁻ Elo⁻	Belas (1994), Belas and Flaherty (1994)
flhA	Tn5 insertion	Swm⁻ Swr⁻ Elo⁻	Gygi et al. (1995a)
gidA	Tn5 insertion	Swm⁺ Swrᶜʳ Elo⁻	Belas et al. (1995)
cld locus	Tn5 insertion	Swm⁺ Swrᶜʳ Elo⁻	Belas et al. (1995)
rfaCD	Tn5 insertion	Swm⁺ Sw₋ᶜʳ Elo⁻	Belas et al. (1995)
fliG	Tn5 insertion	Swm⁺ Swrᶜʳ Elo⁻	Belas et al. (1995)
fliL	Tn5 insertion	Swm⁺ Swrᶜʳ Eloᶜ	Belas et al. (1995)
flgH	Tn5 insertion	Swm⁺ Swrᶜʳ Eloᶜ	Belas et al. (1995)
galU	Tn5 insertion	Swm⁺ Swrᶜʳ Elo⁻	Belas et al. (1995)
dapE	Tn5 insertion	Swm⁺ Swrᶜʳ Elo⁻	Belas et al. (1995)
cpsF	Tn5 insertion	Swm⁺ Swrᶜʳ	C. Hughes (personal communication, 1994)
pepQ	Tn5 insertion	Swm⁺ Swrᶜʳ Elo⁻	Belas et al. (1995)
Urease locus	Northern blot	NA	Allison et al. (1992b)
Hemolysin locus	Northern blot	NA	Allison et al. (1992b)

flagellin species that play a role in antigenic variation of the flagellum. Since flagella and specifically flagellin (H-antigen) are extremely antigenic, changes in flagellin antigenicity may provide the bacteria with an effective means of side stepping the immune response of the host (Brunham et al., 1993).

Three methods have been used to determine the transcriptional regulation of *flaA* and *flaB*. These included plasmid-borne transcriptional fusions, Northern blot DNA:RNA hybridization, and primer extension experiments (Belas, 1994). The results from each confirm that *flaA* is the sole flagellin gene transcribed by wild-type cells. Furthermore, *flaA* is transcribed as a monocistronic message and regulated coordinately with swarmer cell differentiation such that *flaA* transcription increases about eightfold during induction. The transcriptional analyses support the data from previous work using Tn5 mutants that demonstrated that swimmer and swarmer cell flagellin synthesis was interconnected and suggested that only one flagellin gene was involved in the synthesis of both organelles (Belas et al., 1991a).

The function of the gene product of *flaD, flaA,* and *flaB* has been established by constructing mutations in *P. mirabilis*. Mutations in *flaA* completely abolished all motility as well as swarmer cell differentiation, while *flaB* mutations have no demonstrable change in wild-type phenotype. These observations emphasize the central role of FlaA flagellin in swarmer cell differentiation and behavior. Presumably, the cell responds to environmental conditions that prevent the normal rotation of the FlaA flagella (Allison and Hughes, 1991a; Belas, 1992; Belas et al., 1991a). Loss of FlaA filaments resulted inappropriately in an undifferentiated swimmer cell under conditions that should elicit swarmer cell differentiation. This suggests that *flaA* is required for the induction of swarmer cell differentiation, because mutants defective in FlaA do not differentiate. This is contrary to what is observed during swarmer cell differentiation of *V. parahaemolyticus* (McCarter et al., 1988; Sar et al., 1990). In this case, defects causing loss of the polar flagellum (the sensing flagellum) result in a constitutively synthesized swarmer cell. The difference between these two mutants implies that the regulatory mechanisms operating to control swarmer cell differentiation of *P. mirabilis* are probably different from those of *V. parahaemolyticus,* even though the initial signal (inhibition of flagellar rotation) triggering differentiation is the same for both species.

As indicated, FlaA[-] mutants do not synthesize any flagellin species. This observation may be used as an indication that the third copy of *P. mirabilis* flagellin-encoding genes, *flaC,* is not actively transcribed. Although the nucleotide sequence of *flaC* has not been determined or the transcription of this gene assessed, the evidence obtained from FlaA[-] mutants argues that *flaC* is not transcribed in a *flaA* mutation or wild-type cell, or, if it is transcribed, the protein is not a functional flagellin. So, both *flaB* and *flaC* are apparently silent copies of flagellin-encoding genes.

Interestingly, at a low frequency, Mot[+] revertants were found emanating as flares from *flaA* colonies on semisolid Mot agar plates. Moreover, *flaA* Mot[+] revertants are capable of movement through media amended with anti-

FlaA polyvalent antisera. This observation in and of itself suggests that the flagella synthesized by the revertants are antigenically distinct from those produced by wild-type cells. Since the antisera used is polyvalent and produced in response to whole flagella rather than denatured flagellin (Belas et al., 1991a), it may be anticipated that it contains antibodies capable of binding to many different flagellin epitopes. During such binding, bacteria may be tethered together flagellum-to-flagellum, thus preventing them from achieving wild-type swimming or swarming motility. This is exactly what happens when anti-FlaA is added to cultures of wild-type *P. mirabilis:* the bacteria become tethered by flagellum-to-flagellum binding (R. Belas, unpublished data). In contrast, *flaA* Mot$^+$ revertants are only loosely bound when the same antiserum is applied to cell suspensions, indicating that *flaA* Mot$^+$ revertants synthesize flagella containing only a subset of possible FlaA epitopes.

Analyses of Southern blots from *flaA* Mot$^+$ revertants reveal a large deletion of variable size within the *flaAB* locus, which apparently caused the reversion from Mot$^-$ to Mot$^+$. For example, for one such Mot$^+$ revertant, the deletion removes over 50% of *flaA* and most of *flaB*. This observation suggests that the reversion process is a one-way event in which a deletion occurs, permanently removing a portion of DNA downstream from *flaA* and spanning into *flaB* and further into genes downstream from *flaB*. Thus, the mutation may result in a hybrid gene fusion where the 5' end of *flaA* is fused to the 3' end of *flaB*, yielding a functional flagellin as a consequence.

The evidence gathered from SDS-PAGE, immunoblotting of V8 protease-digested flagellin, N-terminal amino acid sequencing, and amino acid composition analyses strongly suggests that FlaA and the revertant flagellin have identical N-terminal ends but different C termini. The hypothetical model used to explain the events producing the revertant flagellin predicts that any splice between the 5' end of *flaA* and either *flaB* or *flaC* that produces a functional flagellum will result in a motile revertant phenotype. Based on this model, the splice site joining FlaA with FlaB is thought to be located at or near *flaA*.

An important question to consider is whether *in vitro* flagellin antigenic variation such as described here can occur *in situ,* such as during colonization of the urinary tract. The inoculation of Mot$^-$ bacterial cells in semisolid media or on the surface of "hard" agar places a strong selection for active motility (Quadling and Stocker, 1957; Silverman and Simon, 1972, 1973, 1974). This is due to the reduction of nutrients around the cells, combined with the accumulation of waste products at the point of inoculation. Spontaneous mutations, occurring at random points around the chromosome, could randomly cause the deletion and genetic fusion seen in the *flaA* Mot$^+$ revertants within the large population of Mot$^-$ cells. Such spontaneous Mot$^+$ mutants would have a tremendous selective advantage compared with the bulk of nonmotile cells due to their enhanced survival by being able to translocate actively to areas of greater nutrients and lower waste products. Such spontaneous Mot$^+$

mutants would also be very easy to spot by a human observer due to the very evident flare of motility they produce.

Such spontaneous mutations giving rise to an antigenically distinct flagellin would be equally likely to occur *in vitro* as *in vivo* and may very well occur in FlaA$^+$ as in Mot$^-$ cells. Such antigenic changes would increase the survivability of a urinary pathogen. For example, urinary pathogens are confronted by secretory IgA as the bacteria attempt to colonize the bladder. Since flagellin is strongly antigenic, a major focus-binding epitope for the immunoglobulins could be the flagellum. Tethering of the bacteria via their flagellum would effectively prevent their motility and impose the same selective pressures as those occurring in vitro. A spontaneous deletion that gave rise to an antigenically distinct flagellum would thus provide the bacteria with an avenue of escape away from immobilization by the immunoglobulins. This would be a successful survival strategy prolonging the chances of colonization by the bacteria.

10.5.2.2. *Flagellar Genes—Other* P. mirabilis *Flagellar Genes*
Inhibition of flagellar rotation is known to cause induction of swarmer cell differentiation (Belas et al., 1986; McCarter et al., 1988). It is therefore not surprising to find that mutations in flagellar genes (other than *flaA*) result in abnormal swarmer cell differentiation. For example, Belas et al. (1995) characterized the Tn*5* insertion sites in *P. mirabilis* mutants defective in swarmer cell elongation. Most of the mutants characterized in that study were in genes associated with flagellar biosynthesis. For example, Tn*5* insertions were identified in *fliL* (the homolog of which in *Caulobacter crescentus* is required for flagellar gene expression and normal cell division [Stephens and Shapiro, 1993]), *fliG* (a component of the flagellar switch), and *flgH* (encoding the basal-body L ring), and they have been shown to play a significant role in swarmer cell elongation (Belas et al., 1995). This suggests that subtle effects to the flagellar motor may cause major perturbations in swarming behavior.

Similarly, Gygi et al. (1995a) analyzed *P. mirabilis* mutants defective in *flhA*. *flhA* is a gene that is necessary for flagellar biosynthesis (Macnab, 1992) and is a member of a newly identified family of putative signal-transducing receptors that have been implicated in diverse cellular processes (Carpenter and Ordal, 1993). Other members of this family include LcrD of *Yersinia pestis* and *Yersinia enterocolitica*, FlbF of *C. crescentus*, FlhA of *B. subtilis* and *E. coli*, MxiA and VirH of *Shigella flexneri*, InvA of *Sal. typhimurium*, HrpC2 of *Xanthomonas campestris*, and Hrp of phytopathogenic bacteria (Carpenter and Ordal, 1993; Dreyfus et al., 1993; Galan et al., 1992; Gough et al., 1993; Lidell and Hutcheson, 1994; Miller et al., 1994; Vogler et al., 1991; Wei and Beer, 1993).

As with *flaA*, *flaB*, and *flaD*, *P. mirabilis flhA* possesses a flagellar-gene-specific σ^{28} promoter (Gygi et al., 1995a). Mutations affecting the function of FlhA result in the lack of flagellar synthesis and the abnormal expression of swarmer cell differentiation (Gygi et al., 1995a). Thus, in general, mutations that prevent the normal function or regulation of flagellar expression in *P. mirabilis* result in the abnormal expression of swarmer cells and point to the central role of flagellar function and filament rotation in maintaining proper sensing of environmental stimuli during swarmer cell differentiation and motile behavior.

10.5.2.3. P. mirabilis *LPS and Capsular Polysaccharide Genes*

Another class of genes found to be important in swarmer cell differentiation and swarming behavior are involved in the synthesis of LPS (O-antigen) and capsular polysaccharide. Belas et al. (1995) have found that the transposon insertion point of some Tn5-generated mutants defective in wild-type swarmer cell elongation is located in an ORF homologous to the hypothetical 43.3 kD protein in the *E. coli* locus responsible for the LPS chain-length determinant. As described in section 10.4.2.1, LPS chain length has been postulated to be important in swarmer cell function (Armitage, 1982; Armitage et al., 1975, 1979). The *cld* locus confers a modal distribution of chain length on the O-antigen component of LPS (Bastin et al., 1993). This protein is thought to have a dehydrogenase activity as indicated by the homology to UDP-glucose dehydrogenase from *Streptococcus pyogenes* and *Streptococcus pneumoniae*, which was confirmed in the *P. mirabilis* homologue by amino acid sequence comparison of the deduced amino acid sequence (Belas et al., 1995). Since *cld* comprises several ORFs, it is possible that such Tn5 insertion mutants of *P. mirabilis* are defective in maintaining the preferred O-antigen chain length. Alternatively, the mutation may be in the *rfb* gene cluster (O-antigen), which is closely linked to *cld* in *E. coli* and *Sal. typhimurium* (Bastin et al., 1993).

In a second *P. mirabilis* mutant, also phenotypically Elo⁻, it was determined that a mutation had occurred in *rfaD*, encoding ADP-L-glycero-D-manno-heptose-6-epimerase (Belas et al., 1995). In *E. coli* and *Sal. typhimurium*, this mutation results in altered heptose (L-glycero-D-mannoheptose) and lipopolysaccharide biosynthesis and increased outer membrane permeability (Pegues et al., 1990). The nucleotide sequence at the distal end of the cloned insert was determined to be highly homologous to *rfaC*, which encodes heptosyltransferase I (Sirisena et al., 1992). In *Sal. typhimurium*, this mutation leads to LPS without heptose and rough colony phenotype (Chen and Coleman, 1993; Sirisena et al., 1992). Such rough mutants have, in other bacteria, been associated with defects in motility, which may explain the Mot⁻ phenotype of this particular mutant. *rfaC* is genetically linked to *rfaD*, supporting the finding that the mutation is in *rfaD* and suggests that the defect affects LPS structure and outer membrane permeability.

Interestingly, all of these mutations appear to affect swarmer cell elongation by preventing normal synthesis and rotation of the flagella. However, while swarming motility is affected by these mutations, swimming is not. Moreover, impairment of the flagella should result in the induction of transcription of the swarmer cell regulon, although each of these mutants is Elo⁻. Thus, there may be a close connection between LPS synthesis and regulation and signal transduction and swarmer cell regulation in *P. mirabilis,* as has been shown by others (Armitage, 1982; Armitage et al., 1975, 1979).

Bacteria often produce extracellular polysaccharides (capsules and "slime") that aid cells in the adhesion to substrates. Many swarming colonies produce a clear extracellular fluid and often have a glistening, mucoid appearance. In *P. mirabilis* this material has been shown to consist of a polysaccharide matrix (Allison and Hughes, 1991a), but its composition in other bacteria has not been determined, nor have defects in swarming been correlated with specific defects in slime production. For example, *Ser. marcescens* excretes a surfactant (Matsuyama et al., 1992) that is not essential for swarming (Harshey, 1994; Harshey and Matsuyama, 1994).

Recently, Gygi et al. (1995b) cloned and determined the nucleotide and deduced amino acid sequence for a genetic locus of *P. mirabilis* responsible for production of capsular polysaccharide that is thought to be important in swarming motility. This region has strong homology to the *Streptococcus pneumoniae* type 19F capsular polysaccharide biosynthesis genes, particularly *cpsF*. The relevance and relationship of capsular polysaccharide biosynthesis swarming migration may now be assessed as *cpsF* mutations are constructed in *P. mirabilis* and the swarming characteristics of the mutants assessed. (Some preliminary results are described in section 10.6.)

10.5.2.4. P. mirabilis *Peptidoglycan and Cell Division Genes*

As part of an analysis of the defects responsible for abnormal swarmer cell elongation, Belas et al. (1995) identified several mutations in genes required for cell wall peptidoglycan biosynthesis and in cell division and chromosome replication. Two of the mutations identified were in *galU,* affecting glucose-1-phosphate uridylyltransferase and cell wall synthesis (Jiang et al., 1991; Liu et al., 1993; Morona et al., 1994; Varon et al., 1993), and in a gene homologous to *dapE* (Bouvier et al., 1992). The latter gene encodes N-succinyl-L-diaminopimelic acid desuccinylase, an enzyme that catalyzes the synthesis of LL-diaminopimelic acid, one of the last steps in the diaminopimelic acid–lysine pathway leading to the development of peptidoglycan (Bouvier et al., 1992). That mutations in cell wall biosynthesis genes produce abnormal swarmer cell elongation is not unusual, as defects in such genes can produce overall effects on the ability of the cell to function.

One of the more interesting mutations that have been found to produce the Elo⁻ Mot⁻ Swrᶜʳ phenotype is in *gidA* (glucose-inhibited division), a nonessential gene in *E. coli* very near *E. coli oriC* (Belas et al., 1995; Ogawa and Okazaki, 1991; von Meyenburg et al., 1982). The homology between *P. mirabilis* GidA and its *E. coli* homologue is one of the strongest observed (Belas et al., 1995). In *E. coli, gidA* mutations are silent on complex media; however, when grown on glucose-containing media, *E. coli gidA* strains produce long filamentous cells (von Meyenburg and Hansen, 1980, 1987; von Meyenburg et al., 1982). Furthermore, there is evidence that *gidA* transcription is regulated by ppGpp and involved in initiation of chromosomal replication (Asai et al.,

1990; Ogawa and Okazaki, 1991). Thus, *gidA* may function to connect glucose metabolism, ribosome function, chromosome replication, and cell division (J. Shapiro, personal communication, 1994). The function of *gidA* in swarmer cell differentiation and elongation is obscure, but evidently essential for swarmer cell elongation and wild-type swarming behavior.

10.5.2.5. Sigma Factors and Other Upstream Regulatory Sequences of P. mirabilis

Nucleotide sequence analysis of cloned swarmer cell differentiation genes from *P. mirabilis* (and other swarming bacteria) has been ongoing for several years. As part of these analyses, regulatory regions of each gene have been scrutinized for evidence suggestive of unique regulatory mechanisms controlling swarmer cell differentiation and behavior. While it is as yet too early to draw conclusions from these analyses, two aspects of the upstream regulatory regions of swarmer cell differentiation genes associated with flagellar synthesis have been found.

The upstream regulatory regions of many flagellar genes from *B. subtilis* (Mirel and Chamberlin, 1989), *E. coli* (Bartlett et al., 1988), and *Sal. typhimurium* (Helmann, 1991) have a unique promoter region for a flagellar-gene-specific RNA polymerase. An alternate σ subunit of RNA polymerase, referred to as σ^{28} (σ^F) in *E. coli* and σ^D in *B. subtilis*, is responsible for this specificity (Helmann, 1991). Thus, it is perhaps not unexpected that the upstream region in front of the start codon to the *P. mirabilis flaA, flaB,* and *flaD* genes (Belas and Flaherty, 1994) and *flhA* gene (Gygi et al., 1995a) also contains a nucleotide sequence that is very similar (if not identical) to the consensus σ^{28} promoter 5'-TAAA-N_{15}-GCCGATAA-3' of *E. coli*. The homology between the *P. mirabilis* sequences and the consensus is very good, with the σ^{28} promoter for *flaA* a direct match to the *E. coli* consensus sequence. The σ^{28} promoter of *flaD* has a single mismatch compared with the consensus, and the *flaB* promoter has two mismatches. This would suggest that these genes (and probably most of the flagellar genes of *P. mirabilis*) are regulated in a manner similar to that demonstrated in *E. coli*.

What then is different about the *P. mirabilis* flagellar gene regulation that might help explain surface-inducible flagellin synthesis in the swarmer cell? The single most notable difference between the *E. coli fliC* regulatory region and those of *flaA* and *flaB* is the presence of a dual, direct tandem repeat (DTR) sequence 5' to the *P. mirabilis* σ^{28} promoters (Belas, 1994; Belas and Flaherty, 1994). This sequence, ATAAAAA repeated twice, is about 90 bp upstream from the midpoint of the −35 region of each σ^{28} promoter. The *flaA* DTR sequence is 5'-ATAAAAATAATATAAAAAAATAA-3', with two overlapping direct repeats, ATAAAAA and ATAAAAATAA. The *flaB* DTR sequence is on the strand opposite the *flaA* DTR and has a nucleotide sequence of 3'-ATAAAAAAAGAGAGGTAGATAAAACAAAAA AGA-5'. The *flaB* DTR has two direct repeats as well. The first of these repeats is identical to *flaA*, ATAAAAA, while the second, AAAAAAGA, does not share sequence identity with the *flaA* DTR. Searches of GenBank bacterial nucleotide sequences failed to find similar sequences in front of any other flagellar gene, but a homologous sequence is found in the upstream regulatory region of *ureR,* the regulatory gene for urease expression (Nicholson et al., 1993). Urease and flagellin are coordinately regulated as part of swarmer cell differentiation (Allison et al., 1992b). While the function of the DTR sequences is currently unknown, it is tempting to speculate, because of sequence conservation and placement, that they may have a function in swarmer-cell-specific gene expression, perhaps as a surface-induced enhancer of transcription (Dingwall et al., 1990; Gober et al., 1991; Gober and Shapiro, 1990, 1992; Kustu et al., 1991).

10.6. MULTICELLULAR MIGRATION

10.6.1. Control of Swarmer Cell Behavior

10.6.1.1. The Migration Cycle: A Model of Extracellular Signals
In contrast to the recent advances at the genetic level characterizing the genes required during swarmer cell differentiation, little is known or understood about the molecular mechanisms controlling swarmer cell motile behavior and, more importantly, the cyclic events of differentiation, migration, dedifferentiation, and consolidation. Mutant phenotypes produced by transposon–insertion mutageneses have implicated a series of multicellular signaling events that presumably regulate the cyclic nature of consolidation in these bacteria, although only limited characterization of these loci has thus far been accomplished. While the role of signals (if they exist at all) is yet to be proven, the following is a brief synopsis of the data and a possible model to explain the observations. Such a model is merely speculative, but will point toward experiments designed to answer the questions regarding multicellular swarming migration.

   One of the most astonishing observations to come from efforts to understand the regulation of swarmer cell differentiation and multicellular swarming behavior in *P. mirabilis* is that many of the mutants defective in wild-type swarming are not null mutants, i.e., those mutants completely lacking the ability to swarm on an agar surface (Fig. 10.6A). Rather, many of the mutants are wild type for the ability to differentiate into swarmer cells and rotate the flagellar filaments in a normal manner (Allison and Hughes, 1991a,b; Belas et al., 1991a). What makes these strains interesting appears to be defects in one or more of the genes that regulate the multicellular interactions associated with the formation of the characteristic "bulls-eye" colony (Fig. 10.6B–D). Such mutants, which have been referred to in my laboratory as *swarming crippled mutants* (Swr^{cr}), have also been constructed in other laboratories (Allison and Hughes, 1991a,b). Swr^{cr} mutants do indeed possess the ability to differentiate and swarm on the appropriate surfaces, yet they are defective in orchestrating the proper series of steps required to form distinct, periodic consolidation zones (Fig. 10.6). The category of crippled mutants is, by its nature, broadly defined and somewhat pleiotropic. Individual strains within the Swr^{cr} group possess a wide variety of mutant phenotypes with the central theme being a defect in the ability to form evenly spaced or well-defined consolidation zones. When compared with a swarming null mutant (Fig. 10.6A), Swr^{cr} strains fall into phenotypic classes that produce (1) nonuniform consolidation zones (Fig. 10.6B); (2) spatially narrow or broad zones (Fig. 10.6C); or (3) indeterminate consolidation zones (Fig. 10.6D). The first two phenotypic classes (nonuniform and narrow or broad spacing strains) appear to have a defect in the spatial or temporal control of the multicellular signaling presumed to control the sequence of events of dedifferentiation, which give rise to the consolidation phase of swarming. It may be that in such mutants the putative signals are either generated more rapidly than in the wild type, giving rise to faster consolidation and narrower zones, or synthesized at a slow rate, resulting in broad zones due to a slower consolidation process. The last class of Swr^{cr} mutants appears to be completely lacking in the signals or, more likely, in the receptors for the signals controling consolidation.

Armed with these observations, a possible hypothetical scenario has been described (Belas, 1992) that may help us understand the unusual Swr^{cr} mutants producing uneven spacing in consolidation zones and thereby extrapolate to the nature of the genetic defects in indeterminate mutant strains. Such a model assumes the presence of a set of extracellular signals (in this particular case three signals, though the number was arbitrarily set). Each signal is assumed to be generated by the swarming colony, increases to a peak level, and then diminishes, perhaps due to breakdown by enzymes or other catalytic processes. As developed, the model requires that each of the three signals is produced, crests, and then diminishes simultaneously with the other signals; hence they are synchronous. Furthermore, according to the model, the cells, having the ability to sense the amplitude of signals, would regulate the differentiation cycle as a response to the sum of all of the incoming signals. These cycles of differentiation and dedifferentiation would therefore be regulated in reference to some maximum and minimum value of the sum of the signals— differentiation commencing at high signal intensity, dedifferentiation occurring at a low signal intensity. Such changes theoretically could be produced by regulatory mutations that affect the rate of production or destruction of one or more of the signal molecules. The threshold signal level for consolidation would thus become arrhythmic. The net outcome would produce a combination of narrow and broad zones, as shown in Figure 10.6B. Using the predictions of this model, it is then feasible to consider the indeterminate consolidation class of mutants (Fig. 10.6D) as lacking all signals or possessing a central defect, such as a mutations in the receptor for these signal molecules.

The models do simulate some of the swarming patterns observed in *P. mirabilis,* but the use of such models is very limited, and interpretation of such depictions of signal intensity may not mirror the real world. There are many different possible explanations that can be used to fit the data presented. Nonetheless, modeling does serve as a way to focus ideas for developing experimental approaches and understanding of the nature of defects in the Swr^{cr} mutants.

10.6.1.2. Genes and Mutant Phenotypes

Several of the genes described in the first section of this chapter are also needed for wild-type multicellular migration, i.e., defects affecting flagellar biosynthesis that result in the failure of the cell to synthesize flagella so that the cell is nonmotile and cannot participate in migration events. However, some of the genes associated with cellular differentiation play slightly different roles in migration.

P. mirabilis swarmer cells secrete substantial amounts of a viscous biofilm defining the leading edge of the migrating population. This substance consists of uncharacterized carbohydrate, which is assumed to assist migration, possibly acting as a surfactant (Stahl et al., 1983). Recently, a mutation that causes frequent consolidation has been characterized and the defect located in a gene, cpsF, involved in synthesis of a major, new capsular polysaccharide, referred to as CPS I (C. Hughes, personal communication, 1994). The mutant fails to polymerize CPS I, and precursors accumulate intracellularly. However, a second, minor capsular polysaccharide (CPS II) still gets produced and forms a thin capsule on the cell surface. Detailed analysis of the swarming cycle shows that the CPS I$^-$ cells differentiate faster than wild type. CPS I$^-$ mutants do not have changes in the temporal control of consolidation pattern formation, but rather are affected by a reduction in the motility of the cells, i.e., spatial placement of the rings is shortened. These data suggest that CPS I is important in reducing friction and acts as a surfactant, perhaps performing a similar function as serrawettin, a surfactant of Ser. marcescens (Matsuyama et al., 1992).

10.6.1.3. The Glutamine Signal: Possible Roles for a Metalloprotease and a Proline Peptidase?

P. mirabilis of diverse types produces an extracellular, EDTA-sensitive metalloprotease that is able to cleave serum and secretory IgA1 and IgA2, IgG, and a number of nonimmunological proteins such as gelatin and casein (Loomes et al., 1990, 1992, 1993; Senior et al., 1987, 1991). The protease activity on IgA is such that the heavy chain of the immunoglobulin is cleaved, which is different from classic microbial IgA proteases (Plaut, 1983). An enzyme with the ability to cleave immunoglobulin molecules has a potential to be a virulence factor and, indeed, is frequently found in the urine of patients with P. mirabilis urinary infections (Senior et al., 1991).

Researchers in my laboratory have recently cloned into E. coli a P. mirabilis genetic locus that confers to the recombinant bacteria all of the properties associated with the metalloprotease described in the wild-type Proteus cells (Wassif et al., 1995). Genetic analysis using Tn5 mutagenesis of the P. mirabilis DNA has indicated that a region of approximately 4.5 kb is required for expression of the metalloprotease in *E. coli*. Using nucleotide primers to IS50L and IS50R regions of the transposon, the nucleotide and deduced amino acid sequences are in the process of being analyzed. Thus far, this analysis has revealed the presence of an operon with three ORFs within the 4.5 kb region. The genes associated with the ORFs correspond to a zinc metalloprotease and two ATP-dependent membrane transport proteins, required for the secretion of the metalloprotease to the external mileau.

It is quite likely that this enzyme functions as a virulence factor during urinary tract infection. Allison et al. (1992a,b, 1994) have demonstrated that the activity of the metalloprotease is coordinately expressed with the cycles of swarmer cell differentiation. But, does this metalloprotease have any other function? One possible role it may play is actually to generate the glutamine signal shown to be required for swarmer cell differentiation and migation (Allison et al., 1993). If this is the case, it represents a significant advance in our understanding of multicellular migration and swarming behavior.

One of the Elo⁻ strains recently examined in my laboratory (Belas et al., 1995) is the result of a mutation in the *P. mirabilis* homologue to *pepQ*. This gene encodes X-proline dipeptidase (Nakahigashi and Inokuchi, 1990) and shows strong identity to other proline dipeptidases. What makes this PepQ⁻ mutant interesting is that, in addition to being defective in swarmer cell elongation (Elo⁻), it also produces abnormal consolidation patterns, similar to that shown in Figure 10.6B. One possible interpretation of these data is that the mutation in *pepQ* functions to generate the signal by cleaving larger polypeptides into transportable peptides. PepQ⁻ defects may prevent normal uptake of these important signal molecules used by *P. mirabilis* to control swarming behavior and consolidation (Belas, 1992). The interaction (if any) between the metalloprotease and PepQ may be significant in the multicellular signaling that occurs during the cyclic events of consolidation in *P. mirabilis* swarming colonies.

10.6.1.4. *Role of Amino Acid Chemotaxis*
Bacteria swim by the rotation of rigid flagellar filaments (Silverman, 1980; Silverman and Simon, 1977). The direction of rotation is reversible and is coordinated by the components associated with chemotaxis behavior (Macnab, 1987). Microorganisms like *E. coli* swim in stretches of smooth runs interrupted by intervals of chaotic motion called *tumbling*. Smooth swimming and tumbling behaviors correspond to opposite directions of flagellar rotation, where counterclockwise (CCW) filament rotation produces swimming and clockwise (CW) rotation results in tumbling (Silverman and Simon, 1973). When bacteria swim up an attractant concentration gradient, the period of smooth swimming is extended, thereby allowing the cell to make progress in the direction of higher concentration of attractant. Conversely, movement away from an attractant source, or toward increasing concentration of a repellant, produces more frequent tumbling response (Adler, 1983).

Historically, the role of chemotactic behavior in swarmer cell differentiation and swarming motility has been disputed. However, over the last several years an overwhelming amount of evidence gathered from *V. parahaemolyticus* (Sar et al., 1990), *P. mirabilis* (Belas et al., 1991a), and *Ser. marcescens* (O'Rear et al., 1992) has clearly identified chemotactic behavior as a major force in controling certain aspects of differentiation and multicellular migration. Moreover, in research demonstrating swarmer cell differentiation and migration in *E. coli* and *Sal. typhimurium*, Harshey and Matsuyama (1994) have unequivocally demonstrated through the use of well-characterized mutants in chemotaxis that chemotactic behavior is critical for swarming in these bacteria.

There are several aspects of chemotactic behavior as it functions in swarming that require further elaboration. One of these is the analysis of the chemotactic genes of *V. parahaemolyticus* (Sar et al., 1990). Examination of the chemotaxis system of *V. parahaemolyticus* is complicated by the existence of two distinct types of motility. Analysis of the two systems was simplified by using mutants capable of only one mode of motility: Fla^+ Laf^- (swimming only) and Fla^- Laf^+ (swarming only) strains (Sar et al., 1990). Thus, the two systems could be evaluated separately. Chemotaxis of swimming bacteria was analyzed using the capillary system of Adler (1973). To measure chemotaxis of swarming bacteria, the viscosity of the assay medium was increased with 5% polyvinylpyrrolidone-360. The flagellar systems responded similarly with respect to attractant compounds and the concentrations that elicited the chemotactic responses. *V. parahaemolyticus* responds to serine and other attractants that in *E. coli* are recognized by the Tsr receptor (Macnab, 1992). A locus required for chemotaxis in a swimmer-only strain of *V. parahaemolyticus* was cloned and mutated with Tn5 in *E. coli*. The transposon-generated defects were transferred to Fla^+ Laf^- and Fla^- Laf^+ strains. Introduction of *che* mutations prevents chemotaxis into capillary tubes and greatly diminishes multicellular swarming and unicellular swimming through semisolid swimming media. Thus, the two flagellar systems of *V. parahaemolyticus*, which consist of distinct motor-propeller organelles, are directed by a common chemosensory control system.

The role of glutamine chemotaxis in *P. mirabilis* swarmer cell differentiation has been described in section 10.5.1.2. *P. mirabilis* grown on a basal medium does not swarm, suggesting that a chemical inducer is missing. Allison et al. (1993) showed that the amino acid glutamine when added to the basal medium induced swarmer cell differentiation and multicellular migration. Glutamine is sensed through a specific transduction mechanism that is independent of the cellular (nutritional) amino acid uptake system. The sensing of glutamine acts in conjunction with the monitoring of flagellar rotation by the cell, and the two signals are processed together in order for swarmer gene–specific expression to be induced. Thus, *P. mirabilis* is different from *V. parahaemolyticus* in that it has a specific requirement for one amino acid, glutamine.

10.7. EVOLUTIONARY SIGNIFICANCE OF SWARMING

Throughout the years, a debate has been waged between those who argue that swarmer cell differentiation and swarming behavior are no more than a laboratory artifact with no significance in the natural environment and proponents who believe that swarming is a naturally occurring phenomenon whose function is surface translocation to a more favorable environment. Indeed, the multicellular swarming response, which allows the bacteria to forage for nutrients on solid surfaces, may provide an adaptive advantage under low nutrient conditions, such as those encountered by marine bacteria. In fact, it has been shown that swarmer cells of *V. parahaemolyticus* actually exclude other bacterial cells from surfaces under nutrient-limiting conditions (Belas and Colwell, 1982). The differentiation of swimmer cells to swarmer cells, with the concomitant multicellular interactions that accompany swarming motility, may also promote pathogenic associations (section 10.2).

While the true function of swarming remains speculative, research on the relationship of swarmer cells and virulence is tilting the argument toward a natural role for swarmer cell differentiation. *P. mirabilis* expresses a set of virulence factors coordinately with differentiation of swimmer to swarmer cells. Thus, the expression of virulence appears to be linked to differentiation, and the swarmer cell type may be a virulence factor (Allison et al., 1992a,b). It has also been suggested that the solid surface provided by urinary catheters and the apparent solidity imposed by the highly viscous mucus that covers urinary epithelial cells could induce differentiation in swarming *Proteus* species, thus facilitating invasion of the urinary tract by the swarmer cell type (Belas, 1992).

In view of the complexity of swarming it also seems reasonable that swarming must occur for a very good reason. Swarming requires an enormous commitment in terms of cellular resources. The cells virtually transform into factories to produce, maintain, and coordinate the movement of thousands of flagella. In the laboratory, swarming is not essential for the livelihood of *P. mirabilis* (or of any other organism), and yet swarming has been maintained in these species, thus providing some advantage to the organism. This is certainly apparent in the reversion of FlaA$^-$ mutants of *P. mirabilis* back to wild-type swarming behavior. Thus, swarming does seem appropriate to life on solid surfaces, and perhaps the main advantage to swarming species is in the colonization of surfaces, especially the cellular surfaces of host tissues.

10.8. A BACTERIAL MODEL OF SENSING AND RESPONSE TO A SURFACE

From the data presented, a model may be derived to describe a possible mechanism used by swarming bacteria for the sensing and ultimate response to a particular surface. As indicated, at the heart of this mechanism is the rotation of the flagellar filament (Fig. 10.7). The flagellar assembly is complex, composed of multiple structural elements and gene products. Central to the rotation of the filament is the functioning of the flagellar basal body composed of the L and P rings, which serve as a bushing in the cell wall; the MS ring, which is believed to function as a sort of mounting plate; and a "transmission

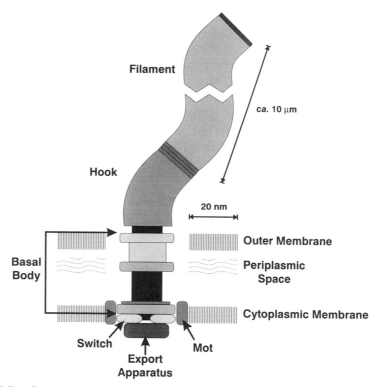

Fig. 10.7. General structure of a gram-negative flagellar assembly. The filament–hook–basal body assembly is a complex structure involving the participation of at least 21 different gene products. The propellor of the bacterial cell is the semirigid filament composed of flagellin (FliC, FlaA, etc.) as well as the filament capping protein (FliD) and two hook–filament junction proteins, FlgK and FlgL. The hook (composed of a single protein, FlgE) functions as a flexible coupling, or universal joint, between the filament and the cell. The hook is attached to a complex structure known as the *basal body,* which is embedded in the cell wall. Several proteins are associated with the basal body. The motor of the flagellum is made up of MotA and MotB proteins that participate in filament rotation through the movement of protons. The switch (FliG, FliM, and FliN) is believed to be responsible for determining the direction of filament rotation and presumably interacts with cytoplasmic chemotaxis proteins (CheY). Flagellar filament protein export is controlled through the action of a group of three proteins (FlhA, FliH, and FliI) located on the cytoplasmic end of the basal body.

shaft" composed of the distal and proximal rods. Two proteins (MotA and MotB) function as proton (Macnab, 1992) or sodium (Atsumi et al., 1992; McCarter, 1994; Tokuda et al., 1988) pumps to power the actual rotation of the filament. The direction of rotation is controlled at the level of three proteins (FliG, FliM, and FliN), which together make up the switch. Thus, the torque produced by the basal body is transmitted through the hook and into the semirigid filament, where it generates force and the resultant cellular movement.

The model to explain the sensory transduction events leading to differentiation of the swarmer cell was first described by McCarter and coworkers (McCarter et al., 1988; McCarter and Silverman, 1990) and is referred to as *reverse chemotaxis* (Fig. 10.8). It relies on our knowledge of the mechanism of bacterial chemotaxis, which has been elucidated through genetic and biochemical studies of *E. coli* and *Sal. typhimurium.* Chemotactic behavior is a consequence of a change in the direction of flagellar rotation in response to external stimuli (Hobson et al., 1982). These stimuli, often small molecules such as amino acids or simple sugars, are gathered by a set of transducer proteins located in the cytoplasmic membrane of gram-negative bacteria. Transmembrane receptors are specific for specific chemoeffector molecules (Adler et al., 1979). When a ligand binds to the receptor, the binding changes the amount of methylation on the cytoplasmic end of the transducer protein. The degree of methylation sends a signal ("X" in Fig. 10.8) that is relayed to other cytoplasmic chemotaxis proteins, most notably CheA and CheY. CheA is an autophosphorylating protein kinase that can transfer phosphate to CheY (Borkovich et al., 1989; Hess et al., 1987, 1988a–c). It is believed that CheY-P binding to the flagellar switch generates tumbling behavior (Bourret et al., 1989, 1990). Thus, in chemotaxis, information is transferred from the outside of the cell, through the membrane transducers, relayed through cytoplasmic chemotaxis protein via phosphorylated intermediates, and ultimately determines the direction of flagellar rotation. (This is a gross oversimplification of chemotaxis. The reader is referred to more thorough reviews for a comprehensive analysis of this subject [Simon, 1992; Simon et al., 1989].)

Reverse chemotaxis makes use of this knowledge but suggests that conditions that inhibit the rotation of the filament have the potential for producing effects within the cell. Inhibition of filament rotation may produce several effects, for example, increased torque may produce changes in the flow rate of protons (or sodium ions) across the basal body. Torque may also cause localized conformational changes in basal body structures or with cell wall and membrane components that interact with the basal body. Such conformational changes could affect the binding affinity of phosphorylated CheY for the switch proteins. The experimental data are not yet available to define this event, and for that reason a question mark has been drawn in Figure 10.8 to indicate that the signal transduction process taking information from the inhibited flagellar filament to the cytoplasm is not known. We do know, however, that this flagellar signal plus signals from glutamine (*P. mirabilis*)

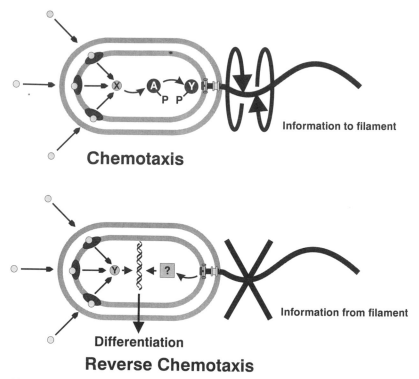

Fig. 10.8. The reverse chemotaxis signal transduction hypothesis of swarmer cell differentiation. Chemotaxis **(top)** results from modulation of the direction of rotation of flagellar filaments in response to extracellular stimuli (Larsen et al., 1974; Silverman and Simon, 1974). Sensory information gathered by a set of transmembrane receptors is integrated and transmitted to the flagellar "switch" (X) by four cytoplasmic proteins, CheA, CheW, CheY, and CheZ (Bourret et al., 1989; Hess et al., 1988b; Simon et al., 1989). Adaptation to environmental stimuli correlates with specific methylation and demethylation of the receptors by cytoplasmic proteins CheR and CheB (Springer et al., 1977, 1979). CheA (A) is an autophosphorylating protein kinase that can transfer phosphate to CheY(Y) (Borkovich et al., 1989; Hess et al., 1988a–c). Phosphorylation of CheY generates a tumbling behavior, presumably through the direct interaction of phosphorylated CheY with the flagellar switch (Bourret et al., 1990).

Swarmer cell differentiation **(bottom)** involves the sensing of the environmental concentration of specific chemical signals through (presumed) transmembrane receptor proteins, for it has been shown that induction of swarmer cell genes is dependent on iron (for *V. parahaemolyticus*) and glutamine (for *P. mirabilis*) concentrations (Allison et al., 1993; McCarter and Silverman, 1989). These signals are insufficient to induce swarmer cell differentiation; instead, the principal signal controlling differentiation in these bacteria is the inhibition of flagellar filament rotation. In reverse chemotaxis, a hypothesis originally described by McCarter and coworkers (1988), it is envisioned that the tethering of the bacterial flagellar filament sends a signal into the cell, essentially in the opposite direction of the signals of chemotaxis. The nature of this signal (?) is unknow, but, combined with the signals from the transmembrane receptors (Y), it induces transcription of a set of genes that ultimately produce the differentiated swarmer cell. It is not known whether the same set of transmembrane proteins involved in chemotaxis also participate in swarmer cell differentiation, nor has the chemical identity of the signal from these proteins been characterized.

and iron (*V. parahaemolyticus*) are assessed before the induction of the swarmer cell differentiation response. The end result is that, unlike chemotaxis, information flows from the filament through the cytoplasm and then outward in the production of the differentiated cell.

We can use our knowledge of chemotactic behavior and employ the hypothesis of reverse chemotactic surface sensing to produce a scenario for how swarming bacteria sense, respond, and adapt to surfaces (Fig. 10.9). In this scheme, cells are attracted to surfaces that have a nutrient-rich boundary layer. Presumably, some of the bound materials are released into the surrounding water where they form a concentration gradient that the cell senses and toward which the cell is drawn (Fig. 10.9A). As the cell gets close to the actual surface, the boundary layer, comprising many denatured molecules forming

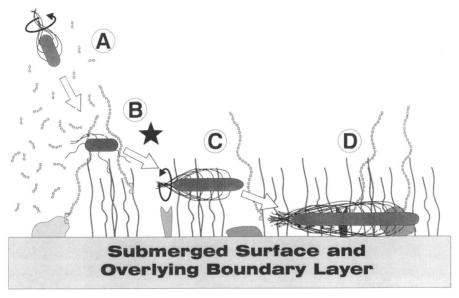

Fig. 10.9. Hypothetical model of swarmer cell differentiation in the sensing and adaptation to life on submerged surfaces. **A:** Motile bacteria may be drawn to surfaces through chemotaxis in response to small attractant molecules (food sources) that emanate from the overlying boundary layer of denatured proteins, carbohydrates, lipids, and other macromolecules. **B:** The boundary layer may present an environment that physically impairs the rotation of the flagella, temporarily trapping the cell in the upper confines of this layer. The inhibition of flagellar rotation, combined with the sensing of specific chemical signals derived from molecular components associated with the surface, produces a specific surface signal (indicated by the star) that induces the expression of specific swarmer-cell-differentiation genes. **C:** Expression of these genes starts the process of swarmer cell differentiation and a resumption of bacterial motility through this viscous surface boundary layer. **D:** Ultimately, the differentiated swarmer cells are able to scavenge nutrients and colonize the surface through adhesion mediated by the many flagellar filaments (Belas and Colwell, 1982).

a viscous microlayer above the surface, impedes the rotation of the bacterial flagella (Fig. 10.9B). As filament rotation becomes further inhibited, a threshold rotation frequency is reached (star) that triggers the transcription of the swarmer-cell-specific genes. At this point, the swimmer cell begins to differentiate into the swarmer cell (Fig. 10.9C). As the cell differentiates, it begins to overcome the viscosity imposed on it by the boundary layer, and its flagella begin to rotate once more. As the flagella rotate, thrust is produced and the now differentiated swarmer cell swarms through the viscous layer (Fig. 10.9D). Motility enables the bacterium to traverse within the boundary layer, thus scavenging nutrients as they are depleted in its immediate environs. As nutrients are metabolized, cell division occurs and a microcolony is formed. Thus, a distinct adaptational advantage is provided to the differentiated cell in its abilities to move and sequester nutrients on submerged surfaces (see also Chapter 3, this volume).

10.9. CONCLUSIONS AND FUTURE DIRECTIONS

For most of the time during which swarming motility has been observed and studied, this form of multicellular prokaryotic behavior has been considered as a curious oddity, usually more of a laboratory artifact and nuisance than a wonder, and certainly not a phenomenon with any connection to bacterial sensing and response to surfaces. These notions are changing as new data reveal the complexity of this sensory phenomenon. What is now becoming apparent is that swarmer cell differentiation and behavior are aspects of a complex signal transduction and genetic regulatory pathway, the mechanisms of which almost certainly have implications to other eubacteria, most notably *E. coli* and *Sal. typhimurium* (Harshey, 1994).

While a model to explain swarmer cell differentiation and behavior is emerging, a number of important questions remain unanswered. We know that inhibition of flagellar rotation is essential for sensing of the environment and inducing the expression of the swarmer-cell-specific genes, yet almost nothing is known about how this signal is transduced into the cell or how it is converted into a regulator of transcription. The recent findings by McCarter (1994) regarding the MotX protein and its function in *V. parahaemolyticus* polar flagellum rotation and the importance of FliG in *P. mirabilis* swarmer cell differentiation (Belas et al., 1995) may be leading to answers to some aspects of this question by directing attention to specific components of the flagellar basal body (Fig. 10.7).

In addition, although we have some limited information regarding the expression and transcriptional control of swarmer-cell-specific genes, e.g., *flaA* (Belas, 1994), most of these genes remain uncharacterized and their regulation unknown. It is intuitive that a complex series of events must occur for the periodic cycles of differentiation and dedifferentiation. Such processes must somehow link all of the cells into one global network of communication so that each cell starts and stops the differentiation cycle at the same time. Extracellular communication via signal molecules is one possible means by which such control could be achieved, but what is the mechanism? Are the metalloprotease and proline peptidase important in swarming migration, and what is the molecular nature of the signals and the regulatory mechanisms employed in controlling the cyclic events of consolidation? We do not have clear answers to any of these questions at present.

Several research groups are actively pursuing such questions using detailed molecular analyses of the genes and mutants affecting multicellular swarming behavior. It is safe to say that many of the answers to the questions raised in this chapter will be found in the next several years. What new questions those data will bring forth and what new mechanisms of signal transduction and genetic regulation will be unveiled can only be surmised at present. Whatever answers are found, however, they are very likely to have major and profound implications on the fields of research focused on bacterial sensing, response, and adaptation to surfaces.

ACKNOWLEDGMENTS

This is contribution No. 1023 from the Center of Marine Biotechnology.

This work was supported by grants from the National Institutes of Health (DK49720 and AI27107) and National Science Foundation (MCB-9206127).

REFERENCES

Adler J (1973): A method for measuring chemotaxis and use of the method to determine optimum conditions for chemotaxis by *Escherichia coli*. J Gen Microbiol 74:77–91.

Adler J (1983): Bacterial chemotaxis and molecular neurobiology. Cold Spring Harbor Symp Quant Biol 2:803–804.

Adler J, Goy MF, Springer MS, Szmelcman S (1979): On the mechanism of sensory transduction in bacterial chemotaxis. Soc Gen Physiol Ser 33:123–137.

Alberti L, Harshey RM (1990): Differentiation of *Serratia marcescens* 274 into swimmer and swarmer cells. J Bacteriol 172:4322–4328.

Allison C, Coleman N, Jones PL, Hughes C (1992a): Ability of *Proteus mirabilis* to invade human urothelial cells is coupled to motility and swarming differentiation. Infect Immun 60:4740–4746.

Allison C, Emody L, Coleman N, Hughes C (1994): The role of swarm cell differentiation and multicellular migration in the uropathogenicity of *Proteus mirabilis*. J Infect Dis 169:1155–1158.

Allison C, Hughes C (1991a): Bacterial swarming: An example of prokaryotic differentiation and multicellular behaviour. Sci Prog 75:403–422.

Allison C, Hughes C (1991b): Closely linked genetic loci required for swarm cell differentiation and multicellular migration by *Proteus mirabilis*. Mol Microbiol 5:1975–1982.

Allison C, Lai HC, Gygi D, Hughes C (1993): Cell differentiation of *Proteus mirabilis* is initiated by glutamine, a specific chemoattractant for swarming cells. Mol Microbiol 8:53–60.

Allison C, Lai HC, Hughes C (1992b): Co-ordinate expression of virulence genes during swarm-cell differentiation and population migration of *Proteus mirabilis*. Mol Microbiol 6:1583–1591.

Alm RA, Guerry P, Power ME, Trust TJ (1992): Variation in antigenicity and molecular weight of *Campylobacter coli* VC167 flagellin in different genetic backgrounds. J Bacteriol 174:4230–4238.

Armitage JP (1981): Changes in metabolic activity of *Proteus mirabilis* during swarming. J Gen Microbiol 125:445–450.

Armitage JP (1982): Changes in the organisation of the outer membrane of *Proteus mirabilis* during swarming: Freeze-fracture structure and membrane fluidity analysis. J Bacteriol 150:900–904.

Armitage JP, Rowbury RJ, Smith DG (1974): The effects of chloramphenicol, nalidixic acid and penicillin on the growth and division of swarming cells of *Proteus mirabilis*. J Med Microbiol 7:459–464.

Armitage JP, Rowbury RJ, Smith DG (1975): Indirect evidence for cell wall and membrane differences between filamentous swarming cells and short non-swarming cells of *Proteus mirabilis*. J Gen Microbiol 89:199–202.

Armitage JP, Smith DG (1978): Flagella development during swarmer differentiation in *Proteus mirabilis*. FEMS Microbiol Lett 4:163–165.

Armitage JP, Smith DG, Rowbury RJ (1979): Alterations in the cell envelope composition of *Proteus mirabilis* during the development of swarmer cells. Biochim Biophys Acta 584:389–397.

Asai T, Takanmi M, Imai M (1990): The AT richness and *gid* transcription determine the left border of the replication origin of the *E. coli* chromosome. EMBO J 9:4065–4072.

Atsumi T, McCarter L, Imae Y (1992): Polar and lateral flagellar motors of marine *Vibrio* are driven by different ion-motive forces. Nature 355:182–184.

Bahrani FK, Johnson DE, Robbins D, Mobley HL (1991): *Proteus mirabilis* flagella and MR/P fimbriae: Isolation, purification, N-terminal analysis, and serum antibody response following experimental urinary tract infection. Infect Immun 59:3574–3580.

Bartlett DH, Frantz BB, Matsumura P (1988): Flagellar transcriptional activators FlbB and FlaI: Gene sequences and 5′ consensus sequences of operons under FlbB and FlaI control. J Bacteriol 170:1575–1581.

Bastin D, Stevenson G, Brown P, Haase A, Reeves P (1993): Repeat unit polysaccharides of bacteria: A model for polymerization resembling that of ribosomes and fatty acid synthetase, with a novel mechanism for determining chain length. Mol Microbiol 7:725–734.

Belas MR, Colwell RR (1982): Adsorption kinetics of laterally and polarly flagellated *Vibrio*. J Bacteriol 151:1568–1580.

Belas R (1992): The swarming phenomenon of *Proteus mirabilis*. Am Soc Microbiol NEWS 58:15–22.

Belas R (1994): Expression of multiple flagellin-encoding genes of *Proteus mirabilis*. J Bacteriol 176:7169–7181.

Belas R, Erskine D, Flaherty D (1991a): *Proteus mirabilis* mutants defective in swarmer cell differentiation and multicellular behavior. J Bacteriol 173:6279–6288.

Belas R, Erskine D, Flaherty D (1991b): Transposon mutagenesis in *Proteus mirabilis*. J Bacteriol 173:6289–6293.

Belas R, Flaherty D (1994): Sequence and genetic analysis of multiple flagellin-encoding genes from *Proteus mirabilis*. Gene 148:33–41.

Belas R, Goldman M, Ashliman K (1995): Genetic analysis of *Proteus mirabilis* mutants defective in swarmer cell elongation. J Bacteriol 177:823–828.

Belas R, Simon M, Silverman M (1986): Regulation of lateral flagella gene transcription in *Vibrio parahaemolyticus*. J Bacteriol 167:210–218.

Bidnenko SI, Bernasovskaia EP, Anisimova I, Barshtein I, Mel'nitskaia EV (1985a): Enteropathogenicity of *Proteus mirabilis* bacteria and pathogenesis of experimental intestinal infections caused by them. Mikrobiol Z 47:81–88.

Bidnenko SI, Mel'nitskaia EV, Rudenko AV, Nazarchuk LV (1985b): Serological diagnosis and immunological aspects of *Proteus* infection. V. Design and trial of polyvalent antigenic preparations. Mikrobiol Epidemiol Immunobiol 2:49–53.

Bisset KA (1973a): The motion of the swarm in *Proteus mirabilis*. J Med Microbiol 6:33–35.

Bisset KA (1973b): The zonation phenomenon and structure of the swarm colony in *Proteus mirabilis*. J Med Microbiol 6:429–433.

Bisset KA, Douglas CW (1976): A continuous study of morphological phase in the swarm of *Proteus*. J Med Microbiol 9:229–231.

Borkovich KA, Kaplan N, Hess JF, Simon MI (1989): Transmembrane signal transduction in bacterial chemotaxis involves ligand-dependent activation of phosphate group transfer. Proc Natl Acad Sci USA 86:1208–1212.

Bourret RB, Hess JF, Borkovich KA, Pakula AA, Simon MI (1989): Protein phosphorylation in chemotaxis and two-component regulatory systems of bacteria. J Biol Chem 264:7085–7088.

Bourret RB, Hess JF, Simon MI (1990): Conserved aspartate residues and phosphorylation in signal transduction by the chemotaxis protein CheY. Proc Natl Acad Sci USA 87:41–45.

Bouvier J, Richaud C, Higgins W, Bogler O, Stragier P (1992): Cloning, characterization, and expression of the *dapE* gene of *Escherichia coli*. J Bacteriol 174:5265–5271.

Brogan TD, Nettleton J, Reid C (1971): The swarming of *Proteus* on semisynthetic media. J Med Microbiol 4:1–11.

Brunham RC, Plummer FA, Stephens RS (1993): Bacterial antigenic variation, host immune response, and pathogen-host coevolution. Infect Immun 61:2273–2276.

Carpenter PB, Ordal GW (1993): Bacillus subtilis FlhA: A flagellar protein related to a new family of signal-transducing receptors. Mol Microbiol 7:735–743.

Chen L, Coleman W (1993): Cloning and characterization of the *Escherichia coli* K-12 *rfa-2* (*rfaC*) gene, a gene required for lipopolysaccharide inner core synthesis. J Bacteriol 175:2534–2540.

Coetzee JN (1972): Genetics of the *Proteus* group. Annu Rev Microbiol 26:23–54.

De Lorenzo M, Herrero M, Jakubzik U, Timmis KN (1990): Mini-Tn5 transposon derivatives for insertion mutagenesis, promoter probing, and chromosomal insertion of cloned DNA in gram-negative eubacteria. J Bacteriol 172:6568–6572.

Dick H, Murray RG, Walmsley S (1985): Swarmer cell differentiation of *Proteus mirabilis* in fluid media. Can J Microbiol 31:1041–1050.

Dienes L (1946): Reproductive processes in *Proteus* cultures. Proc Soc Exp Biol Med 63:265–270.

Dingwall A, Gober JW, Shapiro L (1990): Identification of a *Caulobacter* basal body structural gene and a cis-acting site required for activation of transcription. J Bacteriol 172:6066–6076.

Dlugovitzky DG, Scharovsky OG, Molteni OA, Morini JC, Londner MV (1988): Effect of *Proteus mirabilis* on liver and spleen weight and the hemagglutinin response to SRBC in rats. Rev Latinoam Microbiol 30:31–35.

Douglas CW (1979): Measurement of *Proteus* cell motility during swarming. J Med Microbiol 12:195–199.

Douglas CW, Bisset KA (1976): Development of concentric zones in the *Proteus* swarm colony. J Med Microbiol 9:497–500.

Dreyfus G, Williams AW, Kawagishi I, Macnab RM (1993): Genetic and biochemical analysis of *Salmonella typhimurium* FliI, a flagellar protein related to the catalytic subunit of the F0F1 ATPase and to virulence proteins of mammalian and plant pathogens. J Bacteriol 175:3131–3138.

Ebringer A, Khalafpour S, Wilson C (1989): Rheumatoid arthritis and *Proteus:* A possible aetiological association. Rheumatol Int 9:223–228.

Falkinham JOI, Hoffman PS (1984): Unique developmental characteristics of the swarm and short cells of *Proteus vulgaris* and *Proteus mirabilis*. J Bacteriol 158:1037–1040.

Frolov AF, Parkhomenko LV, Lukach IG (1986): Sensitivity of *Proteus mirabilis* strains to the bactericidal action of human serum. Mikrobiol Z 48:23–26.

Fuscoe FJ (1973): The role of extracellular slime secretion in the swarming of *Proteus*. Med Lab Technol 30:373–380.

Galan JE, Ginocchio C, Costeas P (1992): Molecular and functional characterization of the *Salmonella* invasion gene *invA:* Homology of InvA to members of a new protein family. J Bacteriol 174:4338–4349.

Gmeiner J, Sarnow E, Milde K (1985): Cell cycle parameters of *Proteus mirabilis:* Interdependence of the biosynthetic cell cycle and the interdivision cycle. J Bacteriol 164:741–748.

Gober JW, Champer R, Reuter S, Shapiro L (1991): Expression of positional information during cell differentiation of *Caulobacter*. Cell 64:381–391.

Gober JW, Shapiro L (1990): Integration host factor is required for the activation of developmentally regulated genes in *Caulobacter*. Genes Dev 4:1494–1504.

Gober JW, Shapiro L (1992): A developmentally regulated *Caulobacter* flagellar promoter is activated by 3′ enhancer and IHF binding elements. Mol Biol Cell 3:913–926.

Gough CL, Genin S, Lopes V, Boucher CA (1993): Homology between the HrpO protein of *Pseudomonas solanacearum* and bacterial proteins implicated in a signal peptide-independent secretion mechanism. Mol Gen Genet 239:378–392.

Grabow WOK (1972): Growth-inhibiting metabolites of *Proteus mirabilis*. J Med Microbiol 5:191–204.

Guerry P, Alm RA, Power ME, Logan SM, Trust TJ (1991): Role of two flagellin genes in *Campylobacter* motility. J Bacteriol 173:4757–4764.

Guerry P, Logan SM, Trust TJ (1988): Genomic rearrangement associated with antigenic variation in *Campylobacter coli*. J Bacteriol 170:316–319.

Gygi D, Bailey M, Allison C, Hughes C (1995a): Requirement for FlhA in flagella assembly and swarm-cell differentiation by *Proteus mirabilis*. Mol Microbiol 15:761–769.

Gygi D, Rahmann MM, Lai H, Carlson R, Guard-Petter J, Hughes C (1995b): A cell-surface polysaccharide that facilitates rapid population migration by differentiated swarm cells of *Proteus mirabilis*. Mol Microbiol 17:1167–1175.

Hall PG, Krieg NR (1983): Swarming of *Azospirillum brasilense* on solid media. Can J Microbiol 129:1592–1594.

Harshey R (1994): Bees aren't the only ones: Swarming in gram-negative bacteria. Mol Microbiol 13:389–394.

Harshey R, Matsuyama T (1994): Dimorphic transition in *E. coli* and *S. typhimurium:* Surface-induced differentiation into hyperflagellate swarmer cells. Proc Natl Acad Sci USA 91:8631–8635.

Helmann JD (1991): Alternative sigma factors and the regulation of flagellar gene expression. Mol Microbiol 5:2875–2882.

Henrichsen J (1972): Bacterial surface translocation: A survey and a classification. Bacteriol Rev 36.478–503.

Hess JF, Bourret RB, Oosawa K, Matsumura P, Simon MI (1988a): Protein phosphorylation and bacterial chemotaxis. Cold Spring Harbor Symp Quant Biol 1:41–48.

Hess JF, Bourret RB, Simon MI (1988b): Histidine phosphorylation and phosphoryl group transfer in bacterial chemotaxis. Nature 336:139–143.

Hess JF, Oosawa K, Kaplan N, Simon MI (1988c): Phosphorylation of three proteins in the signaling pathway of bacterial chemotaxis. Cell 53:79–87.

Hess JF, Oosawa K, Matsumura P, Simon MI (1987): Protein phosphorylation is involved in bacterial chemotaxis. Proc Natl Acad Sci USA 84:7609–7613.

Hobson AC, Black RA, Adler J (1982): Control of bacterial motility in chemotaxis. Symp Soc Exp Biol 35:105–121.

Hoeniger JFM (1964): Cellular changes accompanying the swarming of *Proteus mirabilis*. I. Observations on living cultures. Can J Microbiol 10:1–9.

Hoeniger JFM (1965): Development of flagella by *Proteus mirabilis*. J Gen Microbiol 40:29–42.

Hoeniger JFM (1966): Cellular changes accompanying the swarming of *Proteus mirabilis*. II. Observations of stained organisms. Can J Microbiol 12:113–122.

Hoeniger JFM, Cinits EA (1969): Cell wall growth during differentiation of *Proteus* swarmers. J Bacteriol 148:736–738.

Hoffman P, Falkinham JI (1981): Induction of trypotophanase in short cells and swarm cells of *Proteus vulgaris*. J Bacteriol 148:736–738.

Houwink AL, van Iterson W (1950): Electron microscopical observations on bacterial cytology. II. A study of flagellation. Biochim Biophys Acta 5:10–16.

Hughes WH (1956): The structure and development of the induced long forms of bacteria. Symp Soc Gen Microbiol 6:341–360.

Hughes WH (1957): A reconsideration of the swarming of *Proteus vulgaris.* J Gen Microbiol 17:49–58.

Jeffries CD, Rogers HE (1968): Enhancing effect of agar on swarming by *Proteus.* J Bacteriol 95:732–733.

Jiang X, Neal B, Santiago F, Lee S, Romana L, Reeves P (1991): Structure and sequence of the *rfb* (O antigen) gene cluster of *Salmonella typhimurium* (strain LT2). Mol Microbiol 5:695–713.

Jin T, Murray RG (1987): Urease activity related to the growth and differentiation of swarmer cells of *Proteus mirabilis.* Can J Microbiol 33:300–303.

Jin T, Murray RGE (1988): Further studies of swarmer cell differentiation of *Proteus mirabilis* PM23: A requirement for iron and zinc. Can J Microbiol 34:588–593.

Jones HE, Park RWA (1967): The influence of medium composition on the growth and swarming of *Proteus.* J Gen Microbiol 47:369–378.

Kustu S, North AK, Weiss DS (1991): Prokaryotic transcriptional enhancers and enhancer-binding proteins. Trends Biochem Sci 16:397–402.

Kvittingen J (1949a): Studies of the life-cycle of Proteus Hauser. Acta Pathol Microbiol Scand 26:24–40.

Kvittingen J (1949b): Studies of the life-cycle of Proteus Hauser. Part 2. Acta Pathol Microbiol Scand 26:855–878.

Kvittingen J (1953): Studies of the life-cycle of Proteus Hauser. Part 3. Acta Pathol Microbiol Scand 32:170–186.

Larsen SH, Reader RW, Kort EN, Tso WW, Adler J (1974): Change in direction of flagellar rotation is the basis of the chemotactic response in *Escherichia coli.* Nature 249:74–77.

Leifson E, Carhart SR, Fulton M (1955): Morphological characteristics of flagella of *Proteus* and related bacteria. J Bacteriol 69:73–80.

Lidell MC, Hutcheson SW (1994): Characterization of the *hrpJ* and *hrpU* operons of *Pseudomonas syringae* pv. *syringae* Pss61: Similarity with components of enteric bacteria involved in flagellar biogenesis and demonstration of their role in HarpinPss secretion. Mol Plant Microbe Interact 7:488–497.

Liu D, Haase A, Lindqvist L, Lindberg A, Reeves P (1993): Glycosyl transferases of O-antigen biosynthesis in *Salmonella enterica:* Identification and characterization of transferase genes of groups B, C2 and E1. J Bacteriol 175:3408–3413.

Lominski I, Lendrum AC (1947): The mechanism of swarming of *Proteus.* J Pathol Bacteriol 59:688–691.

Loomes LM, Kerr MA, Senior BW (1993): The cleavage of immunoglobulin G in vitro and in vivo by a proteinase secreted by the urinary tract pathogen *Proteus mirabilis.* J Med Microbiol 39:225–232.

Loomes LM, Senior BW, Kerr MA (1990): A proteolytic enzyme secreted by *Proteus mirabilis* degrades immunoglobulins of the immunoglobulin A1 (IgA1), IgA2, and IgG isotypes. Infect Immun 58:1979–1985.

Loomes LM. Senior BW, Kerr MA (1992): Proteinases of *Proteus* spp.: Purification, properties, and detection in urine of infected patients. Infect Immun 60:2267–2273.

Macnab RM (1987): Flagella. In Neidhardt FC, Ingraham JL, Low KB, Magasanik B, Schaechter M, Umbarger HE (eds): *Escherichia coli* and *Salmonella typhimurium:*

Molecular and Cellular Biology. Washington, DC: American Society for Microbiology, pp 70–83.

Macnab RM (1992): Genetics and biogenesis of bacterial flagella. Annu Rev Genetics 26:131–158.

Matsuyama T, Kaneda K, Nakagawa Y, Isa K, Hara-Hotta H, Isuya Y (1992): A novel extracellular cyclic lipopeptide which promotes flagellum-dependent and -independent spreading growth of *Serratia marcescens*. J Bacteriol 174:1769–1776.

McCarter L (1994): MotX, the channel component of the sodium-type flagellar motor. J Bacteriol 176:5988–5998.

McCarter L (1995): Genetic and molecular characterization of the polar flagellum of *Vibrio parahaemolyticus*. J Bacteriol 177:1595–1609.

McCarter L, Hilmen M, Silverman M (1988): Flagellar dynamometer controls swarmer cell differentiation of *V. parahaemolyticus*. Cell 54:345–351.

McCarter L, Silverman M (1989): Iron regulation of swarmer cell differentiation of *Vibrio parahaemolyticus*. J Bacteriol 171:731–736.

McCarter L, Silverman M (1990): Surface-induced swarmer cell differentiation of *Vibrio parahaemolyticus*. Mol Microbiol 4:1057–1062.

McCarter LL, Wright ME (1993): Identification of genes encoding components of the swarmer cell flagellar motor and propeller and a sigma factor controlling differentiation of *Vibrio parahaemolyticus*. J Bacteriol 175:3361–3371.

McLean RJ, Nickel JC, Noakes VC, Costerton JW (1985): An *in vitro* ultrastructural study of infectious kidney stone genesis. Infect Immun 49:805–811.

Miller S, Pesci EC, Pickett CL (1994): Genetic organization of the region upstream from the *Campylobacter jejuni* flagellar gene *flhA*. Gene 146:31–38.

Mirel DB, Chamberlin MJ (1989): The *Bacillus subtilis* flagellin gene (*hag*) is transcribed by the σ^{28} form of RNA polymerase. J Bacteriol 171:3095–3101.

Mobley HLT, Warren JW (1987): Urease-positive bacteriuria and obstruction of long-term urinary catheters. J Clin Microbiol 25:2216–2217.

Morona R, Mavris M, Fallarino A, Manning P (1994): Characterization of the *rfc* region of *Shigella flexneri*. J Bacteriol 176:733–747.

Morrison RB, Scott A (1966): Swarming of *Proteus*. Nature 211:255–257.

Nakahigashi K, Inokuchi H (1990): Nucleotide sequence between *fadB* and *rrnA* from *Escherichia coli*. Nucleic Acids Res 18:6439.

Nicholson EB, Concaugh EA, Foxall PA, Island MD, Mobley HL (1993): *Proteus mirabilis* urease: Transcriptional regulation by UreR. J Bacteriol 175:465–473.

O'Rear J, Alberti L, Harshey RM (1992): Mutations that impair swarming motility in *Serratia marcescens* 274 include but are not limited to those affecting chemotaxis or flagellar function. J Bacteriol 174:6125–6137.

Ochman H, Wilson A (1987): Evolutionary history of enteric bacteria. In Ingraham J, Low B, Schaechter M, Umbarger H (eds): *Escherichia coli* and *Salmonella typhimurium*, Cellular and Molecular Biology. Washington, DC: American Society for Microbiology, pp 1649–1654.

Ogawa T, Okazaki T (1991): Concurrent transcription from the *gid* and *mioC* promoters activates replication of an *Escherichia coli* minichromosome. Mol Gen Genet 230:193–200.

Pegues J, Chen L, Gordon A, Ding L, Coleman W Jr (1990): Cloning, expression, and characterization of the *Escherichia coli* K-12 *rfaD* gene. J Bacteriol 172:4652–4660.

Plaut A (1983): The IgA1 proteases of pathogenic bacteria. Annu Rev Microbiol 37:603–622.

Proom H, Woiwod A (1951): Amine production in the genus *Proteus*. J Gen Microbiol 5:930–938.

Quadling C, Stocker BAD (1957): The occurrence of rare motile bacteria in some non-motile *Salmonella* strains. J Gen Microbiol 17:424–436.

Sar N, McCarter L, Simon M, Silverman M (1990): Chemotactic control of the two flagellar systems of *Vibrio parahaemolyticus*. J Bacteriol 172:334–341.

Scott TN, Simon MI (1982): Genetic analysis of the mechanism of the *Salmonella* phase variation site specific recombination system. Mol Gen Genet 188:313–321.

Senior B, Loomes L, Kerr M (1991): The production and activity *in vivo* of *Proteus mirabilis* IgA protease in infections of the urinary tract. J Med Microbiol 35:203–207.

Senior BW (1983): *Proteus morgani* is less frequently associated with urinary tract infections than *Proteus mirabilis*—An explanation. J Med Microbiol 16:317–322.

Senior BW, Albrechtsen M, Kerr MA (1987): *Proteus mirabilis* strains of diverse type have IgA protease activity. J Med Microbiol 24:175–180.

Silverman M (1980): Building bacterial flagella. Q Rev Biol 55:395–408.

Silverman M, Simon M (1973): Genetic analysis of flagellar mutants in *Escherichia coli*. J Bacteriol 113:105–113.

Silverman M, Simon M (1974): Assembly of hybrid flagellar filaments. J Bacteriol 118:750–752.

Silverman M, Simon MI (1977): Bacterial flagella. Annu Rev Microbiol 31:397–419.

Silverman MR, Simon MI (1972): Flagellar assembly mutants in *Escherichia coli*. J Bacteriol 112:986–993.

Simon MI (1992): Summary: The cell surface regulates information flow, material transport, and cell identity. Cold Spring harbor Symp Quant Biol 57:673–688.

Simon MI, Borkovich KA, Bourret RB, Hess JF (1989): Protein phosphorylation in the bacterial chemotaxis system. Biochimie 71:9–10.

Sirisena D, Brozek K, MacLachlan P, Sanderson K, Raetz C (1992): The *rfaC* gene of *Salmonella typhimurium:* Cloning, sequencing, and enzymatic function in heptose transfer to lipopolysaccharide. J Biol Chem 267:18874–18884.

Springer MS, Goy MF, Adler J (1977): Sensory transduction in *Escherichia coli:* Two complementary pathways of information processing that involve methylated proteins. Proc Natl Acad Sci USA 74:3312–3316.

Springer MS, Goy MF, Adler J (1979): Protein methylation in behavioural control mechanisms and in signal transduction. Nature 280:279–284.

Stahl SJ, Stewart KR, Williams FD (1983): Extracellular slime associated with *Proteus mirabilis* during swarming. J Bacteriol 154:930–937.

Stephens C, Shapiro L (1993): An unusual promoter controls cell-cycle regulation and dependence on DNA replication of the *Caulobacter fliLM* early flagellar operon. Mol Microbiol 9:1169–1179.

Story P (1954): *Proteus* infections in hospitals. J Pathol Bacteriol 68:55–62.

Szekely E, Simon M (1983): DNA sequence adjacent to flagellar genes and evolution of flagellar-phase variation. J Bacteriol 155:74–81.

Tokuda H, Asano M, Shimamura Y, Unemoto T, Sugiyama S, Imae Y (1988): Roles of the respiratory Na⁺ pump in bioenergetics of *Vibrio alginolyticus*. J Biochem 103:650–655.

Toshkov A, Kuiumdzhiev A, Zakharieva S, Georgiev D, Gumpert I (1977): Experimental pyelonephritis from *Proteus mirabilis* L forms in rats. Acta Microbiol Virol Immunol 6:42–48.

Ulitzer S (1975): The mechanism of swarming of *Vibrio alginolyticus*. Arch Microbiol 104:67–71.

VanderMolen G, Williams F (1977): Observation of the swarming of *Proteus mirabilis* with scanning electron microscopy. Can J Microbiol 23:107–112.

Varon D, Boylan S, Okamoto K, Price C (1993): *Bacillus subtilis gtaB* encodes UDP-glucose pyrophosphorylase and is controlled by stationary phase transcription factor sigma-B. J Bacteriol 175:3964–3971.

Vogler AP, Homma M, Irikura VM, Macnab RM (1991): *Salmonella typhimurium* mutants defective in flagellar filament regrowth and sequence similarity of FliI to F0F1, vacuolar, and archaebacterial ATPase subunits. J Bacteriol 173:3564–3572.

von Meyenburg K, Hansen F (1980): The origin of replication, *oriC*, of the *Escherichia coli* chromosome: genes near *oriC* and construction of *oriC* deletion mutants. ICN-UCLA Symp 19:137–159.

von Meyenburg K, Hansen F (1987): Regulation of chromosome replication. In Neidhardt F, Ingraham J, Low K, Magasanik B, Schaechter M, Umbarger H (eds): *Escherichia coli* and *Salmonella typhimurium: Molecular and cellular biology*. Washington, DC: American Society for Microbiology, pp 1555–1577.

von Meyenburg K, Jorgensen B, Neilsen J, Hansen F (1982): Promoters of the *atp* operon coding for the membrane-bound ATP synthase of *Escherichia coli* mapped by Tn*10* insertion mutations. Mol Gen Genet 188:240–248.

Wassif C, Cheek D, Belas R (1995): Molecular analysis of a metalloprotease from *proteus mirabilis*. J Bacteriol 177:5790–5798.

Wei ZM, Beer SV (1993): HrpI of *Erwinia amylovora* functions in secretion of harpin and is a member of a new protein family. J Bacteriol 175:7958–7967.

Williams FD, Anderson DM, Hoffman PS, Schwarzhoff RH, Leonard S (1976): Evidence against the involvement of chemotaxis in swarming of *Proteus mirabilis*. J Bacteriol 127:237–248.

Williams FD, Schwarzhoff RH (1978): Nature of the swarming phenomenon in *Proteus*. Annu Rev Microbiol 32:101–122.

Zieg J, Silverman M, Hilmen M, Simon M (1977): The metabolism of phase variation. In Wilcox G (eds): Molecular Approaches to Eucaryotic Genetic Systems. New York: Academic Press, pp 25–35.

11

MYXOCOCCUS COADHESION AND ROLE IN THE LIFE CYCLE[1]

LAWRENCE J. SHIMKETS

Department of Microbiology, University of Georgia, Athens, Georgia 30602

11.1. INTRODUCTION

Microbes form multicellular consortia in a variety of habitats aided by cell–cell and cell–substrate attachment mechanisms. In the case of biofilms, which cover submerged surfaces in aquatic or aqueous environments, a variety of different species assemble a consortium from a pool of planktonic cells (Characklis and Marshall, 1990). At the other end of the spectrum are species-specific consortia that form for specialized purposes such as mating and development. The myxobacteria can attach to a wide array of substrates and cells. They form biofilms on submerged surfaces (Kuner and Kaiser, 1982) and can attach to unrelated prey bacteria (Burnham et al., 1981, 1984; Daft et al., 1985) or form species-specific attachments during their developmental cycle (Gilmore and White, 1985; Qualls and White, 1982; Shimkets, 1986a,b). Thus the myxobacteria exhibit an extraordinary range of associations.

Physiological and genetic probing has led to identification of the attachment organelle in *Myxococcus xanthus*. This chapter focuses on coadhesion of myxobacterial cells within the context of their life cycle. Work is now underway to determine how the attachment organelle functions and to examine the many roles of attachment in the life cycle. Several reviews (Dworkin, 1996; Hartzell and Youderian, 1995; Reichenbach and Dworkin, 1992; Shimkets, 1990) and

[1] Dedicated to Rich Behmlander 1964–1994 and Jim Dana 1965–1993, who contributed to this field of study.

Bacterial Adhesion: Molecular and Ecological Diversity, pages 333–347
© 1996 Wiley-Liss, Inc.

the book *Myxobacteria II* (Dworkin and Kaiser, 1993) offer insight into other aspects of myxobacteria physiology and genetics.

11.2. THE MYXOBACTERIAL LIFE STRATEGY

The myxobacterial life cycle is most easily depicted in terms of two interlocking circles, one devoted to cellular reproduction and the other to formation of resting cells (Fig. 11.1). In the presence of a continuous supply of nutrients the cells grow and divide indefinitely by binary fission. Development is induced by starvation for carbon, energy, or phosphate (Manoil and Kaiser, 1980). The cells aggregate, form a multicellular fruiting body, and then differentiate dormant spores in all or part of the fruiting body, depending on the species. Coadhesion appears to play a fundamental role in both portions of the life cycle.

In nature myxobacteria feed on other bacterial species and insoluble macro-molecules during the vegetative portion of the life cycle. The predatory nature

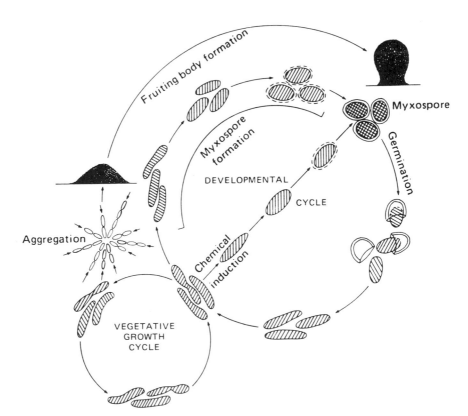

Fig. 11.1. Life cycle of *M. xanthus*. Note that the cells and fruiting bodies are not drawn to scale. (Reproduced from Dworkin, 1985, with permission of the publisher.)

of myxobacteria has been documented in a remarkable series of films by Reichenbach and colleagues (1965) that clearly show the myxobacteria lysing foreign bacteria while rubbing against them. To this end the myxobacteria secrete a variety of extracellular enzymes that hydrolyze peptidoglycan and protein (Sudo and Dworkin, 1972). With most myxobacteria species it is the protein fraction of the prey that serves as the principal carbon, nitrogen, and energy source (Shimkets, 1984). Coculture of *M. xanthus* with the cyanobacterium *Phormidium luridum* results in the formation of multicellular clumps containing both species in which the cyanobacterial cells are lysed (Burnham et al., 1981). When CO_2 and light served as the primary carbon and energy sources for the entire culture, a 9 -day predator–prey population cycle is established (Burnham et al., 1984). The capture of flagellated bacteria, which move 50–200 times faster than myxobacteria, presents a special challenge. *M. xanthus* cells release chemoattractants, most likely amino acids, at concentrations sufficiently high to lure *Escherichia coli* cells to them (Shi and Zusman, 1993a). Once predator–prey contact occurs, the prey are lysed within 30 minutes.

The question of how myxobacteria distinguish self from foreign in predator–prey interactions is not understood. The bacteriolytic enzymes and other lytic factors must be presented to the prey in a manner that renders the predator, which is also sensitive to bacteriolytic enzymes, resistant to enzymatic attack. Furthermore, the gentle rubbing action observed during lysis of the prey is also a common maneuver observed when sibling myxobacterial cells encounter each other with a different end result; the sibling cell is not lysed (Reichenbach et al., 1965).

The aggregation of roughly 10^5 cells and assembly into a fruiting body during development is also dependent on coadhesion. Developing *Stigmatella aurantiaca* cells form single-species clumps, even in the presence of other myxobacterial species (Qualls and White, 1982), suggesting the presence of species-specific coadhesion ligands. Futhermore, *M. xanthus dsp* mutants, which are defective in coadhesion, are unable to aggregate, assemble fruiting bodies, or differentiate spores (Shimkets 1986a,b). Together these experiments point to the importance of the cell surface in morphogenesis, and an organelle involved in coadhesion has now been identified in *M. xanthus*.

11.3. CELL SURFACE FIBRILS MEDIATE COADHESION

Coadhesion is typically quantified using an agglutination assay in which dispersed cells attach to each other forming large clumps that settle out of suspension (Shimkets, 1986a). The divalent cations Ca^{2+} and Mg^{2+} are necessary for agglutination along with an energy source. Cohesion is inhibited by cyanide and azide, which block electron transport; carbonyl cyanide-4-(trifluoromethoxy)phenylhydrozone (FCCP), which uncouples oxidative phosphorylation; and N,N'-dicyclohexylcarbodiimide (DCCD), which prevents ATP synthesis by the membrane bound ATPase (Shimkets, 1986a). Since inhibitors of transcription and translation do not inhibit coadhesion, the energy source is probably necessary for export of an essential component. Protein export in *E. coli* requires a proton motive force and is inhibited by ethanol; similar concentrations of ethanol block agglutination in *M. xanthus* (Shimkets, 1986a). Agglutination is also inhibited by the diazo dye Congo red (Arnold and Shimkets, 1988a,b).

Genes involved in coadhesion were identified through the mapping of mutations in strains with altered coadhesion phenotypes. Two simple assays based on colony morphology and dye binding were used to screen for coadhesion mutants. Colonies of coadhesion mutants are soft to the touch, shiny, tend to spread at a much slower rate than wild-type colonies, and have reduced binding of the dyes Congo red, calcofluor white, and trypan blue (Arnold and Shimkets, 1988a,b; Dana and Shimkets, 1993). Mutant coadhesion was then quantified using the agglutination assay, and the mutants were divided into three groups: those in which agglutination was reduced, those in which agglutination was eliminated, and those in which agglutination was stimulated (Table 11.1). The genes containing these coadhesion mutations have now been placed

TABLE 11.1. Coadhesion Mutants of *M. xanthus*

Gene or allele	Function
Reduced coadhesion	
asgA (Kuspa and Kaiser, 1989)	Histidine kinase/response regulator (Davis et al., 1994)
asgB (Kuspa and Kaiser, 1989)	Transcriptional activator (Plamann et al., 1994)
asgC (Kuspa and Kaiser, 1989)	Sigma factor (Davis et al., 1995)
dsgA (Dana and Shimkets, 1993)	Translation initiation factor IF3 (Cheng et al., 1994)
mglA (Dana and Shimkets, 1993)	GDP/GTP-binding protein (Hartzell and Kaiser, 1991)
sglA (Dana and Shimkets, 1993)	Social motility
sgl-2227 (Dana and Shimkets, 1993)	Social motility
sgl-2234 (Shimkets, 1986a)	Social motility
sgl-3112 (Shimkets, 1986a)	Social motility
sgl-3163 (Shimkets, 1986a)	Social motility
tgl-1 (Shimkets, 1986a)	Social motility
No coadhesion	
dsp-1680 (Shimkets, 1986a)	Social motility
dsp-1693 (Arnold and Shimkets, 1988a)	Social motility
dsp-1694 (Dana and Shimkets, 1993)	Social motility
sgl-3119 (Dana and Shimkets, 1993)	Social motility
Enhanced coadhesion	
stk-1907 (Dana and Shimkets, 1993)	Social motility

on the genetic map of *M. xanthus* and are scattered throughout the chromosome (Fig. 11.2) (Chen et al., 1991; He et al., 1994).

A clue to the nature of the organelle involved in coadhesion was obtained upon microscopic examination of the surfaces of coadhesion mutants and wild-type cells treated with coadhesion inhibitors. Coadhesion mutants failed to produce long thin fibrils that extend outward from the cell surface and attach to other cells (Fig. 11.3) (Arnold and Shimkets, 1988b; Dana and Shimkets, 1993). Inhibition of wild-type cell agglutination with Congo red also inhibited fibril production (Arnold and Shimkets, 1988b). Congo red is thought to inhibit the crystallization of extracellular polysaccharides into microfibrils (Colvin and Witter, 1983; Herth, 1980). A technique for purifying fibrils from wild-type cells has been developed by disrupting cells with sodium dodecyl sulfate (Behmlander and Dworkin, 1991). Addition of extracted, wild-type fibrils to *dsp* mutants restored agglutination (Chang and Dworkin, 1994). Together these results argue that fibrils are the principal coadhesion organelles.

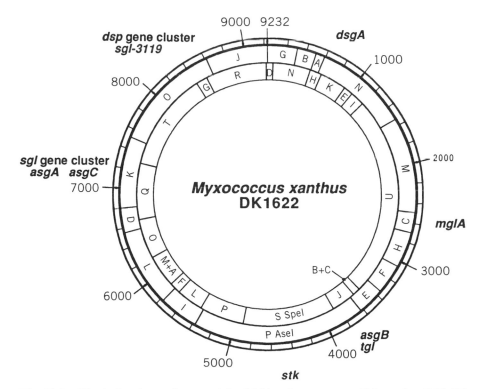

Fig. 11.2. Physical and genetic map of the 9.2 kbp chromosome of *M. xanthus* DK1622 showing the locations of genes involved in coadhesion. Consult Table 11.1 for the functions of the various genes.

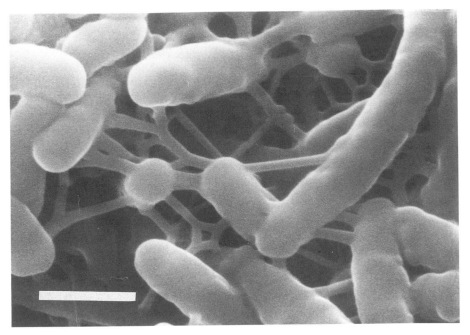

Fig. 11.3. Scanning electron micrograph of agglutinated wild-type *M. xanthus* cells showing the network of fibrils connecting the cells. Bar = 1 μm. (Courtesy of L. Shimkets.)

The fibrils on cells are about 30 nm thick when examined by high voltage scanning electron microscopy (Behmlander and Dworkin, 1994) and about 50 nm thick on negatively stained cells examined with transmission electron microscopy (Arnold and Shimkets, 1988b). Fibrils appear capable of associating along their long axis to form even thicker fibrils and possibly the entire extracellular matrix. Fibrils comprise about 10% of the dry weight of wild-type cells and are composed of a polysaccharide backbone with associated protein in a 1.0:1.2 ratio (Behmlander and Dworkin, 1994). The polysaccharide portion contains galactose, glucosamine, glucose, rhamnose, and xylose. One specific monoclonal antibody raised against cell surface antigens from developing cells, i.e., MAb 2105, reacted with wild-type cells but not fibril-less *dsp* cells (Behmlander and Dworkin, 1991). The antibody bound to protein antigens on the fibrils with a range of apparent molecular sizes on SDS-polyacrylamide gels of 14–90 kDa. There appear to be additional proteins associated with the fibrils that do not react with MAb 2105. The portion(s) of the fibrils involved in coadhesion have not been identified, but the genetic system of *M. xanthus* is sufficiently developed that one could begin making fibrils missing individual proteins or monosaccharides, thereby correlating structure with function.

11.4. COADHESION IS REGULATED BY CELL PROXIMITY

Genetic and physiologic experiments suggest that fibril production is a regulated process. A gene that regulates fibril production has been identified by Tn5 insertion ΩDK1907, which produced the *stk-1907* allele (Dana and Shimkets, 1993). The gene containing ΩDK1907 most likely inhibits coadhesion, since the mutant allele is recessive to the wild-type allele in merodiploids. Strains containing *stk-1907* have constitutive fibril synthesis, making them excessively sticky during all phases of the life cycle. These results suggest that fibril production is a regulated process, and the search for environmental factors that induce fibril production has suggested a relationship with cell proximity. Cells on a solid surface that are not in contact with each other have no fibrils, while 85% of the cells in groups contain fibrils (Behmlander and Dworkin, 1991).

The function of genes known to affect coadhesion was examined by epistasis, a simple genetic test that is helpful in organizing genes into different functional groups. Epistasis involves examining the interaction of nonallelic genes in the formation of a phenotype. In this case the *stk-1907* allele was transduced into the strains with reduced or no coadhesion. Here it stimulated marked improvement of coadhesion in all strains with reduced coadhesion, often to wild-type levels, but failed to improve coadhesion of *dsp* and *sgl-3119* alleles (Dana and Shimkets, 1993). Therefore, *stk-1907* is epistatic to mutations that reduce coadhesion (i.e., masks their phenotype), but hypostatic to mutations that eliminate coadhesion. The simplest interpretation of these results is that *asgA, asgB, asgC, dsgA, mgl, tgl* and most *sgl* mutations produce reduced quantitites of fibrils and that *stk-1907* restores fibril production. This notion was tested by examining these strains with scanning electron microscopy. Indeed, each of the strains had reduced levels of fibrils, and addition of the *stk-1907* allele boosted fibril production. The experiment also suggested that the genes containing the *dsp* and *sgl-3119* mutations encode the fibril structural genes, since *stk-1907* does not restore coadhesion to strains with these mutations. Fibrils have never been detected on these strains with or without *stk-1907*.

11.5. COADHESION IS A PROPERTY OF SOCIAL MOTILITY

The genes involved in coadhesion have a variety of functional roles (Table 11.1). The *asg* genes are involved in the production of an extracellular developmental signal known as A-factor, which is composed of amino acids and peptides (Kuspa et al., 1992a,b). Mutations in any of the three *asg* genes reduced protease secretion and perhaps secretion of other proteins (Kuspa and Kaiser, 1989; Plamann et al., 1992). The *asgA* gene encodes a protein with properties of both a histidine protein kinase and a response regulator (Davis et al., 1994). The *asgB* gene encodes a protein with the properties of a transcriptional activator (Plamann et al., 1994). The *asgC* gene encodes the major sigma factor from RNA polymerase (Davis et al., 1995). The *dsgA* gene encodes translation initiation factor IF3, which is essential for growth (Cheng and Kaiser, 1989; Cheng et al., 1994). Certain point mutations in this gene prevent the secretion of a developmental signal referred to as the D signal. The remaining genes involved in coadhesion are required for social motility.

Myxobacteria move by gliding, a form of motility whose mechanism(s) of propulsion remain unknown. The long, thin, rod-shaped cells typically glide in the direction of their long axis and periodically reverse their direction of movement. Gliding cells often move in large, organized groups known as *swarms* whose members maintain contact with each other during movement. Myxobacterial gliding is restricted to surfaces and is relatively slow, 2–20 μm/ min (Shi and Zusman, 1993b). Gliding cells deposit a slime trail, and other cells preferentially follow that trail (Reichenbach et al., 1965).

The behavior of myxobacterial cells is controlled by two multigene motility systems known as A (adventurous) and S (social), which determine whether cells move as individuals or groups, respectively (Hodgkin and Kaiser, 1979a,b). Mutations in A system genes (A^-S^+) eliminate movement of isolated cells, but movement is still possible using the S sytem, provided that cells are within one cell length of each other (Fig. 11.4) (Hodgkin and Kaiser, 1979a;

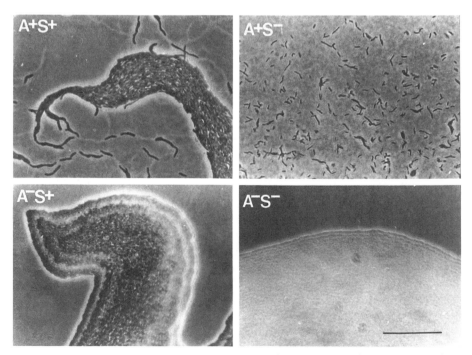

Fig. 11.4. Effect of A (adventurous) and S (social) motility mutations on *M. xanthus* cell behavior. Bar = 50 μm. (Reproduced from Shimkets (1986a with permission of the publisher.)

Kaiser and Crosby, 1983). Conversely, mutations in S system genes (A^+S^-) reduce or eliminate group movement, but cells are still motile via the A system (Hodgkin and Kaiser, 1979b). Therefore, S motility is contact dependent, while A motility is contact independent. Loss of motility occurs only when both systems are disrupted by either mutation of a gene in each system or by a single mutation in the *mglA* gene, the only gene known to be shared by both systems (Hodgkin and Kaiser, 1979a,b). A host of additional genes modify the behavior of gliding cells, such as the *frz* genes, which are similar to the enteric chemotaxis genes (McBride et al., 1989, 1992; McBride and Zusman, 1993; McCleary et al., 1990; McCleary and Zusman, 1990a,b).

Together the A and S systems allow the cells to establish a balance between dispersal, which is useful in searching for new food sources, and contiguity, which is useful in social behaviors. The A system contains at least 22 loci, *aglA* through *aglH, aglJ* through *aglR*, and *cglB* through *cglF* (Hodgkin and Kaiser, 1979a,b; Sodergren and Kaiser, 1983). The S system contains at least 12 loci, *sglA* to *sglH, sgl-3119, sgl-3163, tgl*, and *dsp* (Dana and Shimkets, 1993; Hodgkin & Kaiser, 1979a,b; Shimkets, 1986a,b; Wu and Kaiser, 1995). Additional unnamed A and S loci have recently been described (MacNeil et al., 1994). *mglA*, the gene shared by both systems, appears to encode a cytoplasmic GTP/GDP binding protein (Table 11.1) (Hartzell and Kaiser, 1991).

It is rather striking that mutations in social motility reduce or eliminate coadhesion. The *sgl* and *tgl* mutations reduce agglutination at least fivefold, while *sgl-3119* and *dsp* mutations eliminate agglutination altogether (Table 11.1) (Shimkets, 1986a). The role of coadhesion in social motility is not understood. One possibility is that S-dependent gliding involves selective coadhesion to the surface on which the cell is moving, resulting in treadmilling, a type of motility described for leukocytes (van der Merwe and Barclay, 1994). This possibility was examined by restoring coadhesion to *sgl, tgl*, and *mgl* mutants using the *stk-1907* mutation and assaying cells for restoration of social motility, which was defined as movement in the presence of an A system mutation. Cells with *sgl, tgl*, and *mgl* mutations were not restored for social motility by *stk-1907*, arguing that selective coadhesion does not provide the propulsion for social motility (Dana and Shimkets, 1993). Another possiblity is that coadhesion facilitates multicellular maneuvering in a swarm. This notion is supported by the fact that group movement is restored to A^+ *sgl* and A^+ *tgl* mutants by the *stk-1907* mutation (Dana and Shimkets, 1993). In this situation the propulsion is provided by the A motility system, even though the behavior of the cells is now distinctly group oriented. Coadhesion may place the transmission regulators of adjacent cells in proximity so that cells in a swarm glide in the same direction. Since little is known about the mechanism(s) of propulsion, it is presently not possible to examine this hypothesis in more detail.

A model describing aspects of the regulation of group behavior by S motility genes is shown in Fig. 11.5. The S motility pathway leads to group movement

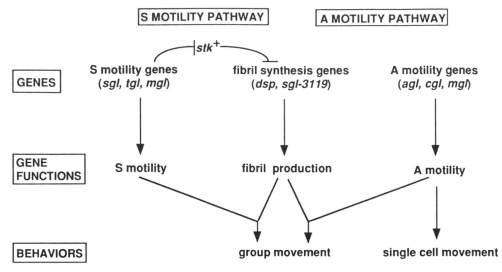

Fig. 11.5. Model outlining the pathways of group and single-cell movement in *M. xanthus*. The social (S) pathway leads to group movement, whereas the adventurous (A) pathway can lead to either single-cell movement or group movement. In the S pathway the social motility genes *sgl*, *tgl*, and *mgl* influence expression of fibril biosynthetic genes *dsp* and *sgl-3119*, possibly by regulating *stk*. The fibrils mediate group movement using either the S motility apparatus (*sgl*, *tgl*, *mgl*) or the A motility apparatus (*agl*, *cgl*, *mgl*). (Reproduced from Dana and Shimkets, 1993, with permission of the publisher.)

only, whereas the A pathway can lead to either single-cell movement or group movement. In the S motility pathway the S motility genes *sgl*, *tgl*, and *mgl* influence expression of fibril biosynthetic genes *dsp* and *sgl-3119* by regulating *stk*. Cell proximity is required for S motility (Kaiser and Crosby, 1983) and may derepress *stk*. Since *sgl* and *tgl* mutants fail to produce pili (Kaiser, 1979), pili are a candidate for the cell proximity monitor (Rosenbluh and Eisenbach, 1992). The Tgl protein may be another candidate for a cell proximity signal, since it appears to be exchanged between cells upon contact (Hodgkin and Kaiser, 1977). Fibrils mediate group movement using either the S propulsion system involving *sgl*, *tgl*, and *mgl* or the A propulsion system involving *agl*, *cgl*, and *mgl*. The next frontier is the identification of the roles these genes play in coadhesion and cell behavior.

11.6. COADHESION IS REQUIRED FOR MYXOBACTERIAL DEVELOPMENT

All the known mutants with reduced fibril levels (Table 11.1) are also defective in fruiting body formation and sporulation (Dana and Shimkets, 1993; Shimkets, 1986b). The *dsp* mutants, which lack fibrils altogether, show little aggregation and produce about 1% wild-type levels of spores. Furthermore, wild-type cells treated with Congo red to prevent fibril formation also fail to form fruiting bodies and have significantly reduced spore yields (Arnold and Shimkets, 1988b). Restoration of fibril production with *stk-1907* restored aggregation in the case of *asgA, dsgA, sglA, sgl-2234,* and *tgl* and restored sporulation to wild-type levels in the case of *dsgA* (Dana and Shimkets, 1993). Furthermore, addition of wild-type fibrils to *dsp* mutants restored fruiting body formation and sporulation in a dose-dependent manner (Chang and Dworkin 1994).

Taken together, these data argue that fibrils play an essential role in both the aggregation and sporulation phases of fruiting body development. However, the precise developmental role(s) they play remain unclear. Fibrils may act as components in tactile or chemosensory systems to facilitate exchange of intercellular signals. While it is unlikely that fibrils are essential for presenting or perceiving the extracellular CsgA protein, which initiates C-signaling during development (Li and Shimkets, 1993), it is possible that other signals are exchanged in this manner or that embedded in the extracellular matrix may be signals to indicate, among other things, the direction to a fruiting body. A second possibility is that cell behavioral changes in the developmental cycle are due, at least in part, to changes in the cohesins. *M. xanthus* coadhesion is likely to consist of transient cell–cell adhesions so that cells may occasionally leave the group. A hallmark of transient intercellular adhesions is stabilization by a large number of weak protein–protein interactions (van der Merwe and Barclay, 1994). Relatively minor changes in the association and dissociation rate constants could result in major changes in cell behavior. This model predicts a series of developmentally regulated changes in the cohesins correlated with the different developmental stages.

11.7. CONCLUSIONS

Coadhesion is important to both the vegetative and developmental modes of the myxobacterial life cycle. The cell surface fibrils appear to be the major coadhesion organelle since production of cells defective in fibrils through the use of chemical dyes such as Congo red (Arnold and Shimkets, 1988a,b) or mutations (Dana and Shimkets, 1993; Shimkets, 1986a,b) eliminates coadhesion. Restoration of fibril production through use of a suppressor mutation restores coadhesion to many coadhesion mutants (Dana and Shimkets, 1993). Finally, addition of fibrils extracted from wild-type cells restores coadhesion of fibril-less *dsp* mutants (Chang and Dworkin, 1994). Together these results present a strong argument that fibrils mediate coadhesion though they do not rule out the possibility that pili or other surface components play an accessory role. All S motility mutants are coadhesion mutants, arguing that coadhesion is essential for the S-dependent social behaviors exhibited during the vegetative and developmental portions of the life cycle. The precise roles coadhesion plays in these behaviors is just starting to be investigated.

The discovery that wild-type fibrils restore coadhesion and development to *dsp* mutants (Chang and Dworkin, 1994) generates a valuable tool for the assay of fibrils *in vitro*. The ability to construct strains with mutations in virtually any locus should enable one to make fibrils with a variety of different monosaccharide and protein compositions. Together these two approaches may enable one to find the cohesin(s) and to determine their role(s) in vegetative and developmental behaviors.

ACKNOWLEDGMENTS

This work was supported by National Science Foundation grant MCB9304083.

REFERENCES

Arnold JW, Shimkets LJ (1988a): Inhibition of cell–cell interactions in *Myxococcus xanthus* by Congo red. J Bacteriol 170:5765–5770.

Arnold JW, Shimkets LJ (1988b): Cell surface properties correlated with cohesion in *Myxococcus xanthus*. J Bacteriol 170:5771–5777.

Behmlander RM, Dworkin M (1991): Extracellular fibrils and contact-mediated cell interactions in *Myxococcus xanthus*. J Bacteriol 173:7810–7821.

Behmlander RM, Dworkin M (1994): Biochemical and structural analyses of the extracellular matrix fibrils of *Myxococcus xanthus*. J Bacteriol 176:6295–6303.

Burnham JC, Collart SA, Highison BW (1981): Entrapment and lysis of the cyanobacterium *Phormidium luridum* by aqueous colonies of *Myxococcus xanthus* PCO_2. Arch Microbiol 129:285–294.

Burnham JC, Collart SA, Daft MJ (1984): Myxococcal predation of the cyanobacterium *Phormidium luridum* in aqueous environments. Arch Microbiol 137:220–225.

Chang B-Y, Dworkin M (1994): Isolated fibrils rescue cohesion and development in the *dsp* mutant of *Myxococcus xanthus*. J Bacteriol 176:7190–7196.

Characklis WG, Marshall KC (1990): *Biofilms.* New York: John Wiley & Sons.

Chen H-W, Kuspa A, Keseler I, Shimkets LJ (1991): Physical map of the *M. xanthus* chromosome. J Bacteriol 173:2109–2115.

Cheng Y, Kaiser D (1989): *dsg,* a gene required for *Myxococcus* development, is necessary for cell viability. J Bacteriol 171:3727–3731.

Cheng Y, Kalman LV, Kaiser D (1994): The *dsg* gene of *Myxococcus xanthus* encodes a protein similar to translation initiation factor IF3. J Bacteriol 176:1427–1433.

Colvin JR, Witter DE (1983): Congo red and calcofluor white inhibition of *Acetobacter xylinum* cell growth and of bacterial cellulose microfibril formation: Isolation and properties of a transient, extracellular glucan related to cellulose. Protoplasma 116:34–70.

Daft MJ, Burnham JC, Yamamoto Y (1985): Lysis of *Phormidium luridum* by *Myxococcus fulvus* in continuous flow cultures. J Appl Bacteriol 59:73–80.

Dana JR, Shimkets LJ (1993): Regulation of cohesion-dependent cell interactions in *Myxococcus xanthus*. J Bacteriol 175:3636–3647.

Davis JM, Li Y, Mayor J, Cantwell B, Plamann L (1994): AsgA and AsgB may function in a signal transduction pathway regulating production of the A-signal. Abstracts of the 21st International Meeting on the Biology of the Myxobacteria.

Davis JM, Mayor J, Plamann L (1995): A missense mutation in *rpoD* results in an A-signalling defect in *Myxococcus xanthus*. Mol Microbiol 18:943–952.

Dworkin M (1985): Developmental Biology of the Bacteria. Menlo Park, CA: Benjamin/Cummings.

Dworkin M (1996): Recent advances in the social and developmental biology of the myxobacteria. Microbiol. Rev. 60:70–102

Dworkin M, Kaiser D (1993): *Myxobacteria II*. Washington, DC: American Society for Microbiology.

Gilmore DF, White D (1985): Energy-dependent cell cohesion in myxobacteria. J Bacteriol 161:113–117.

Hartzell P, Kaiser D (1991): Function of MglA, a 22-kilodalton protein essential for gliding in *Myxococcus xanthus*. J Bacteriol 173:7615–7624.

Hartzell P, Youderian P (1995): Genetics of gliding motility and development in *Myxococcus xanthus*. Arch Microbiol 164:309–323.

He Q, Chen H-W, Kuspa A, Cheng Y, Kaiser D, Shimkets LJ (1994): A physical map of the *Myxococcus xanthus* chromosome. Proc Natl Acad Sci USA 91:9584–9587.

Herth W (1980): Calcofluor white and Congo red inhibit chitin microfibril assembly of *Poterioochromas:* Evidence for a gap between polymerization and microfibril formation. J Cell Biol 87:442–450.

Hodgkin J, Kaiser D (1977): Cell-to-cell stimulation of movement in nonmotile mutants of *Myxococcus*. Proc Natl Acad Sci USA 74:2938–2942.

Hodgkin J, Kaiser D (1979a): Genetics of gliding motility in *Myxococcus xanthus* (Myxobacterales): Genes controlling movement of single cells. Mol Gen Genet 171:167–176.

Hodgkin J, Kaiser D (1979b): Genetics of gliding motility in *Myxococcus xanthus* (Myxobacterales): Two gene systems control movement. Mol Gen Genet 171:177–191.

Li S-F, Shimkets LJ (1993): Effects of *dsp* mutations on cell–cell transmission of CsgA in *Myxococcus xanthus*. J Bacteriol 175:3648–3652.

Kaiser D (1979): Social gliding is correlated with the presence of pili in *Myxococcus xanthus*. Proc Natl Acad Sci USA 76:5952–5956.

Kaiser D, Crosby C (1983): Cell movement and its coordination in swarms of *Myxococcus xanthus*. Cell Motil 3:227–245.

Kuner JM, Kaiser D (1982): Fruiting body morphogenesis in submerged cultures of *Myxococcus xanthus*. J Bacteriol 151:458–461.

Kuspa A, Kaiser D (1989): Genes required for developmental signalling in *Myxococcus xanthus:* Three *asg* loci. J Bacteriol 171:2762–2772.

Kuspa A, Plamann L, Kaiser D (1992a): Identification of heat stable A factor from *Myxococcus xanthus*. J Bacteriol 174:3319–3326.

Kuspa A, Plamann L, Kaiser D (1992b): A signalling and the cell density requirement for *Myxococcus xanthus* development. J Bacteriol 174:7360–7369.

MacNeil SD, Calara F, Hartzell PL (1994): New clusters of genes required for gliding motility in *Myxococcus xanthus*. Mol Microbiol 14:61–71.

Manoil C, Kaiser D (1980): Guanosine pentaphosphate and guanosine tetraphosphate accumulation and induction of *Myxococcus xanthus* fruiting body development. J Bacteriol 141:305–315.

McBride MJ, Weinberg RA, Zusman DR (1989): Frizzy aggregation genes of the gliding bacterium *Myxococcus xanthus* show sequence similarities to the chemotaxis genes of enteric bacteria. Proc Natl Acad Sci USA 86:424–428.

McBride MJ, Zusman DR (1993): FrzCD, a methyl-accepting taxis protein from *Myxococcus xanthus*, shows modulated methylation during fruiting body formation. J Bacteriol 175:4936–4940.

McBride MJ, Kohler T, Zusman DR (1992): Methylation of FrzCD, a methyl-accepting taxis protein of *Myxococcus xanthus*, is correlated with factors affecting cell behavior. J Bacteriol 174:4246–4257.

McCleary W, Zusman D (1990a): FrzE of *M. xanthus* is homologous to both CheA and CheY of *Salmonella typhimurium*. Proc Natl Acad Sci USA 87:5898–5902.

McCleary W, Zusman D (1990b): Purification and characterization of the *M. xanthus* FrzE protein shows that it has autophosphorylation activity. J Bacteriol 172:6661–6668.

McCleary W, McBride M, Zusman D (1990): Developmental sensory transduction in *M. xanthus* involves methylation and demethylation of FrzCD. J Bacteriol 172:4877–4887.

Plamann L, Kuspa A, Kaiser D (1992): Proteins that rescue A-signal–defective mutants of *Myxococcus xanthus*. J Bacteriol 174:3311–3318.

Plamann L, Davis JM, Cantwell B, Mayor J (1994): Evidence that *asgB* encodes a DNA-binding protein essential for growth and development of *Myxococcus xanthus*. J Bacteriol 176:2013–2020.

Qualls GT, White D (1982): Developmental cell cohesion in *Stigmatella aurantiaca*. Arch Microbiol 131:334–337.

Reichenbach H, Dworkin M (1992): The Myxobacteria. In Balows A, Truper HG, Dworkin M, Harder W, Schleifer K-H, (eds): the Prokaryotes. 2nd Ed. New York: Springer-Verlag, pp 3416–3487.

Reichenbach H, Heunert H, Kuczka H (1965): *Myxococcus* spp. (Myxobacterales)-Schwarmentwicklung und bildung von protocysten. Inst Wissen Gottingen, Film E779.

Rosenbluh A, Eisenbach M (1992): Effect of mechanical removal of pili on gliding motility of *Myxococcus xanthus*. J Bacteriol 174:770–777.

Shi W, Zusman DR (1993a): Fatal attraction. Nature 366:414–415.

Shi W, Zusman DR (1993b): The two motility systems of *Myxococcus xanthus* show different selective advantages on various surfaces. Proc Natl Acad Sci USA 90:3378–3382.

Shimkets L (1984): Nutrition, metabolism, and initiation of development. In Rosenberg E (ed): Myxobacteria: Development and Cell Interactions. New York: Springer-Verlag, pp 92–107.

Shimkets LJ (1986a): Correlation of energy-dependent cell cohesion with social motility in *Myxococcus xanthus*. J Bacteriol 166:837–841.

Shimkets LJ (1986b): Role of cell cohesion in *Myxococcus xanthus* fruiting body formation. J Bacteriol 166:842–848.

Shimkets L (1990): Social and developmental biology of the myxobacteria. Microbiol Rev 54:473–501.

Sodergren E, Kaiser D (1983): Insertions of Tn5 near genes that govern stimulatable cell motility in *Myxococcus*. J Mol Biol 167:295–310.

Sudo S, Dworkin M (1972): Bacteriolytic enzymes produced by *Myxococcus xanthus*. J Bacteriol 110:236–245.

van der Merwe AP, Barclay AN (1994): Transient intracellular adhesion: The importance of weak protein–protein interactions. Trends Biochem Sci 19:354–358.

Wu SS, Kaiser D (1995): Genetic and functional evidence that type IV pili are required for social gliding motility in *Myxococcus xanthus*. Mol Microbiol 18:547–558.

INDEX

Page numbers followed by the letter "t" designate tables.

349

gt9510b@prism.gatech.edu

DATE DUE

100297			
100297			
GAYLORD			PRINTED IN U.S.A.